SURVIVAL AT SEA
THE LIFEBOAT AND LIFERAFT

BY

C. H. WRIGHT
NATIONAL SEA TRAINING SCHOOLS
Liverpool, Retired

A COMPANION BOOK TO: THE EFFICIENT DECKHAND

ANGUS ALEXANDER MACDOUGALL

8 GARRYGALL
CASTLEBAY
ISLE OF BARRA
PA80 - 5UH

GLASGOW
BROWN, SON & FERGUSON, LTD.
4-10 Darnley Street

First Edition	1971
Second Edition	1977 (Revised)
Third Edition	1986 (Revised)
Fourth Edition	1988 (Revised)

ISBN 0 85174 540 7

Made and Printed in Great Britain

By Capt. C. H. Wright, Principal, National Sea Training School. Liverpool (Retired).

THE EFFICIENT DECKHAND

This up-to-date book has been specially written to assist candidates who are attempting the Department of Transport Examinations for Efficient Deckhand Certificate or A.B. Certificate and for pre-sea training. Contains detailed information on the use of various bends and hitches, splicing hawser laid, multiplait and wire ropes and all the latest equipment including fire fighting, steel hatch covers and precautions to be taken when carrying oils and chemicals, heavy lifting gear, container ships and roll-on roll-off ships.

THE COLLISION REGULATIONS 1972 AS AMENDED 1981 Fully Explained.

Written for both amateur and professional seamen it contains a copy of the Final Act of the International Conference of Revision of the International Regulations for Preventing Collisions at Sea, 1972 as amended 1981 and which came into force on 1st June 1983. Fully illustrated with coloured diagrams, suitably explained.

A MUST FOR EVERY NAVIGATOR.

SURVIVAL FOR YACHTSMEN

Originally "Know your Liferaft" completely re-written and brought up to date with all the latest regulations. Specifically for all yachtsmen, small boat owners and fishermen. A masterpiece of practical information on survival afloat, in the event of a disaster.

PROFICIENCY IN SURVIVAL CRAFT CERTIFICATES

Specifically written for candidates attempting to pass Certificates of Proficiency in Survival Craft. Contains all the necessary information.

Obtainable from the Publishers
Messrs. Brown, Son & Ferguson, Ltd., 4-10 Darnley Street, Glasgow G41 2SD and from Nautical Booksellers.

ACKNOWLEDGEMENTS

The Author would like to record his appreciation and thanks to all those who have so kindly contributed to this text book, both with advice and material assistance and in particular to:–

Commander C. G. Cuthbertson D.S.C., R.N.R., Nautical Surveyor
Captain H. C. Large, National Sea Training Centre, Liverpool.
Mr. H. Hambling. The British Shipping Fedaration Ltd.
Mrs. L. Shaw. The British Shipping Federation Ltd.
Mr. C. E. J. Sheather. C.Eng., M.I.Mech.E., A.M.I.E.D., Walter Tangen Ltd.
Mr. C. M. Wright. M.C.M.S.
Captain R. Ogilvy. National Sea Training Centre, Liverpool.
G. M. Fitz Gibbon. M.D., F.R.C.S.
Captain W. Y. Higgs. F.N.I., M.S.N.A.M.E.
Captain R. W. Joughin. B.Sc.(Hons), F.N.I. College of Nautical Studies, Warsash.

Beaufort Air-Sea Equipment Ltd.
The R.F.D. Company Ltd.
The Dunlop Company Ltd.
The Permutit Company Ltd.
E. P. Barrus (Concessionaires) Ltd.
Hugh McLean and Sons Ltd.
Watercraft Ltd.
The Welin Davit and Engineering Company Ltd.
Dowty Hydraulic Units Ltd.
Schat Davits Ltd.
William Cubbin Ltd.
Pains Wessex Ltd.
William Mills (Sunderland) Ltd.
Automatic Rescue Equipment Ltd.
F. P. T. Industries Ltd.
Mashford Bros. Ltd.
Salter Bros. Ltd.
G. R. Woodford Ltd.
Firdell Multiflectors Ltd.
Damage Control (U.K.) Ltd.
Lifeguard Equipment Ltd.
Hawker Siddley Marine.
R. Perry and Co. Ltd.
Radcom Ltd.
Whittaker Survival Systems.
Ameeco Oceanographic and Marine Division.
The Marconi International Marine Company Ltd.
Whittaker Survival Systems
For permission to reproduce copyright:
'The Lifeboat Glossary' 'Safety at Sea International'.
'Yachting World'. 'Department of Transport'.
'Siglingamal'. 'National Maritime Institute'.
The Nautical Institute. 'Nautical Review'.
'Seaways'.
Berwin Engineering Ltd.

PREFACE

My object in writing this book is to provide not only a reliable and useful guide for candidates attempting the Department of Trade Examination for Certificated Lifeboatmen, but also a description of the latest aids available which contribute towards safety of life at sea, together with the means whereby they can be used to the best advantage, and which can be understood by all who sail the sea.

While only the regulations concerning foreign-going British merchant ships have been quoted, these regulations incorporate the life-saving regulations applicable to most other vessels. Aids not required by these regulations have also been included in an effort to make the book fully comprehensive.

In addition to merchant seamen, I trust that fishermen, trawlermen, yachtsmen and air-crews will find much useful material included, be they sailors, engineers, radio-officers or members of the catering department. The period of qualifying service may be served in any type of sea-going vessel.

All whose business takes them either on or over the oceans should possess a knowledge of the means available to them to assist in saving lives in an emergency, and how to use them. For if they do not know beforehand, they will most certainly not know when sudden disaster unexpectedly strikes.

C. H. WRIGHT, December 1970

PREFACE TO THIRD EDITION

Since writing the first edition of "Survival at Sea - The Lifeboat and Liferaft" in 1970, many of the regulations have been revised and much of the equipment has been improved. It has therefore become necessary to revise and rewrite the book, in order to bring it once again fully up-to-date. The U.K. Department of Transport 1983 requirements for Fishing Vessels and Pleasure Yachts have been inserted, together with much useful information, in order to improve the information available to yacht owners and crews.

Attention has been paid to all the recommendations of the International Maritime Organization made at the conventions held in 1981, 1982 and 1983 to amend the recommendations of the 1974 SOLAS Convention.

The Governments (Administrations) of the majority of sea-going nations are parties to the Organization and therefore will in due course and not later than 1st July 1986 ratify the recommendations. Indeed a number of authorities have already implemented a number of these new recommendations as a requirement on their own shipping.

The main requirements under the original 1974 convention have been retained for the benefit of seamen serving on vessels which were built prior to 1st July 1986 and which will therefore only be required to implement some of the new recommendations by 1st July 1991.

C. H. WRIGHT, December 1983

CONTENTS

DEFINITIONS.

BMT.	British Maritime Technology Ltd. (Formerly NMI).
Buoyant apparatus.	flotation equipment other than lifebuoys and lifejackets.
Cargo vessel.	any vessel that is not a passenger vessel, pleasure yacht or fishing vessel.
Certificated person.	a person who holds a Certificate of Proficiency in Survival Craft or a Certificate of Efficiency as Lifeboatman issued prior to 28th April 1984.
DoT.	UK Department of Transport.
GCBS.	General Council of British Shipping. (Formerly The British Shipping Federation).
IMO.	International Maritime Organization. (Formerly IMCO).
Inflatable boat or raft.	Requires to be inflated before use.
Inflated boat or raft.	Kept inflated at all times.
MERSAR.	Merchant Ship Search and Rescue Manual.
Mile.	1852 metres. 6080 feet.
MNTB.	Merchant Navy Training Board.
NI.	Nautical Institute.

Passenger ship.	any ship carrying more than 12 passengers.
Person practiced in the handling and operation of liferafts.	a person who has attended a basic sea survival course and has at least two months sea service.
Person.	A person over the age of one year.
SAR.	Search and Rescue.
SI.	Statutory Instrument.
SOLAS.	Safety of Life at Sea.
Thermal Protective Aid.	A bag or suit made of waterproof material with low thermal conductivity.
U.K.	United Kingdom.

The I.M.O. Regulations unless expressly provided otherwise, do not apply to:-

(a) Ships of war and troopships.
(b) Cargo ships of less than 500 tons, gross tonnage.
(c) Ships not propelled by mechanical means.
(d) Wooden ships of primitive build.
(e) Pleasure yachts not engaged in trade.
(f) Fishing vessels.

THE "L" TICKET

Lifeboat efficiency means a sufficiency
of knowledge of things great and small
pertaining to boats, equipment and ropes,
and practice in using them all:

Gunnwales and cringles, earrings and gymbals;
the boat's liquid compass to box;
the tack and the clew, and what not to do,
when detailed to act as the cox:

The luff of the lug, the head and the throat,
the foot and the yard and the peak;
the issue of stores and drink in each boat,
and how one should cope with a leak;

To know how to lower, without or with power;
to slacken the guys, or to trim;
to let go of the falls (and keep clear of the walls)
to hoist up the boat and launch in;

To heave a sea-anchor or toggle a painter;
to set the sails right on the mast;
to turn up a cleat with the slack of a sheet
but never, no NEVER make fast;

To row round the harbour with rhythm and ardour,
the Board of Trade Captain to show,
how, if one were coxswain, to keep off the rocks and
steer the boat home, one should know;

If this you can do (and he thinks so too!)
and you land him ashore fairly dry;
you'll find you have made the B of T Grade
if not – it's well worth it to try.

D. E. W.

(Compiled by an anonymous candidate
for the original Lifeboat Certificate).

SURVIVAL

Issued by the Council of the Nautical Institute as a guide to survivors, a basic text for instruction and for discussion during boat drill.

ON HEARING THE EMERGENCY SIGNAL.
Go to your station.
If possible, collect warm *(woollen)* clothing, waterproof clothing, *(including a close fitting hat)* and lifejacket from their stowage.
Have a good drink of water.

IF ORDERED TO ABANDON SHIP
Put on warm clothing, waterproof clothing, *(a close fitting hat)* and life-jacket, loosen neckwear.
Assist loading extra water, provisions (not protein foods) and blankets if time allows.
Secure painters as appropriate, launch lifeboats/liferafts **to orders** and board dry if possible.
Avoid unnecessary swimming if you have to enter the water.
Release lifeboat or cut liferaft painter **to orders.**
Assist clearing the ship's side and danger area.
Stream the sea-anchor if required.
Help protect yourself from the environment by:

a. in liferafts taking anti-seasickness pills, inflating floor and using doors to regulate conditions.
b. in open lifeboats, erecting canopy, taking anti-seasickness pills, keep as warm and dry as possible.

Follow the instructions contained in the survival equipment.

SURVIVAL AFTER SHIPWRECK.
The senior fit survivor must assume command. Decisions and action must be taken concerning:–
*The distribution of survivors (full complements in rafts in cold climates, as few as possible in the tropics).
Take any unoccupied survival craft in tow, use them for stores, sleeping accommodation and as a back-up craft for use in future emergency.
*Congregation of rafts and boats to make a better target for air search.
*Navigation if thought more prudent.
*Fair rationing.
*Allocation of duties, lookouts and routine.
You are unlikely to drown in a liferaft but you can easily die of cold or injury so your first priority is:–
PROTECTION. You should:
*Bail out the boat/raft, rig canopy (boats), close entrances and prevent reswamping. Keep head to wind (boats) and deploy sea anchors, check for leaks and make repairs. Collect useful flotsam.
*Apply first-aid (do this while others bail-out).
*Treat asphyxia (clear airways, commence mouth-to-mouth).
*Stop bleeding (apply pad and bandages).
*Treat for shock (keep warm, encourage, relieve pain).
*Treat fractures (immobilise with pads, bandages and splints).
*Prevent sea-sickness (everyone to take a pill if available).
*Keep clothing on - wet clothing is much better than no clothing. Wind-

proof clothing is valuable, if necessary strip the dead.
*Insulate yourself from wind and water if possible. In liferafts inflate floors/canopy.
Evacuate the bowels and urinate.
Your second priority is **LOCATION**
Normally the search will start in the vicinity of the wreck so stay where you are if practicable. Meanwhile prepare your signalling devices and mount a lookout. Put the emergency transmitter to use. Do not forget the earth wire which must be put into the sea.
*Read survival literature in the liferaft survival pack.
Your third priority is **WATER**
You should institute strict rationing at once.
Normally drink nothing for the first 24 hours, then issue not less than one pint or ½ litre per person per day (more in the tropics) and be prepared to catch any rain water *(wash salt off catchment area with rain, first).* **Do not drink any sea water, birds blood, urine or alcohol.** (Note injured persons need more water.)
Finally your fourth priority is **FOOD.**
You don't need food for long periods so, unless water is plentiful, only eat carbohydrates such as sweets, glucose, etc.
Do not eat protein, it will dehydrate you. Do not worry about bowel motion *(Once at the outset is sufficient).*
Cigarettes will also dehydrate you so no smoking is best.

IN COLD CLIMATES.
*Protect face, ears and hands from frostbite, keep each other warm and watch one another for tell tale white patches; warm affected parts with palm of a warm hand and cover it with cotton wool or similar material.

DO NOT MASSAGE FROSTBITE
*Keep feet dry as possible. Move the fingers and toes, move at ankles and knees, clench fists and stretch limbs, wrinkle face and nose, manipulate ears. This keeps the blood circulating.
*Put feet up for at least 5 minutes in every hour (boats).
*Keep weather cover closed except for small opening (about 150mm (6in) diameter) to ensure ventilation.

IN HOT CLIMATES.
*Minimise sweating.
*Protect skin against sunburn.
*Adjust weather cover to provide maximum through draught by day; close it at night except for a small opening for ventilation.
*Keep outside of weather cover wet with sea water throughout daylight the evaporation lowers the inside temperature.
*Deflate floor of raft by day, soak shirt in sea water and put it on wet. Rinse clothes before sundown and squeeze out salt. Clothing and floor of raft should be dry by sundown.
*Do not swim, it wastes energy.
You have survived so far. You can continue to survive if you don't make too many mistakes. **SO READ THIS ADVICE AGAIN.**

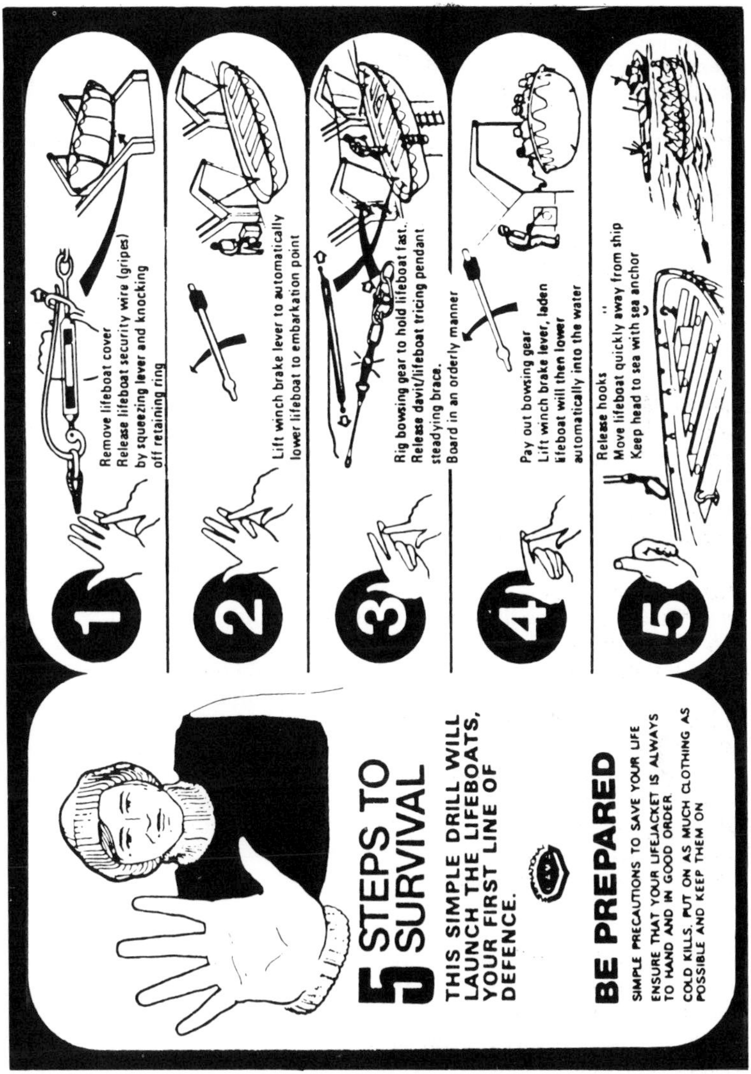

A simplified ships lifeboat launching poster
Issued by:– The International Life-saving Manufacturers' Association
Reproduced from:– 'Safety at Sea International'. March 1976.

IMO. RECOMMENDATIONS ON TRAINING OF SEAFARERS IN PERSONAL SURVIVAL TECHNIQUES

Every prospective seafarer should, before being employed in a sea-going ship, receive approved training in personal survival techniques. In respect of such training, the following recommendations are made.

1. Every prospective seafarer should be instructed in the following:
   a. Types of emergencies which may occur, such as collision, fire and foundering;
   b. types of life-saving appliances normally carried on ships;
   c. need to adhere to the principles of survival;
   d. value of training and drills;
   e. need to be ready for any emergency and to be constantly aware of the information in the muster list, in particular;
      (i) his specific duties in an emergency;
      (ii) his own survival craft station;
      (iii) the signals calling all crew to their survival craft or fire stations;
      location of his own and spare lifejackets;
      location of fire alarm controls;
      means of escape;
      consequences of panic;
   f. actions to be taken when called to survival craft stations, including
      (i) putting on suitable clothing;
      (ii) donning a lifejacket;
      (iii) collecting additional protection such as blankets, time permitting;
   g. actions to be taken when required to abandon ship, such as;
      (i) how to board survival craft from ship and water;
      (ii) how to jump into the sea from a height and reduce the risk of injury when entering the water;
   h. actions to be taken when in the water, such as, how to survive in circumstances of;
      (i) fire or oil on the water;
      (ii) cold conditions;
      (iii) shark infested waters;
      (iv) how to right a capsized survival craft;
   i. actions to be taken when aboard a survival craft, such as;
      (i) getting the survival craft quickly clear of the ship;
      (ii) protection against cold or extreme heat;
      (iii) using a drogue or sea-anchor;
      (iv) keeping a look-out;
      (v) recovering and caring for survivors;
      (vi) facilitating detection by others;
      (vii) checking equipment available for use in the survival craft, and using it correctly;
      (viii) remaining so far as possible in the vicinity;
   j. main dangers to the survivors and the general principles of survival including;
      (i) precautions to be taken in cold climates;

(ii) precautions to be taken in tropical climates;
(iii) exposure to sun, wind, rain and sea;
(iv) importance of wearing suitable clothing;
(v) protective measures in survival craft;
(vi) effects of immersion in water and of hypothermia;
(vii) importance of preserving body fluids;
(viii) protection against seasickness;
(ix) proper use of fresh water and food;
(x) effects of drinking sea water;
(xi) means available for facilitating detection by others;
(xii) importance of maintaining morale.

2. Every prospective seafarer should be given practical instruction in at least the following;
   a. wearing a lifejacket correctly;
   b. entering the water from a height wearing a lifejacket;
   c. swimming while wearing a lifejacket;
   d. keeping afloat without a lifejacket;
   e. boarding liferafts from ship and water while wearing a lifejacket;
   f. assisting others to board a survival craft;
   g. operation of survival craft equipment including basic operation of portable radio equipment;
   h. streaming a drogue or sea anchor.

IMO. MANDATORY MINIMUM REQUIREMENTS FOR THE ISSUE OF CERTIFICATES OF PROFICIENCY IN SURVIVAL CRAFT

Every seafarer to be issued with a certificate of proficiency in survival craft shall:–

a. be not less than 17½ years of age;
b. satisfy his Administration as to medical fitness;
c. have approved sea-going experience of not less than 12 months or have attended an approved training course and have approved sea-going service of not less than nine months;
d. satisfy his Administration by examination or continuous assessment during an approved training course that he possesses knowledge of the contents of the appendix to this Regulation;
e. demonstrate to the satisfaction of his Administration by examination or by continuous assessment during an approved training course that he possesses the ability to:
   (i) don a lifejacket correctly; safely jump from a height into the water; board a survival craft from the water while wearing a lifejacket;
   (ii) right an inverted liferaft while wearing a lifejacket;
   (iii) interpret the markings on survival craft with respect to the number of persons they are permitted to carry;
   (iv) make the correct commands required for launching and boarding the survival craft, clearing the ship and handling and disembarking from the survival craft;
   (v) prepare and launch survival craft safely into the water and clear the ship's side quickly;
   (vi) deal with injured persons both during and after abandonment;
   (vii) row and steer, erect the mast, set the sails, manage a boat under sail and steer a boat by compass;
   (viii) use signalling equipment, including pyrotechnics;
   (ix) use portable radio equipment for survival craft;

APPENDIX

MINIMUM KNOWLEDGE REQUIRED FOR THE ISSUE OF CERTIFICATES OF PROFICIENCY IN SURVIVAL CRAFT.

1. Types of emergency situations which may occur, such as collision, fire and foundering.
2. Principles of survival including:
   a. value of training and drills;
   b. need to be ready for any emergency;
   c. actions to be taken when called to survival craft stations;
   d. actions to be taken when required to abandon ship;
   e. actions to be taken when in the water;
   f. actions to be taken when aboard a survival craft;
   g. main dangers to survivors.
3. Special duties assigned to each crew member as indicated in the list, including differences between the signals calling all crew to survival craft and to fire stations.
4. Types of life-saving appliances normally carried on board ships.

5. Construction and outfit of survival craft and individual items of their equipment.
6. Particular characteristics and facilities of survival craft.
7. Various types of devices used for launching survival craft.
8. Methods of launching a survival craft into a rough sea.
9. Action to be taken after leaving the ship.
10. Handling survival craft in rough weather.
11. Use of painter, sea-anchor and all the other equipment.
12. Apportionment of food and water in survival craft.
13. Methods of helicopter rescue.
14. Use of first-aid kit and resuscitation techniques.
15. Radio devices carried in survival craft, including emergency position-indicating radio beacons (E P I R B's).
16. Effects of hypothermia and its prevention; use of protective covers and protective garments.
17. Methods of starting and operating a survival craft engine and its accessories together with the use of fire extinguishers provided.
18. Use of emergency boats and motor lifeboats for marshalling life-rafts and rescue of survivors and persons in the sea.
19. Beaching a survival craft.

IMO. REQUIREMENTS FOR ALL SHIPS COMMENCING 1st JULY 1986

TRAINING MANUAL.

Every ship shall carry a training manual, which may comprise several volumes, shall contain instructions and information, in easily understood terms illustrated wherever possible, on the life-saving appliances provided in the ship and on the best methods of survival. Any part of such information may be provided in the form of audio visual aids in lieu of the manual. The following shall be explained in detail.

1. donning of lifejackets and immersion suits, as appropriate;
2. muster at the assigned stations;
3. boarding, launching and clearing the survival craft and rescue boats;
4. method of launching from within the survival craft;
5. release from launching appliances;
6. methods and use of devices for protection in launching areas where appropriate;
7. illumination in launching areas;
8. use of all survival equipment;
9. use of all detection equipment;
10. with the assistance of illustrations, the use of radio life-saving appliances;
11. use of drogues;
12. use of engines and accessories;
13. recovery of survival craft and rescue boats including stowage and securing;
14. hazards of exposure and the need for warm clothing;
15. best use of survival craft facilities in order to survive;
16. methods of retrieval, including use of helicopter rescue gear, (slings, baskets, stretchers), breeches-buoy and shore life-saving apparatus and ship's line throwing apparatus;
17. all other functions contained in the muster list and emergency instructions;
18. instructions for emergency repair of the life-saving appliances.

Instructions for on-board maintenance of life-saving appliances shall be easily understood, illustrated wherever possible, and, as appropriate, shall include the following for each appliance:

1. a check list for use when carrying out the inspections required by these Regulations;
2. maintenance and repair instructions;
3. schedule of periodic maintenance;
4. diagram of lubrication points with the recommended lubricants;
5. list of replaceable parts;
6. list of sources of spare parts;
7. log for records of inspection and maintenance.

IMO. MUSTER LIST AND EMERGENCY INSTRUCTIONS

1. **This Regulation applies to all ships. Commencing 1st July 1986.**
2. Clear instructions to be followed in the event of an emergency shall be provided for every person on board.
3. Muster lists complying with the requirements set out below shall be exhibited in conspicuous places throughout the ship including the navigating bridge, engine room and crew accommodation spaces.
4. Illustrations and instructions in appropriate languages shall be posted in passenger cabins and conspicuously displayed at muster stations and other passengers spaces to inform passengers of;
   (i) their muster station;
   (ii) the essential actions they must take in an emergency;
   (iii) the method of donning lifejackets.
5. The muster list shall specify details of the general emergency alarm signal prescribed by these Regulations, and also actions to be taken by crew and passengers when this alarm is sounded. The muster list shall also specify how the order to abandon ship will be given.
6. The muster list shall show the duties assigned to the different members of the crew including the:
   (i) closing of the watertight doors, fire doors, valves, scuppers, sidescuttles, skylights, portholes and other similar openings in the ship;
   (ii) equipping of the survival craft and other life-saving appliances;
   (iii) preparation and launching of survival craft;
   (iv) general preparation of other life-saving appliances;
   (v) muster of passengers;
   (vi) use of communication equipment;
   (vii) manning of fire parties assigned to deal with fires;
   (viii) special duties assigned in respect to the use of fire-fighting equipment and installations.
7. The muster list shall specify which officers are assigned to ensure that life-saving and fire appliances are maintained in good condition and ready for immediate use.
8. The muster list shall specify substitutes for key persons who may become disabled, taking into account that different emergencies may call for different actions.
9. The muster list shall show the duties assigned to members of the crew in relation to passengers in case of emergency. These duties shall include:
   (i) warning the passengers;
   (ii) seeing that they are suitably clad and have donned their lifejackets correctly;
   (iii) assembling passengers at muster stations;
   (iv) keeping order in the passageways and on the stairways and generally controlling the movements of the passengers;
   (v) ensuring that a supply of blankets is taken to the survival craft.
10. The muster list shall be prepared before the ship proceeds to sea. After the muster list has been prepared, if any change takes place in the crew which necessitates an alteration in the muster list, the master shall either revise the list or prepare a new list.
11. The format of the muster list used on the passenger ships shall be approved.

GENERAL EMERGENCY ALARM SYSTEM.

The general emergency alarm system shall be capable of sounding the general emergency alarm signal consisting of **seven or more short blasts followed by one long blast** on the ship's whistle or siren and additionally on an electrically operated bell or klaxon or other equivalent warning system, which shall be powered from the ship's main supply and the emergency source of electrical power required by these Regulations. The system shall be capable of operation from the ship's bridge and except for the ship's whistle, also from other strategic points. The system shall be audible throughout all the accommodation and normal crew working spaces.

ABANDON SHIP TRAINING AND DRILLS.

1. **This Regulation applies to all ships. Commencing 1st July 1986.**
2. A training manual shall be provided in each crew messroom and recreation room or in each crew cabin.
3. **Practice musters and drills.**
   - (i) Each member of the crew shall participate in at least one abandon ship drill and one fire drill every month. The drills of the crew shall take place within 24 hours of the ship leaving port if more than 25% of the crew have not participated in abandon ship and fire drills on board that particular ship in the previous month. The Administration may accept other arrangements that are at least equivalent for those classes of ship for which this is impracticable.
   - (ii) On a ship engaged on an international voyage which is not a short international voyage, musters of the passengers shall take place within 24 hours after their embarkation. Passengers shall be instructed in the use of the lifejackets and the action to take in an emergency. If only a small number of passengers embark at a port after the muster has been held it shall be sufficient, instead of holding another muster, to draw the attention of these passengers to the emergency instructions required by these Regulations.
   - (iii) On a ship engaged on a short international voyage, if a muster of the passengers is not held on departure, the attention of the passengers shall be drawn to the emergency instructions required by these Regulations.
   - (iv) Each abandon ship drill shall include:
     - a. summoning of the passengers and crew to muster stations with the General emergency alarm system, and ensuring that they are made aware of the order to abandon ship specified in the muster list;
     - b. reporting to stations and preparing for the duties described in the muster list;
     - c. checking that the passengers and crew are suitably dressed;
     - d. checking that lifejackets are correctly donned;
     - e. lowering of at least one lifeboat after any necessary preparation for launching;
     - f. starting and operating the lifeboat engine;
     - g. operation of davits used for launching liferafts;

(v) Different lifeboats shall, as far as practicable, be lowered in compliance with the requirements of paragraph 3, (iv), (e) at successive drills.

(vi) Drills shall, as far as practicable, be conducted as if there were an actual emergency, *(and should include the rigging of exposure covers or the covers on partially enclosed lifeboats).*

(vii) Each lifeboat shall be launched with its assigned operating crew aboard and manoeuvred in the water at least once every three months during an abandon ship drill. The Administration may allow ships operating on short international voyages not to launch the lifeboats on one side if their berthing arrangements in port and their trading patterns do not permit launching of lifeboats on that side. However, all such lifeboats shall be lowered at least once every 3 months and launched at least annually.

(viii) As far as is reasonable and practicable, rescue boats other than lifeboats which are also rescue boats, shall be launched each month with their assigned crew aboard and manoeuvred in the water. In all cases this requirement shall be complied with at least once every 3 months.

(ix) If lifeboat and rescue boat launching drills are carried out with the ship making headway, such drills shall, because of the dangers involved, be practiced in sheltered waters only and under the supervision of an officer experienced in such drills.

(x) Emergency lighting for mustering and abandonment shall be tested at each abandon ship drill.

4. **On-board training and instructions.**

(i) On-board training in the use of the ship's life-saving appliances, including survival craft equipment, shall be given as soon as possible but not later than 2 weeks after a crew member joins the ship. However, if the crew member is on a regularly scheduled rotating assignment to the ship, such training shall be given not later than 2 weeks after the time of first joining the ship.

(ii) Instruction in the use of the ship's life-saving appliances and in survival at sea shall be given at the same interval as the drills. Individual instruction may cover different parts of the ship's life-saving system, but all the ship's life-saving equipment and appliances shall be covered within any period of 2 months. Each member of the crew shall be given instructions which shall include but not necessarily be limited to:

a. operation and use of the ship's inflatable liferafts;

b. problems of hypothermia, first-aid treatment for hypothermia and other appropriate first-aid procedures;

c. special instructions necessary for use of the ship's life-saving appliances in severe weather and severe sea conditions.

(iii) On-board training in the use of davit launched liferafts shall take place at intervals of not more than 4 months on every ship fitted with such appliances. Whenever practicable this shall include the inflation and lowering of a liferaft. This liferaft may be a special liferaft intended for training purposes only, which is not part of the ship's lifesaving equipment; such a special liferaft shall be conspicuously marked.

5. **Records.**

The date when musters are held, details of abandon ship drills and fire drills, drills of other life-saving appliances and on-board training shall be recorded in such log book as may be prescribed by the Administration. If a full muster, drill or training session is not held at the appointed time, an entry shall be made in the log book stating the circumstances and the extent of the muster, drill or training session held.

On all passenger ships an abandon ship drill and fire drill shall take place weekly.

NO SHIP IS TO BE ABANDONED, EXCEPT BY ORDER OF THE MASTER.

As apart from the General emergency alarm signal, the master of the ship will designate a special signal for "abandon ship".

There is to be a separate signal for the practice of boat and fire drills; the letter "B" _ . . . is commonly used.

There is to be a special signal for **"Fire stations"** and the rapid ringing of a gong or bell is commonly used.

OPERATIONAL READINESS, MAINTENANCE AND INSPECTIONS.

1. **This Regulation applies to all ships as from 1st July 1986.** The requirements of paragraphs 3 and 6 (ii) shall be complied with, as far as is practicable, on ships constructed before 1st July 1986.
2. **Operational Readiness**

   Before the ship leaves port and at all times during the voyage, all lifesaving appliances shall be in working order and ready for immediate use.
3. **Maintenance**
   (i) Instructions for on-board maintenance of life-saving appliances complying with the requirements of the Training Manual shall be provided and maintenance shall be carried out accordingly.
   (ii) The Administration may accept, in lieu of the instructions in paragraph 3 (i), a shipboard planned maintenance programme which includes the requirements of the Training Manual.
4. **Maintenance of falls**

   Falls, used in launching shall be turned end-for-end at intervals of not more than 30 months, and be renewed when necessary due to deterioration of the falls or at intervals of not more than 5 years, whichever is the earlier.
5. **Spares and repair equipment** shall be provided for life-saving appliances and their components which are subject to excessive wear or consumption and need to be replaced regularly.
6. **Weekly Inspection.**

   The following tests and inspections shall be carried out weekly:
   (i) All survival craft, rescue boats and launching appliances shall be visually inspected, to ensure that they are ready for use;
   (ii) All engines in lifeboats and rescue boats shall be run ahead and astern for a total period of not less than 3 minutes provided the ambient temperature is above the minimum temperature required for starting the engine. In special cases the Admin-

istration may waive this requirement for ships constructed before 1st July 1986;

(iii) The general emergency alarm system shall be tested.

7. **Monthly Inspections.**

Inspection of the life-saving appliances, including lifeboat equipment, shall be carried out monthly using the checklist required by the Training Manual to ensure that they are complete and in good order. A report of the inspection shall be entered in the log book.

8. **Servicing of inflatable liferafts, inflatable lifejackets and inflated rescue boats.**

(i) Every inflatable liferaft and every inflatable lifejacket shall be serviced;

a. at intervals not exceeding 12 months. However, in cases where it appears proper and reasonable, the Administration may extend this period to 17 months;

b. at an approved servicing station which is competent to service them, maintain proper servicing facilities and uses only properly trained personnel.

(ii) All repairs and maintenance of inflated rescue boats shall be carried out in accordance with the manufacturer's instructions. Emergency repairs may be carried out on board the ship, however, permanent repairs shall be effected at an approved servicing station.

9. **Periodic servicing of hydrostatic releases.**

(i) at intervals not exceeding 12 months. However, in cases where it appears proper and reasonable, the Administration may extend this period to 17 months;

(ii) at a servicing station which is competent to service it, maintains proper servicing facilities and uses only properly trained personnel.

Extract from:-

U.K. Statutory Instrument 1986 No. 1071.

Penalties

9.-(1) If the master of a ship fails to perform any of the obligations imposed upon him under the "Abandon ship—Training and Drills" paragraph 5, or if there is any breach of the requirements of Regulations 3(vi) and (vii) the master of the ship shall be guilty of an offence and liable on summary conviction to a fine not exceeding One Thousand Pounds or, on conviction or indicment, to imprisonment for a term not exceeding two years and a fine.

(2) Any person who fails to carry out the duty assigned to him under the Muster List and Emergency Instructions paragraphs 6, 7 and 9 shall be guilty of an offence and liable on summary conviction to a fine not exceeding Five Hundred Pounds.

(3) If the master of a ship fails to comply with any of the requirements of the Abandon Ship Training Drills paragraph 5 he shall be guilty of an offence and liable on summary conviction to a fine not exceeding Fifty Pounds.

(In ships not required to keep an official logbook a record of each matter specified in the Abandon Ship Training Drills paragraph 5 shall be made by the master and shall be retained on board for a period of not less than 12 months).

(4) It shall be a good defence to a charge under this Regulation to prove that the person charged took all reasonable steps to avoid commission of the offence.

PART ONE — SURVIVAL AT SEA

CHAPTER ONE — EMERGENCY STATIONS

On joining a vessel, or as soon thereafter as may be reasonable and practicable, read the muster list. Ascertain the various signals for boat drill, fire drill, **fire** and **abandon ship.** Ascertain to which boat you have been allocated and for what task you are responsible in the event of having to abandon ship, also your fire station and your duties in the event of fire. Locate the stowed position of your boat, your muster station and the place from which the boat will be embarked. Locate the stowed positions of all the liferafts on board and their embarkation positions. Locate the position of all the fire alarm points and the positions of all the portable fire extinguishers, hoses and fire hydrants. Locate the stowage of both your own and the spare lifejackets. Decide where you will keep your warm and waterproof clothing and keep it there when not in use, so that it will always be available if you should need it. Ascertain the shortest route from your cabin to the muster station. If you are responsible for placing stores, an E P I R B or portable radio in your survival craft, ascertain how you can most easily and quickly obtain them. Keep a good strong gas-tight torch with your own warm clothing.

Everyone is confident that they will never have to abandon ship and most people never have to. However, in the event of an unexpected emergency, and unfortunately such emergencies do occur when they are least expected, to the people who least expect them, perhaps in the middle of a very dark night and with bad weather to boot. You will have no time to go searching for warm clothing, your lifejacket, or anything else, you will certainly not have time to read the muster list. It is absolutely essential to know what to do and how to do it, beforehand. Under no circumstances ever attempt to take alcohol of any description in the survival craft. On the other hand a knife, needles and cotton, some nylon string, spare clothing and polythene bags could all be valuable assets. If you smoke, cigarettes or tobacco and a lighter may help soothe both your own and other survivors nerves at the outset. Smoke only at the door of an inflatable liferaft and watch the lighted end of your cigarette. Do not puncture the liferaft with ash or that lighted end. As various different types of portable fire extinguishers are to be found on ships. Read the instructions on the portable fire extinguishers on the vessel you have joined, so that in an emergency, you are able to use one quickly and effectively.

When the signal for "Emergency stations" is sounded, it is a signal for everyone to put on warm clothes and a lifejacket and then go to their emergency station, wearing their immersion suit or taking their thermal aid if they have one. Thermal aids are normally fragile, be very careful not to tear it. Members of the crew will carry out the duty allocated to them in the muster list, then wait for orders.

Make a habit of always taking a long drink of fresh water whenever the signal for boat drill or emergency stations is sounded.

When you put on a lifejacket, always remember to tie the neck tapes (where these are provided) tightly and first, otherwise some designs of lifejacket may allow your head to slip through the hole, you will then hang from the lifejacket by your waist and drown. The loop at the back must be left hanging free, as this is provided for your rescuers to catch hold of you.

A properly adjusted lifejacket will turn you onto your back and keep your mouth and nose out of the water, even if you should lose consciousness. On at least one occasion a survivor, who had not tied his tapes tightly (if at all), slid from his lifejacket and drowned when rescuers attempted to lift him out of the water by means of the loop on the back of his lifejacket.

All float-free life-saving equipment shall be stowed in such a manner that it cannot be obstructed from surfacing by either superstructure or any other means.

Never paint lifeboat or liferaft disengaging hooks or hydrostatic release mechanism. Maintain them as instructed, paint can cause them to fail to function.

When permanently attaching an inflatable liferaft painter to the ship, on no account is any painter to be pulled out of the container. The painter has been treated at its point of entry into the container, to prevent water wicking up the cord. Pulling any of the painter out of the container will remove the treated portion and allow water to wick up the cord into the container. Some containers are provided with a rubber link connecting the painter inside the container, with the cord for securing the liferaft outside the container.

Never stand on any liferaft container after it has been stowed or at any other time. Cases have ocurred where a GRP container has cracked with the weight of a person standing on it, so that water has entered the container and rendered the liferaft completely unserviceable. Never roll a liferaft container along the deck, it could cause the liferaft to shift inside the container.

Lifeboats are numbered from forward to aft. Odd numbers to starboard, even numbers to port.

On the Sounding of the General Emergency Alarm.

Every coxswain and boat's crew will go to their boat where the coxswain will have the following duties carried out:–

1. The cover taken off the boat, folded and placed in the boat.
2. Two men in the boat, one forward to pass out the toggle painter, and one aft to ship the plug and report "Plug shipped".
3. Send the toggle painter as far forward as possible, inboard of the falls, and outside everything else, the slack picked up, and the painter made fast.
4. Have all the crew and passengers and their lifejackets and warm clothing checked.
5. Stand-by for orders.

On ships that have a cover and strongback fitted to the boats, the strong back and cover should be placed on the outboard side-benches, and on a ship fitted with luffing davits and manila rope falls, it will be necessary to set tight the falls.

Particular attention must be paid to the toggle painter. If this is not taken outside the davit and everything else except the falls, and sufficiently far forward, the boat will be hung up on it before reaching the water. On the other hand, if too much slack is left and the ship has any way on her, the boat will fall astern and be towed by the falls, possibly making it impossible to "Let-go Falls", added to which, if there is another boat aft of your own boat's position, the after boat may well be lowered on top of the forward one, causing a major tragedy, or if the boat is lowered from the poop, it may get caught either under the counter by a rising sea, or in the propeller.

The gripes should not be "let-go", or the boat lowered to the embarkation deck, unless instructions have been issued by the Master previously that this is to be done when the signal for emergency stations is sounded.

If it is obvious that the ship is in great danger of sinking, but that there is still a little time in hand before it will become necessary to abandon ship, a coxswain would do well to have some extra gear put in the boat, always remembering that the more extra gear put in the boat, the less room will be available for survivors. A suitable collection of extra gear might be summed up as follows:—

Blankets, tinned milk, milk tablets, fruit, biscuits and sweets. Notebook, pencil and a waterproof watch, torches, batteries and bulbs, palm, needles and sail twine, the ship's pyrotechnics if unused, plastic bags and a small pocket radio receiver. To take extra water, fuel and lubricating oil, three quarter fill clean bottle-necked containers, then cork, float and tow them. Take a grapnel and line if one is available (in a boat only).

Remember that the ship herself is always your number one life-saving unit, the ship's damage control and firefighting organisation should be efficient, so as to be able to overcome any emergency, and you will be trained by frequent drills to learn to make the best possible use of the ship's equipment. Many lives have been lost by premature and unnecessary abandonment of ships.

Should you have to abandon ship, **try and keep dry**. If you have to go into the water, **never go in without a lifejacket** and an immersion suit or thermal aid, if you have one. However, warm clothing will trap air and air provides warmth. You cannot swim far in heavy clothing, neither can you swim far in a lifejacket. In any case you cannot swim the Atlantic Ocean. Do not try to swim unnecessarily, it uses vital energy and assists hypothermia to set in. Try and take something buoyant with you into the water to assist you to keep afloat. If you are only scantily clothed, you will certainly die of exposure either in or out of the water. Wet clothing is far better than no clothing. In cold weather in a survival craft, remove and wring out the top layer of wet clothing and put it on again as quickly as possible.

SURVIVAL

Do not rely on a quick rescue, it may not happen.

If possible, launch all the ship's survival craft and rescue boats. Spare survival craft will provide additional stores, some extra sleeping accommodation, help avoid over-crowding in hot weather and provide back up craft that may be used in the case of a further unforeseen emergency. Inflatable liferafts must never be inflated on deck. Inflation must always be overside.

As soon as possible after the survival craft are launched, tie all the survival craft together, to assist rescue, i.e., when one is found all are found.

If you have to go into the water from a survival craft, perhaps to help rescue another survivor. Be sure to take a line with you. A survival craft will drift far faster than you can swim, without a line to help you get back to the survival craft, you may well find yourself unable to get back to safety.

Survivors in the water should hang onto lifeboats, liferafts and buoyant apparatus by putting their arms through the loops of the lifelines, rather than hold onto them, for the hands get numb and let go.

In order to handle a lifeboat more easily, the skates should be unshipped, but do not dump them until the ship herself has sunk. If you are able to return to the ship you will want to recover the boats and having done so, reship the skates.

Tie boats together stem to stern with the full length of the permanent painter between them. Tie liferafts together by means of doubled rescue lines, allowing at least 30 to 50 feet (9 to 15m.) between them. When inflatable liferafts are tied together any closer than this in a seaway they tend to snatch and allow the wind to get beneath them, this can cause them to capsize.

In an inflatable liferaft, in heavy weather with the sea-anchor streamed, sit the majority of the survivors on the same side of the liferaft as that to which the sea-anchor is attached, this will help to prevent the raft capsizing.

If the vessel carried a life-saving Emergency position-indicating radio buoy and this is afloat, moor the survival craft to it, to avoid drift and so aid rescue.

If, having launched a liferaft, you are on the weather side of the ship, your sea-anchor will soon hold you, and the ship will drift away from you, but if you are on the lee side of the ship you may have great difficulty in getting a survival craft away particularly if there is a strong wind blowing. Therefore, it is sometimes better, if conditions permit, to abandon ship on the weather side.

In a liferaft on the lee side or when there is a calm, you will have difficulty in getting away from the ship's side with the paddles, (circular rafts tend to rotate when the paddles are used). So, in addition to paddling, tie a weight (a boot or shoe will do) to the sea-anchor, throw the sea-anchor as far as you can in the direction you wish to go, allow the sea-anchor to sink and then pull the liferaft towards the sea-anchor. Repeat this as many times as may be necessary.

If you are trying to join other liferafts to leeward of you, take in your sea-anchor and drift down to them, while they keep their sea-anchors streamed. Get across wind by using your sea-anchor with a weight tied to it, in the same way as you would to get away from the ship.

Search for possible survivors in the water; at night, use the torch to help. but do not waste the battery. Rescue survivors by picking them up over the weather bow or stern of the boat. The boat would be liable to drift over the top of a survivor on the lee side. Throw the rescue line and quoit from a raft and pull them in. Take injured survivors on board the raft by putting your hands under their armpits and sliding them in on their backs.

In an inflatable liferaft, when the relief valves have stopped blowing off after the initial inflation, fit the plugs provided into the valves, to prevent salt water and dirt getting in. Remember to remove them occasionally during the day in hot weather and whenever the buoyancy tubes are being topped-up. When the liferaft is first inflated and whenever the relief valves blow-off and in the event of a leak occuring in one of the tubes it is essential to ventilate the liferaft.

As soon as possible examine all the equipment and read the instructions on it and also the survival instructions. Anti-seasickness pills should be issued to all hands as soon as may be practicable. The most hardened and experienced seaman is prone to seasickness in a survival craft and the consequent loss of moisture and distress caused by vomiting will reduce vitality.

A first-aid kit is provided to help assist any injured survivors. Read the instructions. Do not try to do too much, warmth and sleep are nature's finest cures and should be encouraged. Massage can often be injurious.

If possible, board the survival craft without entering the water. Do not jump into the water unless it is essential; use an overside ladder or, if necessary, lower yourself by means of a rope or fire hose. Unless it is unavoidable, never jump from higher than 20 feet (6 metres) into the water. If it is necessary to enter the water, choose a suitable place from which to leave the ship, bearing in mind (a) the ship may drift down on you faster than you can swim away. (b) Survival craft may drift far more quickly than you can swim. If there is no survival craft available it may be preferable to abandon ship from the bow or stern or weather side in order to get clear of the ship with more certainty. There may be difficulty in getting clear of the ship's side from amidships on the lee side. (c) Get clear of the ship as quickly as possible, the danger of being struck from below by surfacing wreckage is greater than that from the suction caused by the ship sinking.

Points to bear in mind before jumping overboard:–

a. Have your lifejacket securely tied on and hold it down by crossing the arms over the chest; blocking off the nose and mouth with one hand.
b. Keep your feet together, check that it is all clear below; look straight ahead; jump feet first.
c. Do not look down when jumping as it makes you unstable and likely to fall forward.
d. Wearing a lifejacket and possibly an immersion suit or thermal plastic aid and certainly wearing heavy clothing. It may be easier when swimming to a survival craft, to swim on your back. Do not swim or tread water unnecessarily; It wastes valuable body heat and energy.

In a survival craft wearing clothes that are wet with salt water, you will be much more prone to the debilitating effects of salt water boils and bed sores, all of which will quickly turn into open wounds. Woollen clothing next to the skin will aggravate the condition. While wool is probably the best type of warm clothing you can wear. A long sleeved cotton shirt and cotton underpants next to the skin will reduce the aggravation. Do not urinate in your clothes, it will quickly cause a most painful nappy rash.

Provided you have been castaway in a temperate zone and encounter a heavy rainstorm and you are not suffering badly from hypothermia. Undress, wash the salt and sweat from your body and the salt out of your clothing. Having wrung out your clothing, put it on again. Clean wet clothes and a clean body will be far more comfortable than dry salt laden clothes and body heat will dry your clothes. Unfortunately, sitting for long periods in a survival craft is bound to cause the formation of bed sores on the back, buttocks and heels, the presence of salt in your clothes adds to the inflammation and increases the resulting pain.

Never jump into a lifeboat, abandoning ship is hardly the right time to choose to deliberately break an arm or a leg or both.

If an inflatable liferaft has a fabric valise and you are able to recover it, do so. It will help to provide warmth for a scantily clad person or provide a makeshift stretcher for an injured survivor.

Sit in a liferaft with your feet towards the centre, putting your arms through the handlines secured around the inside of the raft to give you some support. Space survivors evenly, except in severe weather when there is a danger of a capsize, in which case the majority should sit on the side to which the sea-anchor is attached.

In an inflatable liferaft have a personal search to make sure that no one is in possession of anything that might puncture the raft. In the event of a puncture use one of the composition plugs as a temporary stopper. Repair with a patch as soon as possible.

An inflatable liferaft when launched should always inflate the right way up, for the weight of the gas bottle turns the container or valise to the correct attitude and the liferaft is packed to emerge the right way up. However, research indicates that an inflatable liferaft is at its greatest risk of capsize immediately on launching. The ballast pockets beneath the raft have not had time to fill. There is no crew aboard to act as ballast and the raft has been inflated hard by the gas and is therefore rigid, it can lift to the sea exposing its underside to the wind which can then capsize the liferaft. The same result can occur if the wind gets beneath the canopy.

The single most likely factor to cause a loaded liferaft to capsize occurs when the sea-anchor fails to operate correctly and snatches the liferaft, causing it to tilt and allow the wind to get beneath the liferaft. The new BMT sea-anchor should correct this. The next most likely factor is when liferafts are tied too closely to each other and again snatching causes the liferaft to lift and allows wind to get beneath it. If a loaded liferaft capsizes, do not panic, there will be ample air space inside the liferaft and it will not hurt or harm you. Climb out and hold on to the grablines until the liferaft is rerighted, then pull yourself round to the entrance and climb back in again. **Never let go of a liferaft once you have hold of it.**

To right a capsized liferaft, **turn the canopy into the wind** and standing on the gas bottle, pull yourself up the righting strap and lie back on it. As the raft comes over, keep clear from underneath. Provided that they use the wind to help them, any one person can right a capsized liferaft easily, although on the larger liferafts it will help if the uppermost water pockets are emptied first. However, it will require two persons to right a 42 man liferaft. Remember that if any survivor lets go of the liferaft, it will drift away from him, faster than he can swim.

As from 1st January, 1984 all new inflatable liferafts must be fitted with BMT sea anchors and new water ballast systems. Existing liferafts are to be modified at the next survey after 1st January, 1985.

In a liferaft, in the tropics, rig the entrance curtains with the paddles to form a wind scoop and stream both sea-anchors (on different lengths of line, to avoid fouling) to increase the draught.

It is impossible to navigate a liferaft, survivors can only choose between taking in the sea-anchor(s), emptying the water pockets and allowing the raft to dirft and reducing the drift by filling the water pockets and streaming both sea-anchors (on different lengths of line).

In the event of having to abandon ship under conditions where the chances of rescue are remote and if there is a lifeboat on the ship fitted with a mast and sails, the survivors might well be advised to take that boat and tow the liferaft. Given a fair wind, there is no limit to the distance it is possible to travel in a lifeboat. Voyages lasting 28 days and more have been made in lifeboats. They have also withstood the most severe weather conditions. On the other hand, more people lose their lives from exposure, than from any other single cause. The protection of survivors from exposure, especially under arctic conditions, in a liferaft, is far superior to the protection provided by an open lifeboat. Again should survivors in an inflatable liferaft be washed up on a rocky shore, they will have a fair chance of survival, whereas a boat would be certain to capsize and break up, giving very little opportunity for survival to the occupants.

In the event of a sudden calamity it takes only seconds to launch a liferaft against the ten minutes or more required to launch a lifeboat and liferafts fitted with hydrostatic release gear will in fact launch themselves, if the ship sinks before they can be launched by the crew. Liferafts fitted with electrical release can be launched from a remote position by the push of a button.

When abandoning ship by means of an inflatable liferaft, it is well to remember that when these liferafts were first approved, they were approved for use on fishing vessels only. These were vessels with a very low freeboard and crews were encouraged to jump onto the rafts in an emergency, for the jump would only be of a few feet. However, the rafts were so successful that they were soon approved for use on all ships, ships in fact with a very high freeboard. On one occasion during a practice drill with an inflated liferaft on a large high sided cargo ship, an apprentice was instructed to jump from the boat deck onto the canopy of the floating liferaft. The result was that the apprentice suffered very serious back injuries. On another occasion when abandoning a high sided vessel, a heavy crew member jumped onto the canopy; he tore the canopy rendering it useless, and severely injured a survivor who was beneath it. Panic jumping either into the entrance or onto the canopy can cause severe injury, possibly loss of life and damage to the liferaft. **Never jump from a height of more than about 2 metres (6ft. 6in.) and then only into an empty entrance, never onto the canopy.**
Discard you footwear before jumping.

It is essential, when abandoning ship in a liferaft, to cut the painter. Although if the painter remains uncut, the probability is that either the painter will break or the painter patch pull off. Before this can occur, the liferaft will have been capsized and almost totally submerged if the ship sinks with the liferaft still attached by the painter. When the "Nesam" was lost 9 survivors were in the liferaft, and with the ship rapidly sinking, they were unable to locate the knife with which to cut the painter. Two survivors jumped into the sea and were lost. Three more were thrown into the sea when the painter parted. Of these, one was lost and the other two were 4 hours in the water before being rescued. Apart from the knife located at one entrance, on a raft with a central pillar remember that a second knife is usually located in a pocket on the central pillar.

When an inflatable liferaft is cast overboard, it is essential that it is inflated by means of the painter, from the ship's deck. It is almost an impossibility for a survivor in the water to inflate a floating unopened liferaft.

Warm clothing too, is essential. On one occasion of shipwreck in Icelandic waters, three men, the mate and two-engine room hands, abandoned ship in an inflatable liferaft. The mate was wearing warm clothing (having just been employed on deck) the engine-room hands were in singlet and trousers. The engine-room hands died of hypothermia. The mate survived and after three days, drifted ashore and was still able to walk two miles to the nearest habitation.

The importance of keeping the interior of the liferaft dry cannot be overstressed. Any water in the liferaft should be bailed out either with the bailer or by using the pump or bellows in reverse. The floor being dried by means of the sponges provided. Some liferafts are provided with an integral self-bailer.

A double floor which can be inflated by survivors, by means of the pump or bellows, for insulation against cold from the sea, should never be pumped up hard, for it has no relief valve, and damage to the interior air space may result. A soft cushion inflation is sufficient. Deflation keys or valves are provided but once inflated, it is impossible to completely deflate the floor. In cold weather, after ventilation, close the door curtains and inflate the floor. Body temperature will soon raise the temperature of the interior of the liferaft to a comfortable level.

Although the carbon-dioxide used to inflate a liferaft is non-toxic, it can cause asphyxia if the raft is not ventilated properly every half hour and whenever there is an escape of gas into the liferaft, i.e., when the safety valves are blowing, immediately after inflation, on the expansion of the gas due to a hot sunny day or when a leak occurs. Lives have been lost when survivors have failed to ventilate the liferaft thoroughly before closing the entrances.

A towing patch is fitted to every liferaft and this patch must always be used when tying two or more rafts together or when being towed. The smaller patches may pull off.

On some liferafts a transparent panel is let into either the canopy or a door curtain of the liferaft, so that it is possible to keep a look-out with the door curtain closed. However, owing to salt spray and condensation it is not always completely successful. On some other liferafts an observation porthole is provided in the canopy through which a lookout may thrust his head.

Nothing, except the skates is ever to be thrown away from a lifeboat. In an inflatable liferaft, jettison all knives, except the safety knives provided, brooches and anything else that could puncture the liferaft. Footwear that could damage the liferaft by friction on the floor or lower buoyancy tube, should also be discarded.

In all survival craft endeavour to get well clear of a vessel before she sinks. Flotsam will surface with considerable force and may well damage the survival craft beyond repair. After the ship has sunk, collect any useful flotsam that you can get hold of.

An emergency position-indicating radio beacon (EPIRB) should be energised as soon as possible after abandonment. Its function is to alert passing aircraft. Owing to the curvature of the earth it is unlikely to be heard by a ship any distance away and it will not be using a wavelength likely to be monitored by shipping.

After their yacht had been sunk by a whale, the Bailey family survived

118 days afloat in an Avon inflatable liferaft.

Keep the survival craft bailed out and dry, wet feet and legs can be most painful. In bad cases of immersion foot it sometimes becomes necessary to amputate a limb after rescue. To bail out a lifeboat, bring all the survivors to one end of the boat, this will lift the other end and make it easier to bail the boat out.

To conserve both energy and moisture in the body and prevent urine retention, everyone should be encouraged to urinate and have a bowel motion within the first few hours. There should be no necessity for any subsequent bowel motions. Do not worry about constipation. A plastic bag may be used for urinating.

In every survival craft keep **all the gear lashed** and water containers tightly bunged all the time, so that in the event of an unexpected capsize. nothing is lost

In the tropics, do not go swimming. You waste valuable energy and swallow sea water. Wet your clothes by day to help reduce body temperature and perspiration, but dry out before sun-down as the nights are chilly. Keep in the shade of the boat cover or canopy to avoid sunburn. Keep the cover or canopy wetted with sea water to help reduce the interior temperature. In an inflatable liferaft, leave the floor uninflated to obtain the coolness of the water through the floor.

If a radar reflector has not been supplied, keep the canopy or cover wetted to aid radar reflection whenever you think rescuers are in the vicinity. In a boat, a wet cover or sail will give better radar reflection than a bucket upside down on the masthead.

Remember the aids you have to attract attention:–

1. All hands to shout (bang buckets etc., in a boat)
2. Pyrotechnics.
3. Daylight signalling mirror.
4. Whistle
5. Torch
6. Highly coloured cover or canopy.

and in addition in an open lifeboat:–

7. Oil lamp dipped in and out of a bucket to signal S.O.S.
8. Orange sails

Remember the morse signals:–

S.O.S. . . . – – – . . . **HELP**

U . . – You are standing into danger.

The spoken word "MAYDAY" on the radiotelephone.

Never waste your pyrotechnics, oil or torch batteries. Pyrotechnics should only be let off with the permission of the coxswain and when there is a reasonable chance of their being seen. Use the whistles and shout in thick weather, when help is thought to be within one mile.

In a lifeboat have both axes passed aft and if a survivor should lose his reason and become violent, dump the axes.

As soon as everything in the survival craft has settled down and any injured survivors have been made as comfortable as possible, anti-seasickness tablets issued, an E.P.I.R.B. energised, the equipment checked and stowed and the survival craft has been bailed out and dried and in a **lifeboat**, the covers erected. In order to maintain both discipline and **morale**, the coxswain should organise all the survivors in watches, giving

B

everyone a job to do, no matter how small. Two hourly watches or even less in very cold weather, would seem to be the best. A constant look-out should be kept on all four points of the compass 24 hours a day. Someone trustworthy and reliable should be placed in charge of the water and rations. In an inflatable liferaft the buoyancy tubes must be closely watched for any slight leaks and repaired. Make someone responsible for keeping the buoyancy tubes topped-up. In all survival craft, survivors should be made responsible for bailing out, catching rain water, catching birds and fish, organising a little physical exercise, tending the sea-anchor and so on. In a boat, helmsmen will need to be appointed, the boat cover tended, sleeping quarters organised and when sailing, the sheets will need to be tended. Also operators for the emergency radio if aboard.

For exercise:— Move fingers, toes, ankles, wrists and knees, clench fists, stretch limbs, shrug shoulders, wrinkle face and nose, manipulate the ears, if possible, lie on the back and do cycling exercises with the legs for short periods daily. Do deep breathing exercises. All this helps to keep the blood circulating.

Catching rain water:— The catchment area must be washed free of all salt with the rain, before collecting rain water. Taste all water before storing, to insure that it is salt free. Condensation may also prove to be a useful source of fresh water. In an inflatable liferaft, try and keep one sponge salt free, in order that it can be used to mop up and store condensation. Use the plastic bags which have contained items of equipment, for additional water storage. Some liferafts supply sealing lids for water tins.

To catch fish:— Use the tinfoil from the food rations or brightly coloured cloth (red is best) for bait. Fishing by night with a torch can be effective. Generally speaking, a fish with scales is safe to eat. Fish that blow up and the points of spikes and spines and the teeth of all fish are poisonous. Be very careful not to allow yourself to be pricked or bitten. Additional water must be available if fish are to be eaten.

To keep up morale:— Community singing, yarn spinning and word games are all helpful. A small pocket radio receiver can help.

To provide sleeping quarters in a lifeboat:— Lift the bottom boards and lay them fore and aft on the thwarts. Pad with spare clothing and life-jackets.

To right a capsized lifeboat:— Hang onto the loops of the keel grablines and rock the boat, gaining a little each time. Three men have succeeded in righting a lifeboat by this means.

Another method in a heavy sea is to lie across the keel, with the hands grasping the weather gunwale. As the boat lifts on a sea, pull together as the boat reaches the crest of a wave, it may be possible to have enough weight on the lee side, which together with the roll, will right her.

Do not issue any food or water for the first 24 hours, except to a person who has either bled or vomited considerably, as there is ample moisture in the body. After 24 hours issue one pint (½ litre of water) per person per day, two ounces at a time at intervals, (20 ounces in all). Do not go below this ration while you have water (it may rain tomorrow). Provided every effort is made to preserve the moisture in the body by avoiding perspiration, it is possible to survive up to 14 days or more without water. Survival craft rations consist of non-thirst provoking foods; barley sugar and condensed milk will both act to reduce the body's need for water; even them out and eat them with the water ration. Apart from the survival craft rations, eat only when you have sufficient water to allow 2 pints (1 litre) per person per day. This applies particularly to any fish, plankton or birds you may have caught. Do not under any circumstances drink sea water, do not even use it to eke out the fresh water available. Statistics show that on voyages in survival craft lasting 3 days or more, 40% of the deaths are due to the drinking of sea water. The best way to alleviate thirst is to suck a button or similar object.

To catch plankton:— Provided sufficient fresh water to provide a ration of at least 2 pints (1 litre) per person, per day, is available, nourishment sufficient to sustain life can normally be obtained from the sea in the form of plankton. Plankton is to be found in large areas of the oceans, and is not normally visible to the naked eye. To obtain plankton trail a fine mesh net. A suitable net could be made from a shirt sleeve, open at one end and tied at the other. In a plankton area, when it is hauled in, a sludge will be found in the end of the net. Sort out and discard the small shell fish and eggs, from the jelly and vegetable matter. The vegetable matter will provide nourishment and tastes somewhat like fishpaste, the jelly is best returned to the sea. If at first no plankton is found in the net, do not be discouraged for you will probably soon drift into an area where it is in plentiful supply.

Apart from rain and condensation, water may be obtained in Arctic waters from either fresh water ice or old salt water ice. Old salt water ice contains virtually no salt and can be distinguished by its blueness and splintery nature. Fresh salt water ice is totally unsuitable.

Do not throw spoilt provisions, offal from birds or fish you have caught or any other food into the sea, unless you are drifting reasonably fast. Foods attracts fish, and apart from the danger of sharks. Schools of porpoises have a tendency to rub their backs on the underside of an inflatable liferaft, with possibly disastrous consequences.

In a motor lifeboat always endeavour to keep sufficient fuel for a few hours in hand, to meet a possible vital emergency such as drifting onto an inhospitable lee shore, or making a weather shore. Beach in surf in the same way as in an ordinary lifeboat, keeping the engine running slow ahead.

If you find yourself drifting ashore in an inflatable liferaft, stream both sea-anchors (on different lengths of line) and pump up the floor. Secure all gear especially lines and open entrance doors to avoid being trapped. The raft will have a tendency to bounce off rocks and wash over the top of them. In either breakers or surf you will have a far better chance of survival in the liferaft, than you would out of it. When you are washed ashore, take the raft's rescue line with you. Do not lose the raft.

Should you be stranded on a deserted beach, haul your raft to an exposed position above the high water mark, peg it down with stones and use it as a tent. It will be highly visible to searching aircraft. Safe but somewhat brackish water can normally be obtained by digging shallow holes a little above the high water mark.

When survivors from a survival craft are rescued, the survival craft should be either taken aboard the rescuing vessel or sunk. For if left afloat, it will cause undue alarm each time it is sighted. If an inflatable liferaft is taken aboard, it should be kept inflated and ready for any emergency until it can be sent to a service station for repacking.

Some of the survivors who have been adrift in inflatable liferafts in severe weather conditions, have found it more favourable to allow the liferaft to drift without a sea-anchor. Others preferred to use the sea-anchor. Whether or not to stream a sea-anchor from an inflatable liferaft in heavy weather would seem to depend on the make of raft, the sighting of the sea-anchor hawser patch on the raft and the type of sea-anchor provided. Different types of severe weather may also be a factor. Nevertheless, a raft will drift considerably faster and further, without the restraint of a sea-anchor, which will make it harder for the rescurers to locate.

Pyrotechnics.

Always fire pyrotechnics from the lee side of the survival craft.

Rockets are for use at night and on a clear night they may be seen at a distance of up to 30 miles. Rockets tend to climb into the wind, so always fire a rocket a little to leeward, to enable it to gain maximum height.

Flares are for use by day or night and are used to pinpoint your position when help is near at hand, they may be seen at distances of up to 5 miles. They are better than smoke floats for attracting the attention of surface vessels. If you do not have tight cuffs on your wrists, roll up your sleeve. A small piece of burning composition falling into your sleeve will result in a very severe burn. If you have a glove, wear it.

Smoke Floats are to attract the attention of aircraft by day. However, a flare is more easily seen in rough weather.

Always read the instructions before attempting to ignite a pyrotechnic.

Keep rockets and flares in their canister with the lid screwed tight.

It is no use igniting a smoke float to attract an aircraft you cannot see, even though you may be able to hear it.

Inflatable liferafts may be constructed in either a circular, oval, square or polygon shape. Some have an inflated thwart, on most the canopy is supported by an arch, but in others by a central pillar.

The raft itself is made of coated materials and is required to have an even number of buoyancy chambers, half of which must be capable of supporting all the persons the raft is certified to accommodate. This however, does not mean that a 10 person liferaft will accommodate 20 persons, there is just not the room. If in an emergency, some of the survival craft are unavailable for one reason or another, the remaining survival craft may have no option but to suffer overcrowding. In an inflatable liferaft, in these circumstances, it may help if survivors remove their lifejackets and sit on them, so increasing the space available per person.

During the Falklands invasion, 47 men crowded on to a 20 man inflatable liferaft and were able to get the short distance to the beach in one voyage.

Once in a survival craft, disconnect your lifejacket light if possible and so save the battery for later use if necessary.

When afloat, circular inflatable liferafts have a tendency to spin when not restrained by a sea-anchor.

Lifejackets make good seats if it is safe to take them off, for the base of the spine is the area most at risk to injury, i.e. bed sores.

Liferafts constructed with an arch, seem to lie best to the sea anchor. When the sea anchor is attached to the liferaft in such a way that the arch is across wind. This method appears to reduce the yawning of the raft and risk of capsize in heavy weather. Care should be taken to try and avoid wind getting into and billowing out the canopy as this can cause capsize and may tear the canopy.

THE M. V. LOVAT DISASTER.

On 25th January 1975 off the South-West coast of England, the weather though bad, was not unduly so, when owing to a shift of cargo, the M.V. "Lovat" capsized. Of a crew of 13, eleven were lost. The ensuing Court of inquiry was unable to find any single cause for this loss of life but a number of factors all contributed their share.

The inflatable liferaft, contained in a valise, was of the now condemned rubber coated cotton fabric type, stowed in a box. It had not been serviced for 13 months and a service was therefore just overdue.

Until July 1974, the liferaft had been stowed on the bridge but because it had become very wet in this stowage it was transferred to a box in the funnel casing. Undoubtedly, the wetness had had some effect on the condition of the raft and may have caused the sea cell to discharge. However, the sea cell was lost so the question remains unanswered.

The liferaft was then stowed in the funnel casing where the heat would undoubtedly have had some effect on its condition. It was stowed in a box that was too large and allowed movement of the valise, the resultant friction caused by movement occasioned by the ships motion probably helped cause damage to the liferaft, particularly in the vicinity of the gas bottle. The box had been used to stow anchor lights and other equipment on top of the valise which may have helped cause damage by copper oxidation due to the close proximity of the brass fittings. There had also been considerable damage caused by fungal attack to which this type of raft was prone. There was also a smell of oil but this may have been the result of spreading oil during rescue attempts.

When the liferaft inflated, the lower chamber failed to inflate owing to damaged material allowing the gas to escape, the upper chamber was soft and some of the crew, fearing it was sinking, went into the water, another climbed onto the canopy.

Of the thirteen crew members, only the two who were wearing warm clothing survived the two and a half hour ordeal. The remainder in thin clothing lost their lives.

The ten-man liferaft should easily have supported the thirteen men had it been in good condition, as it was, with the lower chamber useless no light with which to be able to see, unable to close the doors either because of overcrowding or cold hands, unable to open the equipment container owing to cold hands and overcrowding and so unable to either bail out or pump up the liferaft, it soon became flooded and helped to add to the survivors discomfort. Those who lost their lives died from drowning but there is no doubt that their deaths were greatly assisted by hypothermia.

It seems safe to conclude that had the raft been safely stowed in a dry unheated position and in a box that prevented movement and in which other gear could not be stowed and had the survivors been warmly clad, all might well have survived, fungal attack on the liferaft notwithstanding.

CHAPTER 2 ABANDON SHIP

No ship is to be abandoned except by order of the master.

The coxswain will normally be the senior crew member in the survival craft and should use his personality to make it clear that he is in command, that his orders are to be obeyed and that he intends to keep discipline. In order to give confidence to the survivors, he should never hesitate or countermand his own orders. The safety of the survivors must at all times be his prime thought.

GRAVITY DAVITS

To abandon a ship equipped with totally enclosed lifeboats:–

1. Coxswain ensures that the harbour safety pins are out.*
2. Coxswain to open entry door and enter lifeboat.
3. Embark passengers and crew, second-in-command last.
4. The gripes are normally automatically released when the boat is lowered. However, if they are not of the automatic release type, the second-in-command will release them prior to boarding the lifeboat.
5. Second-in-command to ensure that everyone who should be on board, is on board, seated and with safety belt secured.
6. Coxswain to start engine, water spray and air support systems. (Water spray will start when the boat is waterborne).
7. Second-in-command secures the entry door and ensures all vents are closed. Secures his own seat belt and reports to the coxswain "All Secure".
8. On receiving the report "All secure", the coxswain will pull the lowering control wire to release the gripes and lower the boat.
9. As soon as the boat reaches the water, the coxswain will pull the hook release handle and a crew member will slip the painter.**
10 As hooks and painter are released, the coxswain will open wide the throttle and proceed at speed away from the ship's side to a safe distance.

   He will then heave-to and endeavour to rendezvous with all other survival craft and search for any survivors in the water.

   Ventilators should not be opened until the lifeboat is well clear of all oil and fire on the surface of the water and any toxic fumes in the vicinity.

Whenever a risk of breathing toxic fumes is a possibility or oil on the surface of the water is on fire (or likely to catch fire), the coxswain should endeavour to get the lifeboat to windward of the stricken vessel as quickly as possible and before it becomes necessary to open the ventilators.

**The coxswain should also ensure that it is all clear over side and safe to lower the boat.*

***Vessels equipped with totally enclosed lifeboats will normally keep the painters run out and ready for use at all times when the vessel is at sea.*

To abandon a ship equipped with open or partially enclosed lifeboats.
The crew should consist of a coxswain and at least 5 men.

1. Check that the harbour safety pins are out.Two men to be in the boat.
2. Let-go gripes (check that the triggers have fallen).
3. Lower to the embarkation deck.
4. Make fast bowsing-in tackles.

5. Let-go tricing pendants.
6. Embark passengers and crew (women and children first, and all to be seated as low as possible).
7. Ease off and let-go bowsing-in tackles. (In a motor boat start the engine).
8. Ensure that it is all clear below. Lower to the crest of a wave.
9. Unhook falls (after fall first, if no dis-engaging gear). Ship the tiller.
10. Embark the launching party.
11. Spring off and let-go the toggle painter.
12. Proceed under power or row for a quarter of a mile, stream the sea-anchor and wait.

The two men who let-go the gripes should ensure that the triggers fall, and the two men in the boat should clear the gripe wires and lifelines and hold onto the lifelines while the boat is being lowered or get out of the boat and rejoin it at the embarkation deck. The winchman should lower the boat on the brake until the tricing pendants have brought the boat close alongside. The full weight of the boat should not be allowed to come on the tricing pendants. The two men in the boat, hook the bowsing-in tackles onto the floating blocks, and onto the ring bolts provided on the ship's side, making sure that the blocks with the hauling parts coming from them are the ones that are hooked onto the ship's side. They then haul tight the bowsing-in tackles and make them fast in the boat. Tricing pendants must be released before persons are allowed to embark, in order to ensure that the full weight of the boat is taken by the falls. If the link on the senhouse slip is difficult to dislodge, use the back of an axe to knock it off.

Passengers, crew and coxswain now board the boat, women and children first, all being seated as low as possible, **with the hands and elbows off the gunwale**. Having first ensured that the water overside is clear and that it is safe to lower the boat, the coxswain will order the bowsing-in tackles to be eased up and let-go, the winchman will lower the boat. The boat should be landed on the crest of a wave and allowed to drop into the trough while afloat, this will overhaul the falls so that when the boat rises to the next crest the coxswain will order "Let-go falls." Care must be taken, if dis-engaging gear is not fitted, that the falls are unhooked together. Otherwise if the ship has any way at all on her, it is better to let-go the after fall first, rather than risk the boat being towed by the after fall.

In a flat calm, let the boat go the last ten feet (3m) with a run, when she should plunge and overhaul the falls sufficiently for them to be un-hooked.

When unhooking the bowsing-in tackles, unhook them from in the boat, and leave them hanging on the ship's side. Never hook the bowsing-in tackle onto the lifeboat fall, and never have it hooked onto the fall while the boat is being lowered, for cases have ocurred where the bowsing-in tackle has been hooked onto the running part of the fall, cut through the fall as the boat was being lowered, and allowed one end of a loaded boat to fall with tragic consequences. If it is considered necessary to bowse the falls into the ship's side while the boat is being lowered, perhaps because the ship is rolling, this should be done by passing a heavy fibre rope (the lifelines will do, if no other heavy ropes are handy) several times round the falls and a stanchion, keeping the turns as low as possible, hauling tight and making fast before letting-go the bowsing-in tackles. As the boat is being lowered the falls will run through the rope which will keep them close to the ship's side, and if anything should carry away, it will be the bowsing-in rope and not the fall.

As soon as the falls are unhooked, the tiller is to be shipped.

When the boat has been lowered into the water, care must be taken that the blocks of the falls do not injure anyone as they swing about after they have been let-go, heaving lines made fast to them would allow the winchman to raise them, or if the ship still has emergency power on the winch, the winchman could hoist the blocks out of harms way as soon as they have both been released. (Do not do this on boat drills, you will be unable to get the falls back). Alternatively, the boat could be pulled a little way forward by means of the toggle painter, while waiting for the winchman to come down a ladder. As soon as the winchman and any other persons who may have been left on the ship, have boarded the lifeboat, the coxswain will put the tiller towards the ship's side to spring off the boat; as soon as it is clear have the toggle painter let go. If there is no way on the ship, it will be best if the toggle painter is passed aft, to be hauled upon, while the bowman bears off with the boathook. Once clear of the ship, proceed under power or man the oars and row for a quarter of a mile, either stream the sea-anchor and wait or help round up the other survival craft and pick up survivors from the water.

When a ship has headway, boats should clear one at a time from aft.

LUFFING DAVITS

To abandon a ship equipped with luffing davits.

The crew should consist of a coxswain and at least 7 men.

1. Let-go gripes and turn down chocks. Two men to be in the boat.
2. Turn out the davits.
3. Lower to the embarkation deck.
4. Turn in the davits and make fast the bowsing-in tackles.
5. Embark passengers and crew (women and children first and all to be seated as low as possible).
6. Ease off and let-go bowsing-in tackles and turn out the davits.
7. Ensure that it is all clear below. Lower to the crest of a wave.
8. Unhook the falls (after fall first) and ship the tiller.
9. Embark the launching party. Spring off with the toggle painter.
10. Row for a quarter of a mile, stream the anchor and wait.

Ensure that all hands and arms are kept off the gunwales.

The two men in the boat should clear the lifelines and hold onto the lifelines while the boat is being turned out and lowered. If the rudder has had to be unshipped in order to stow the boat, it should be shipped by the man in the stern of the boat as soon as the boat has been turned out.

There are no tricing pendants and if the boat is not alongside when it is lowered to the embarkation deck, turning in the davits will bring it alongside. The boat should be held steady for the embarkation of persons by either bowsing-in tackles or by having a bowsing-in rope taken round each of the falls.

After embarkation of the survivors, the bowsing-in tackles must be let-go and the davits turned fully out again. Boats must not be lowered with the davits only partially turned out. The procedure is now exactly the same as for gravity davits.

If the boats are equipped with manila rope falls, two men are to attend each fall (a second man to 'back-up' the first), and a reliable person should remain to give directions to the lowerer's, as the boat is lowered to the water in order to keep it on an even keel. Orders for lowering would be "Turns for lowering", "Start the falls", and "Lower away". If the order "Still" is given, everyone will stop and keep silent, the lowerer's taking an extra turn of the fall on the bitts, until the fault is corrected and the order to continue lowering the boat is given. Manila boat falls must be set tight before the davits are turned out.

RADIAL DAVITS

To abandon ship equipped with radial davits.

Radial davits are now obsolete but may still be found on a very old vessel. They will be equipped with manila rope falls.
The crew should consist of a coxswain and at least 9 men.

1. Set tight the falls. Let-go gripes and turn down chocks. Two men to be in the boat.
2. Haul away on the after guy. Slack away on the forward guy. Launch the boat aft.
3. Haul away on the forward guy. Slack away on the after guy. Launch the boat forward and outboard.
4. Haul away on the after guy. Slack away on the forward guy to centre the boat. Ship the rudder.
5. Lower to the embarkation deck.
6. Haul away on the forward guy. Slack away on the after guy to bring the boat alongside. Make fast bowsing-in tackles.
7. Embark passengers and crew. Hands and elbows kept off the gunwale
8. Let-go bowsing-in tackles. Haul away on the after guy. Slack away on the forward guy to centre the boat. Ensure all is clear below.
9. Lower to the crest of a wave.
10. Unhook the falls (after fall first) and embark the launching party.
11. Spring-off with the toggle painter, row for a quarter of a mile. Stream the sea-anchor and wait. Pick up survivors and join other survival craft.
    Boats may be launched outboard either bow first or stern first, which ever may be the most suitable.

When launching or recovering a lifeboat that is not equipped with dis-engaging gear. Hold the floating blocks by the cheeks. Keep your fingers out of hooks, links, falls and sheaves.

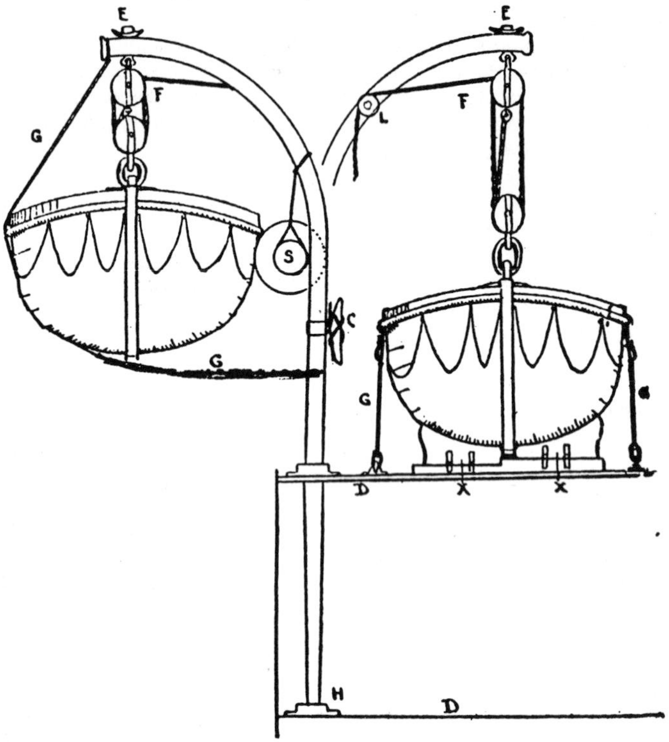

The lifeboat on the right shows how a lifeboat is stowed under radial davits while the boat on the left shows how it is sometimes stowed at sea griped against a griping spar and turned-out ready for instant use.

C. Cleat for making fast the boat's fall.
D. Deck.
E. Davit head eye-plate to which the guy and span are shackled.
F. The boats fall.
G. Gripes fitted with senhouse slips.
H. Socket for the heel of the davit.
S. Griping spar.
X. Chocks. (Note: step on the outboard side which enables the boat to be swung out without lifting, after the chocks have been turned down.

SINGLE ARM DAVITS

To abandon a ship equipped with a single arm davit.
The crew should consist of a coxswain and at least 4 men.
Not more than two men are to be in the boat while it is being swung outboard and lowered into the water.
Two men wll be required to turn the davit against an adverse list.

1. Let-go gripes. Set tight the fall or hoist the boat if required.
2. Turn down chocks and remove portable rail if necessary.
3. Turn the davit outboard, ensure that it is all clear below and lower the boat to the crest of a wave.
4. Unhook the sling and allow survivors to board by ladder.
5. Spring off with the toggle painter, row for a quarter of a mile, stream the sea-anchor and wait.

At the time of going to press, it is understood that single arm davits are being developed from which it will be possible to lower a fully loaded lifeboat or rescue boat by lowering on the fall, the lowering being controlled from within the boat.

FREE–FALL

If a totally enclosed lifeboat is to be launched by "Free-Fall" it is essential that before launching, the coxswain ensures that all doors, ports and ventilators are properly closed and that all the survivors are seated with their safety belts correctly and tightly adjusted. He should ensure that it is all clear below and give ample warning that he is about to let the boat fall and allow sufficient time for an unready survivor to object, before actually launching the boat.

Boats, however, should never be launched by free-fall from davits except in dire emergency such as when an explosion is imminent.

On at least one occasion, when a North Sea oil-rig was being evacuated by means of a Whittaker Capsule adapted for free-fall, the hooks were released prior to all the survivors being seated and belted-in. The end result was that some of the survivors lost their lives many others were gravely injured.

If an inflated rescue boat or D o T inflatable boat is to be launched by free-fall from a tilt launcher. The painter should first be well secured. The lashings are then released and the boat launched. The rescue crew or survivors should embark after the boat has been launched and is alongside.

LIFERAFTS

To abandon ship by means of a rigid liferaft.

1. Let-go straps and remove the cover.
2. Check that the painter is well secured to a strong point.
3. Make sure that it is all-clear overside and launch the liferaft.
4. Pull the remainder of the painter out of the liferaft and tug it hard. The canopies will erect themselves.
5. Board the liferaft from a ladder, a rope or from the sea.
6. When everyone is aboard, cut the painter with the knife provided.
7. Endeavour to remain in the vicinity with the other rafts and boats. Stream the sea-anchor.

Rigid liferafts may be launched from a single arm davit fully loaded by suspending the raft from the hook in strops.

To abandon ship by means of a throw-overboard inflatable liferaft.

1. Check that the painter is well secured to a strong point.
2. Check that all is clear overside. Let-go lashings and take the raft to the ship's side, remove a portable rail if necessary.
3. Launch the liferaft overboard in its container.
4. Pull the remainder of the painter out of the floating container and tug it hard to fire the gas bottle.
5. The liferaft will take 20 to 30 seconds to inflate.
6. Board the liferaft from a side ladder, a rope or from the sea.
7. Jettison all shoes and sharp objects (knives, diamond rings, brooches, etc.)
8. When everyone is aboard, cut the painter with the safety knife.
9. In cold weather inflate the double floor.
10. Ventilate the raft well before closing entrances. Stream the sea-anchor
11. Endeavour to remain in the vicinity with the other rafts and boats.

Should the liferaft fail to inflate or only half inflate, give the painter another hard tug.

Should the liferaft not be close alongside, provided that the entrance is clear and that you are not wearing heavy footwear, provided also that the raft is not more than 6 feet (2m) away from you, jump form the ladder into the liferaft entrance. **On no account jump into the liferaft from a height of more than 6 feet (2m) and do not jump onto the canopy or into the entrance while another person is there.**

Avoid snatching the raft when hauling it alongside with the painter. Coax it into position.

On some liferafts, one sea-anchor will stream itself automatically.

To abandon ship in a davit launched inflatable liferaft.

Unfortunately, the various makes of liferafts vary slightly in their construction, and consequently in their makers' drills for the safe launching of davit launched liferafts. However, it is considered that should a member of the crew be untrained in the particular make of liferaft fitted to his ship, any of the several makes of British liferafts could be safely launched by using the following drill.

1. Remove the portable rail and turn out the davit.
2. Bring a raft in its valise or container, to the ship's side, having the bowsing lines inboard.
3. Break out the bowsing lines and make them fast to the deck cleats provided, leaving some slack. Break out the painter.
4. Hook the davit fall onto the raft.
5. Heave away on the davit fall (to the pre-set mark, if the fall is marked) taking the raft overside.
6. If the liferaft does not inflate automatically, inflate by pulling out the rest of the painter and tugging it hard.
7. When inflation is complete, bowse the liferaft into the ship's side and inspect the interior for any defects.
8. With two men tending the entrance, board the passengers, seating them in the raft feet towards the centre, on alternate sides,

commencing outboard and working inboard, having first removed all footwear, brooches, etc.

9. When the liferaft is loaded, make sure that it is all clear below, release the bowsing lines and **throw both bowsing lines and the painter into the raft,** this is most important.
10. Lower away. The raftsman pulls the red lanyard to release the hook safety catch as the raft reaches the water.
11. The hook will release itself when the raft is waterborne. Get the raft away from the ship's side as quickly as possible and in cold weather inflate the floor. Do not close the entrances before the raft has been well ventilated. In cold weather, inflate the floor and endeavour to remain in the vicinity with the other boats and rafts. Stream the sea-anchor. One man trained in the use of liferafts is to go in each raft. On some liferafts, one sea-anchor will stream itself automatically. On board, the crew will raise the fall, recovering it by means of the tricing line, and continue with the next raft.

Remember to always throw the bowsing lines and painter into the raft before lowering the raft, because if one of these lines should become foul on deck as the liferaft is being lowered, it could cause a nasty tear to the raft and possibly render it useless.

On Dunlop liferafts contained in a valise, the painter is incorporated in the left hand bowsing line which should be pulled out to the mark before being made fast.

Every ship will have an emergency drill suited to the particular needs of that ship, and which may conflict with the drills given in this book. Nevertheless the same actions will need to be carried out in order to launch the lifeboats and liferafts. It is essential to understand what has to be done to safely abandon a sinking ship, in order to avoid any panic and prevent the unnecessary loss of lives.

Single arm davits supplied for the launching of liferafts are required to have a hook attached to the fall that will automatically release the raft when the raft is waterborne, provided that the safety catch has been taken off.

When the hook is hooked onto the liferaft a locking lever is closed by the operator. The hook cannot be closed without at the same time closing the lever and so locking the safety catch on. Once the liferaft has been hoisted, and the weight of the raft is on the hook, a red lanyard attached to the locking lever on the hook may be pulled and the safety catch will be taken off. The hook however, will not become unhooked until the weight of the raft is removed from the hook, as happens when the raft becomes waterborne, as soon as the weight of the raft is removed, the hook disengages automatically. To avoid the possibility of an accident by the weight of the raft becoming supported on the ship's side, particularly if the ship is rolling. **The tripping of the red lanyard should be delayed until the raft is within 3 feet or 1 metre from the water,** for as long as the safety catch is in the "on" position, the hook cannot release the liferaft. These hooks are polished and it is most important that they are not painted.

Once an inflatable liferaft has been inflated, either by design or accident, it is to be kept inflated until it can be sent ashore for servicing and packing.

ATLAS AUTOMATIC RELEASE/ RECOVERY HOOKS

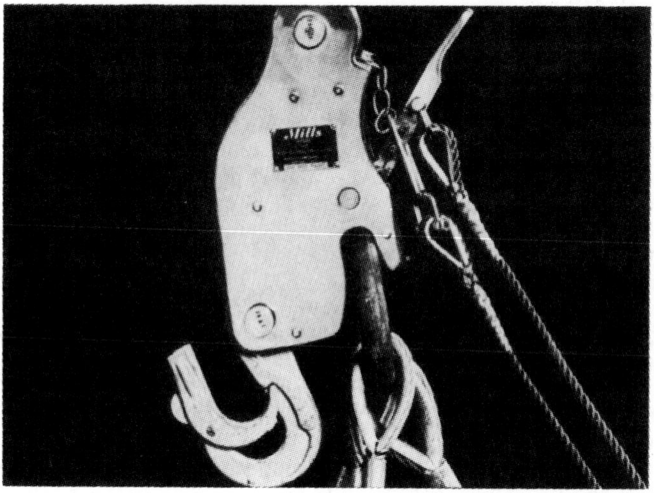

The "Atlas" Automatic Release Hook is for use with liferafts or small boats where a quick, simple and automatic disengagement of the craft from the lowering gear is required. The hook itself is basically of the 'Fail-safe' or off-load release type, which means that it cannot be released whilst laden, but it has an automatic device which releases as soon as the craft is waterborne.

The 'fail-safe' characteristic is achieved by the geometry of the hook, which is so shaped that in order to release the craft before it is waterborne the entire weight of the craft would have to be lifted; this is due to the peculiar curve of the bill of the hook. As there is no force to swivel the hook, it is clearly impossible to effect a release until the craft is settled on the water.

When the craft is waterborne, a small spring automatically opens the hook and so releases the craft.

The launching procedure is as follows:—

1. **The "Atlas" hook is engaged with the craft's lowering link or shackle by lifting the setting (or cocking) lever upwards, so engaging the hook.**

2. **When the raft or boat nears the water, this lever is pulled downwards by means of a red coloured lanyard from within the craft, allowing the automatic hook release spring to operate when the craft is waterborne.**

3. **When this has happened the weight is released from the hook, the small release spring automatically opens the hook and disengages the hook from the craft. If for any reason the hook fails to disengage, another lanyard, coloured green, may be pulled which directly releases the hook. The green lanyard is a secondary safety device which can only be operated by reaching up outside the craft; it does not extend so far downwards into the craft as the red lanyard, and therefore there is no possibility of confusion between the two lanyards, even in the dark. (The green lanyard, like the red one, will not release the hock until the craft is waterborne, even in the extremely unlikely event of its being pulled by mistake.)**

The lack of corrosion means that the hook should always function properly, even after long periods without use.

No maintenance is required, apart from possibly an inspection from time to time to prevent the accumulation of dirt in the mechanism or to check the lanyards.

The added facility of a recovery hook means that the raft may be retrieved easily from the water by just snapping the sling lift ring into the automatic self closing recovery hook attached to the lower part of the hook body.
The raft may therefore be lifted back on board ship and stowed before changing back into the launching mode.

William Mills (Sunderland) Ltd.

THE BEAUFORT LIFERAFT

LAUNCHING AND BOARDING INSTRUCTIONS

How to launch and board Container-stowed, Davit-Type Liferafts.

1. Undo straps.
2. Remove lid.
3. Release base retaining straps. Attach Davit hook to shackle.
4. Hold securing lines and painter. Swing outboard. Tug painter to inflate.
5. Secure alongside. Remove shoes and board.
6. Free liferaft and lower to water.
7. Cock Davit hook for automatic release.

To launch without Davit

1. Secure painter.
2. Throw overboard.
3. Tug and continue tugging.
4. If inverted, correct as shown.

LAUNCHING SEQUENCE
R.F.D. INFLATABLE LIFERAFTS TYPE M.C.

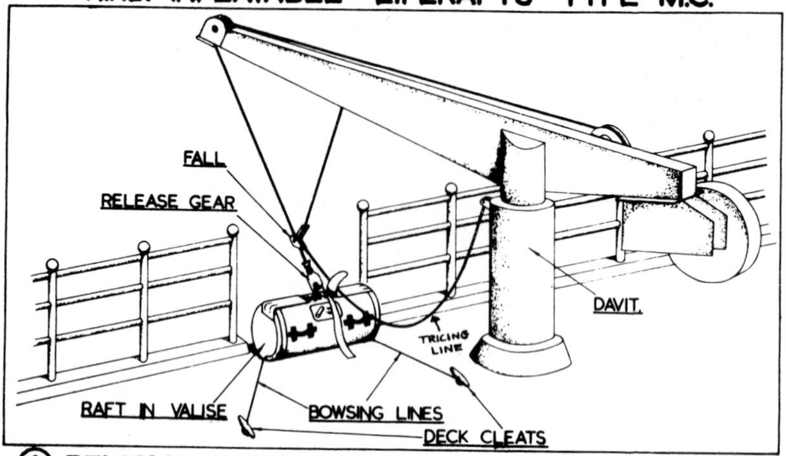

① REMOVE GUARD RAIL, SLEW DAVIT OUTBOARD.
② POSITION RAFT, ATTACH BOWSING LINES, ATTACH RELEASE GEAR.
③ RAISE FALL, INFLATE RAFT.
④ BOARD RAFT.
⑤ UNDO BOWSING LINES, LOWER AWAY.

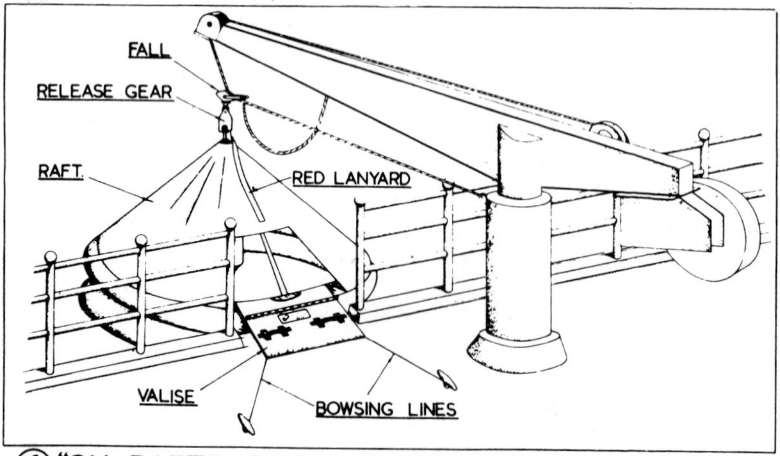

⑥ "ON RAFT." COCK RELEASE BY PULLING RED LANYARD DURING DESCENT.

RFD PT No 6298/258

To abandon a ship equipped with the R.F.D. Marine Evacuation System.

A crew of 9 men is required.

1. No. 1 crewman will operate the outer door release handle and fire the ejection bag, thus deploying the slide and platform.
2. The Escape Slide and boarding platform deploys and inflates.
3. No. 1 crewman now releases the inner door and takes up a position at the safety curtain edge, so as to be in a position to control passenger evacuation.
4. Nos. 2, 3, 4, 6, 7, 8 and 9 crewmen now descend to the boarding platform.
5. No. 5 crewman releases the first two liferaft containers (containing four liferafts) as soon as the boarding platform is deployed.
6. Crewmen on the boarding platform, secure the liferaft containers alongside and deploy the liferafts.
7. No. 1 crewman then despatches the passengers down the slide, one at a time, at one second intervals, having first ensured that all footwear, knives, brooches, etc., have been discarded.
8. From the foot of the escape slide the passengers go directly to the nearest liferaft and board. Each liferaft will take 41 passengers and one crewman making a total of 42 survivors in each liferaft.
9. When the first two liferafts are full, they are cut free and the next two filled.
10. When the first four liferafts have been filled and cleared away from the boarding platform, No. 5 crewman will release the two remaining containers, leave his station and go down the slide to assist on the platform.
11. The procedure is now repeated using the second lot of four liferafts.
12. No. 1 crewman after despatching all the passengers, will leave his station, go down the slide and either join the last liferaft or take charge of the platform if any survivors remain on it.

Four containers, containing eight liferafts are carried in two tiers on a rack at the ship's side. There are two blue release handles, one for each tier of containers which are provided to release the liferafts when the boarding platform is deployed.

Under the most severe weather conditions it may prove necessary to evacuate the ship by using the ladders rather than the slide. Provision has been made for deploying the liferafts independently. Two black release handles are provided to allow the containers to be deployed separately if required. The twin track system is designed to remain operational on one track if the other half is damaged.

The 2 x 42 liferafts are laced together but a slip line can be pulled to separate them if required e.g. for towing.

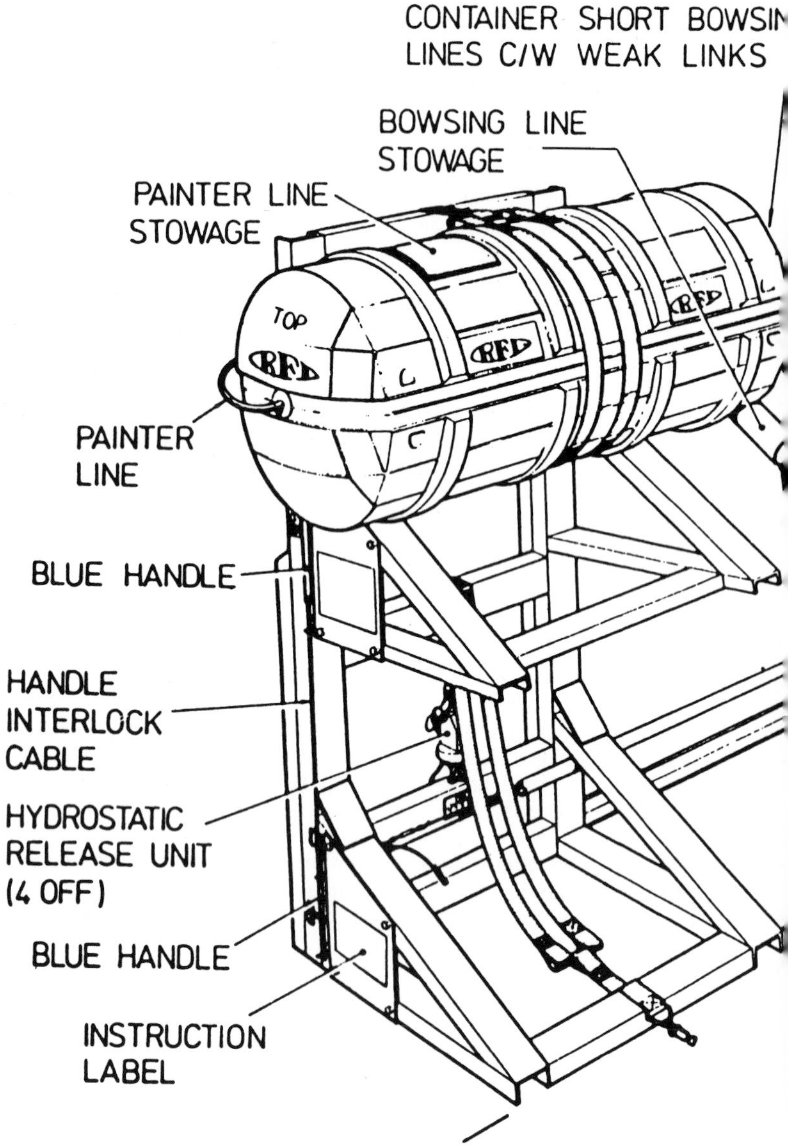

THE RFD. MES SYSTEM –
LIFERAFTS STOWED ON SPECIAL RACKS

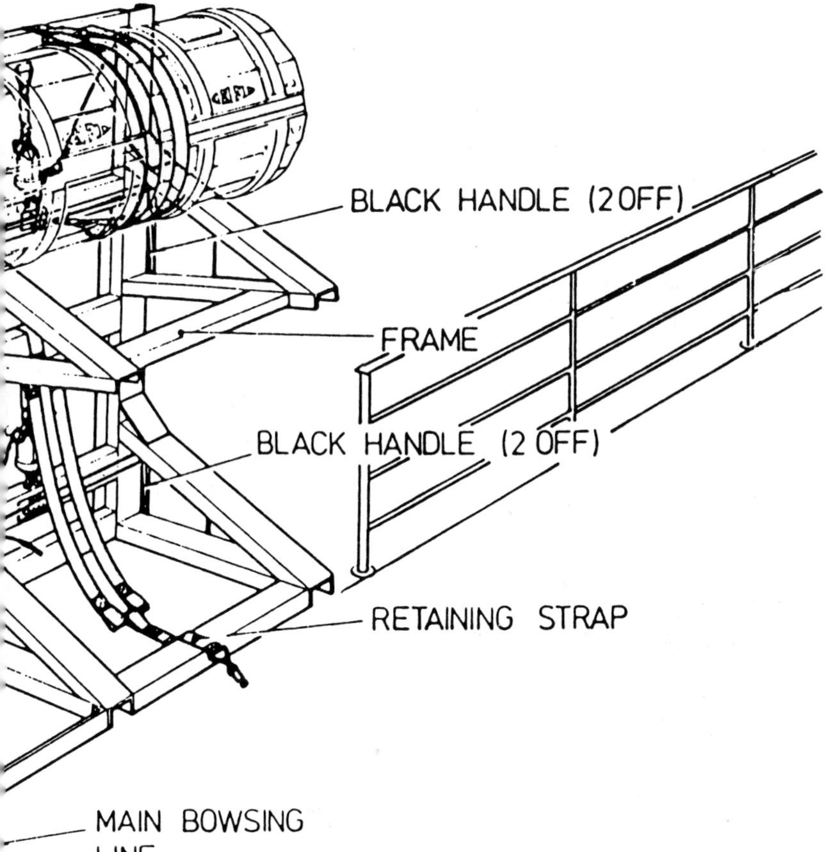
BLACK HANDLE (2OFF)
FRAME
BLACK HANDLE (2 OFF)
RETAINING STRAP
MAIN BOWSING
LINE
NOTE: LOWER CONTAINERS
DEPLOYED

DUNLOP

INFLATABLE LIFERAFT

LAUNCHING AND BOARDING PROCEDURE FOR DAVIT TYPE LIFERAFT

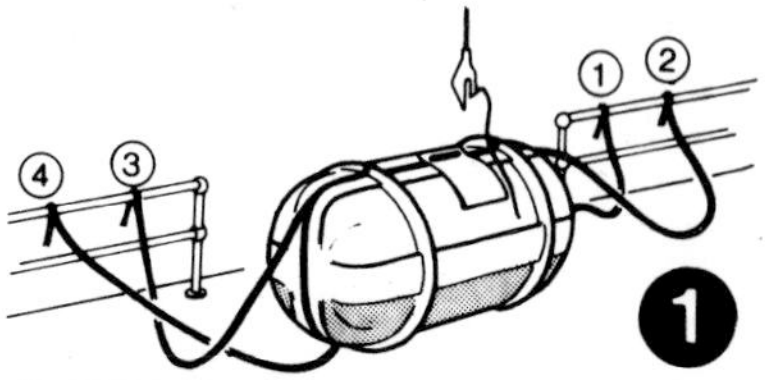

1

BRING CANISTER FROM STOWAGE POINT AND POSITION AS SHOWN. HAUL OUT APPROX 3 METRES OF PAINTER ① AND SECURE TO STRONG POINT. HAUL OUT FULLY THE BOWSING LINES ② AND ③, CANISTER RETAINING LINE ④ AND SECURE TO STRONG POINTS AS ILLUSTRATED IN DIAGRAM.

2

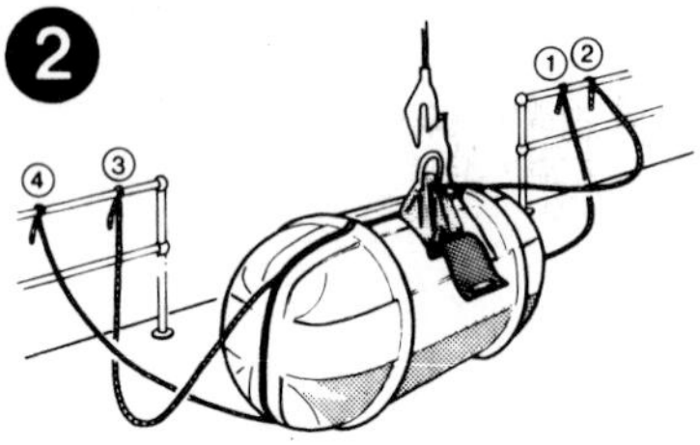

PEEL OFF RUBBER PATCH. RELEASE SHACKLE BY PULLING STRAP IN DIRECTION OF ARROW. ATTACH QUICK RELEASE HOOK TO SHACKLE AND LOCK IN POSITION.

3

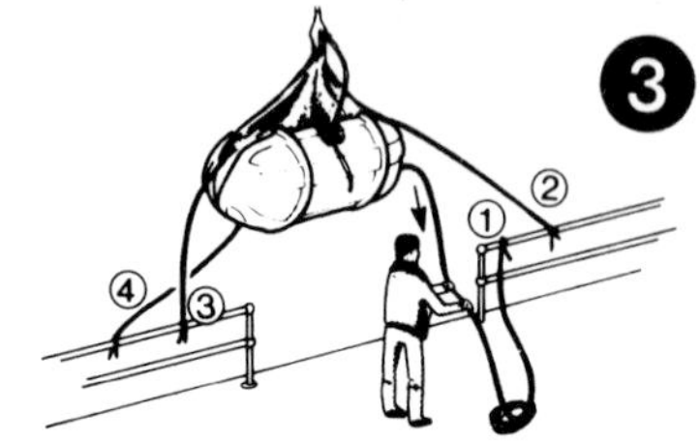

HOIST AND SWING OUTBOARD. STEADY CANISTER WITH BOWSING LINES. PULL OUT PAINTER TO MAXIMUM EXTENT THEN GIVE A SHARP JERK TO INFLATE RAFT.

4

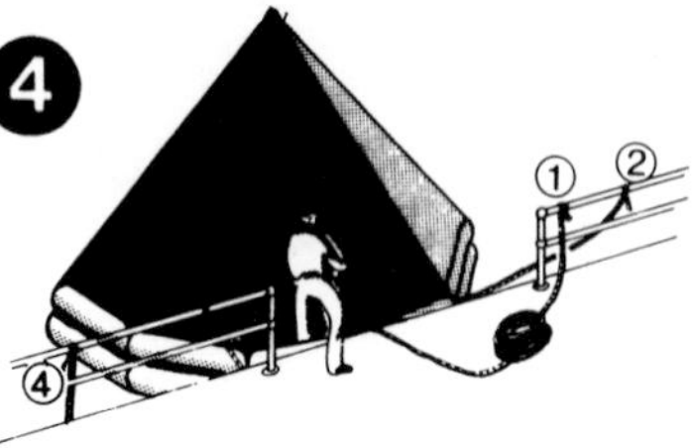

INFLATED RAFT IS STEADIED ALONGSIDE AT SILL LEVEL BY MEANS OF BOWSING LINES. THE RAFT CAN NOW BE BOARDED.

5

RELEASE BOWSING LINES AND PAINTER AND THROW THEM INTO RAFT. ENSURE TRIP LINE FOR QUICK RELEASE HOOK IS ACCESSIBLE. BEGIN TO LOWER RAFT. RETRIEVE OR MOVE SPENT CANISTER FROM LAUNCHING AREAS.

6

ACTIVATE QUICK-RELEASE HOOK AS RAFT APPROACHES SEA LEVEL. THE RAFT WILL AUTOMATICALLY RELEASE ON TOUCHING THE WATER.
GET AWAY FROM SIDE OF SHIP LEAVING AREA CLEAR FOR NEXT LAUNCHING.

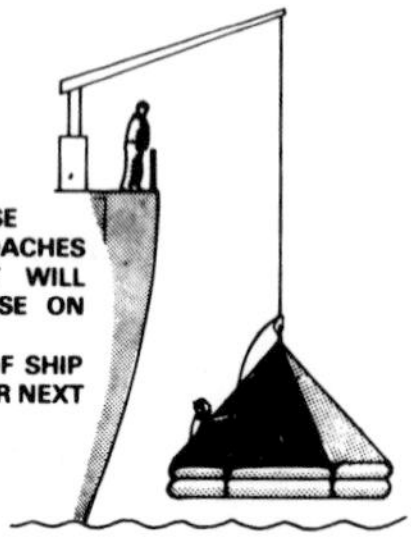

TO LAUNCH WITHOUT DAVIT

1. SECURE PAINTER ONLY TO STRONG POINT.
2. PUSH CANISTER OVERBOARD.
3. PULL OUT PAINTER TO FULL EXTENT AND GIVE A SHARP TUG TO INFLATE RAFT.

To abandon a ship equipped with the R.F.D. Marine Slide Raft.

The R.F.D. Marine Slide Raft is a smaller version of the Evacuation System and is suitable for small vessels and ferries whose freeboard does not exceed 6m (19.5ft) and can be deployed by one person.

1. Container is released from the cradle and descends into the water.
2. Inflation of liferaft, subsequent to positioning at exit.
3. During inflation, the bowsing line is used to manoeuvre the liferaft into position. As the head of the slide is also attached to the bowsing line, this manoeuvre also positions the head of the slide at the exit.
4. During the above operation, the separate cylinder attached to the slide is triggered to inflate the slide and its support system.
5. Subsequent to use, the liferaft may be released from the slide by the use of the quick release system at the foot of the slide.

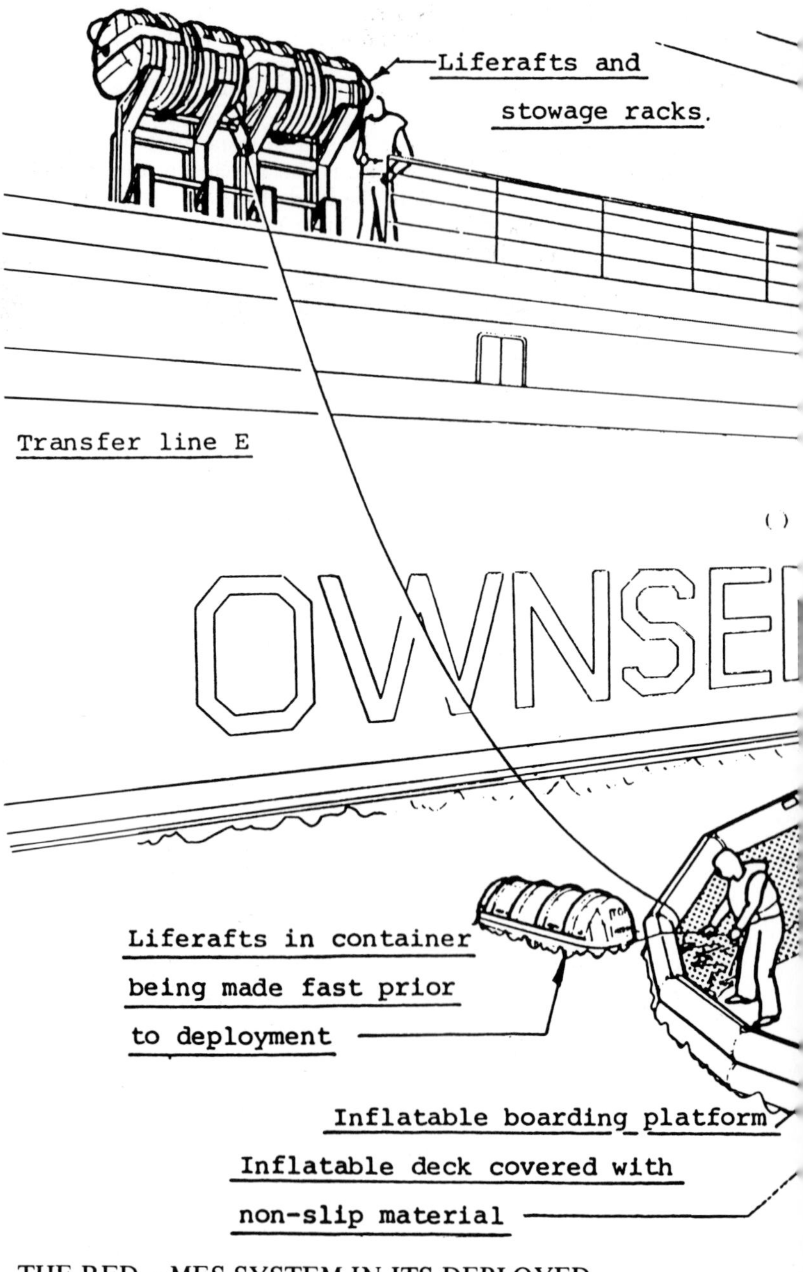

THE RFD. - MES SYSTEM IN ITS DEPLOYED CONDITION ON A TYPICAL FERRY

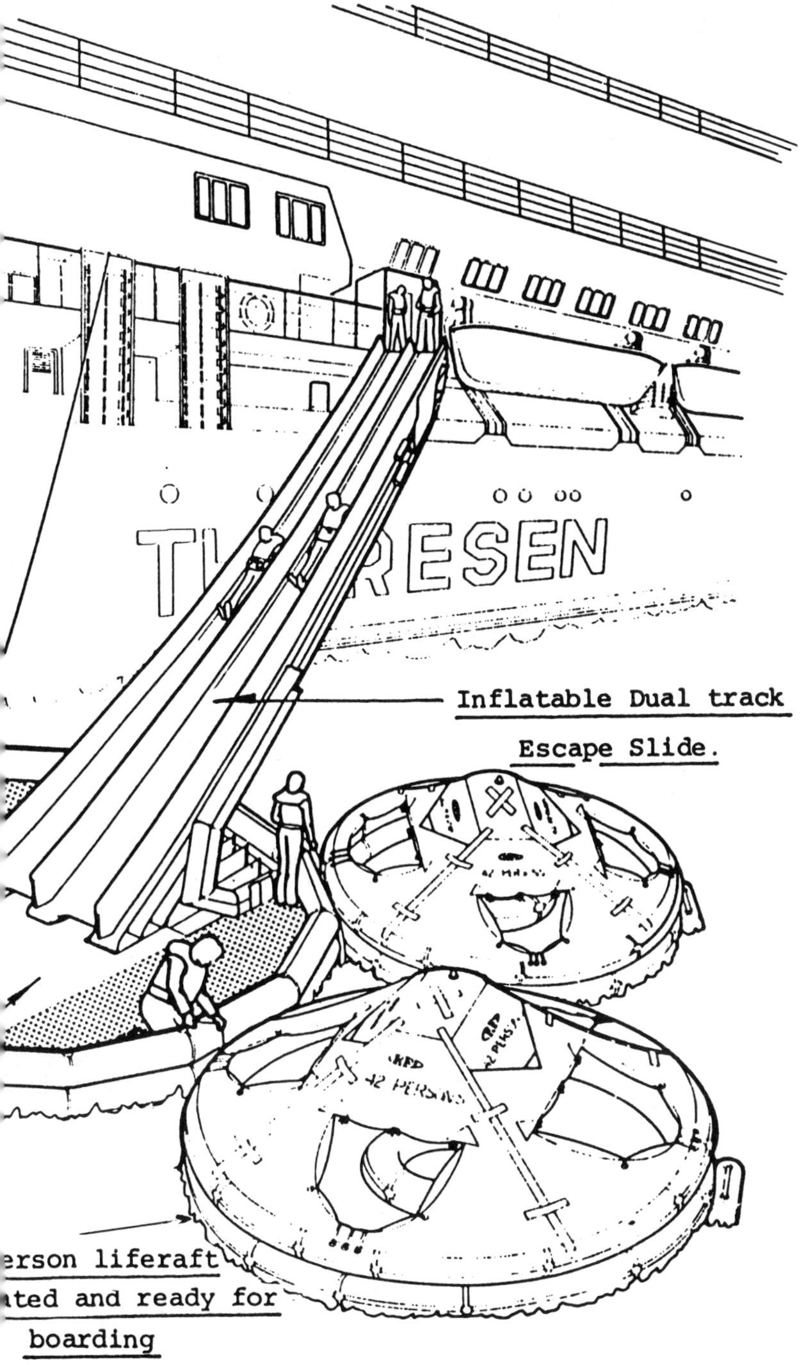
Inflatable Dual track
Escape Slide.
erson liferaft
ted and ready for
boarding

THE BREECHES BUOY

If a ship runs ashore under high ground or onto rocks or a sand bank, within reasonable distance of the shore, attempts may be made by the Coastguard to rescue those on board by means of the breeches buoy.

Due to heavy weather a vessel may well run aground on a lee shore or even in fine weather drift ashore for reasons beyond her control, such as engine failure. Provided that a vessel is not more than 250 yards (200m) from the shore and having sent up distress signals is aware that the Coastguard have arrived at the nearest point. A line carrying rocket should be fired towards the shore. Failing this the Coastguard will endeavour to fire a line carrying rocket towards the ship. While a Coastguard rocket is considerably more powerful than a ship's rocket, it must be remembered that the Coastguard will in all probability have to fire against the wind and that the ship makes a far smaller target than the shore

As soon as contact has been established between the ship and the Coastguard and the rocket line is in possession of both parties, the Coastguard will signal the ship to haul on the line.

When finally a tail block is hauled on board the ship, it should be taken as high as possible and well secured to a very strong point, such as the crosstrees.

When the ship signals the Coastguard that the tail block is secure, the Coastguard will haul out a heavier line by means of an endless fall rove through the tail block on the ship. When this line known as the "Jackstay" is received on board, it must be made very securely fast about 2 to 3 feet (60 to 90cm) above the tail block. On occasions, the tail block and endless fall are used without a jackstay.

When the Coastguard receives a signal to say this has been done, they will place a traveller in the shape of a block on the jackstay and connect it to the endless fall rove through the tail block on the ship. Then suspending a lifebuoy connected to a pair of canvas breeches, from the traveller, they will haul the breeches buoy out to the ship, by means of the endless fall. On being signalled that a survivor is in the breeches buoy, the Coastguard will haul the survivor ashore and repeat the operation until all the survivors are rescued.

There is a steadying line rigged from the breeches buoy to the endless fall, being made fast on the shore side near the traveller. Survivors travelling in the buoy should face the steadying line and use it to help maintain their balance.

It is most important to remember that a great deal of strain will be placed on both the tail block and the jackstay, so that they must be well secured to good strong points. They must also be as high as possible to avoid the survivor being dragged through the sea and very possibly washed out of the buoy.

Signals for use with the Breeches Buoy

Vertical waving of the arms with a flag by day and a light by night. Yes or affirmative.

Horizontal waving of the arms with a flag by day and a light by night. No or negative.

On a vessel carrying highly inflammable cargo, when it is unsafe to fire a rocket, a red flag should be flown by day and a red light exhibited by night.

ROCKET LIFE-SAVING APPARATUS

SIMPLIFIED DIAGRAM OF CENTRAL RIG

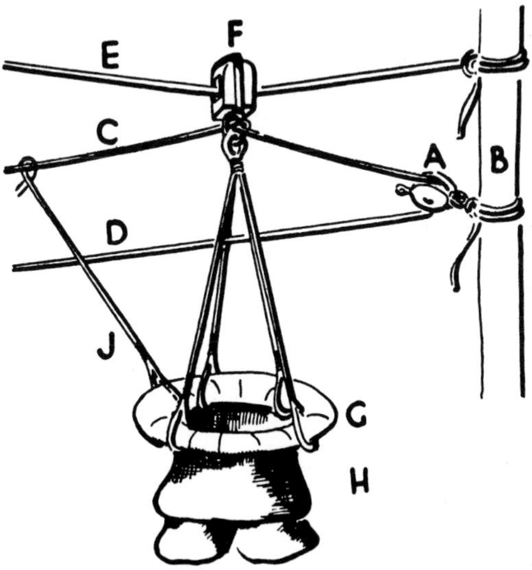

DETAILS OF RIG OF JACKSTAY, WHIP, AND BOUY

A. Tail block
B. Ship's superstructure
C. Shore side of endless whip
D. Return part of endless whip
E. Jackstay
F. Traveller
G. Breeches buoy
H. Breeches
J. Steadying line

SIGNALS TO BE EMPLOYED IN CONNECTION WITH THE USE OF SHORE LIFE-SAVING APPARATUS.

SIGNAL		SIGNIFICATION
By day—Vertical motion of a white flag or the arms or firing of a green star-signal. By night—Vertical motion of a white light or flare or firing of a green star-signal.		In general—"Affirmative." Specifically:- "Rocket line is held." "Tail block is made fast." "Hawser is made fast." "Man is in the breeches buoy." "Haul away."
By day—Horizontal motion of a white flag or arms extended horizontally or firing of a red star-signal. By night—Horizontal motion of a white light or flare or firing of a red star-signal.		In general—"Negative" Specifically:- "Slack away." "Avast hauling."

Helicopters are frequently used for maritime rescues. For further information on helicopter rescue operations see Annual Notice No. 4 of the annual summary of Admiralty Notices to Mariners.

For signals used by aircraft engaged on search and rescue operations. See "Life-Saving Signals. SOLAS. No. 2. on pages 313 to 316. See also M.S. Notice No. 1204 "Helicopter Assistance at Sea" on page 333.

The signal GU may be made by sound in morse code.
At night a ship may light a flare when the tail block is fast and show a white light when the jackstay is fast.
Whenever possible contact by morse code should be established, either by light or sound.
It is important that a man should be detailed as a signalman. He should stand well away from the main party, in a conspicuous position and fully understand the meaning of all the signals.

A capable seaman should be the first person sent ashore to liaise with the Coastguard.

AIR SEA RESCUE

Air-sea rescue is in actual fact a part of Her Majesty's Armed Forces. Nevertheless they are in contact with both police and Coastguard throughout the United Kingdom, if the rescue of civilians in distress is not one of their statutory duties, there is no disguising the fact that they have indeed unhesitatingly gone to the aid of countless civilians in dire need of assistance both ashore and afloat, saving many lives that could never have been saved without their help. There is no doubt whatsoever that they will continue to render this valuable service.

Rescue by helicopter of survivors at sea is a job that requires both courage and skill, especially in either fog or gale lashed seas. Horizons can disappear and the pilot may have great difficulty in estimating his height above the water. Winds under cliff faces can change direction and force in a most haphazard manner. Yet the air-sea rescue crews will cheerfully brave all this and in spite of the elements achieve the rescue of survivors against all odds, often when it appears to the helicopter crews themselves to be humanly impossible.

Owing to the danger to the rotors, a helicopter requires a space on the ship to be clear of masts and other obstructions. Preferably, a raft, dinghy or boat drifted astern of the vessel, with the survivors on board. The usual approach is for the helicopter to lower a member of it's crew on a length of steel wire. He will have with him a helicopter strop, which is slipped over the survivor's shoulders and under his armpits. Held by the crewman, the survivor is then hoisted up to the helicopter. On occasion the helicopter may lower the wire, which will be weighted, without the crew member but with helicopter strop attached. The survivor should slip the strop over his shoulders and under his armpits, signal the crew to hoist and hold on.

Injured survivors can be lifted aboard, having first been strapped in a "Neil Robertson" stretcher, should one be available. Indeed, there may well be one already on the helicopter.

When survivors are out of helicopter range, it is possible to drop inflatable liferafts and supplies to them from aircraft, who will then direct a surface vessel to the rescue.

It is important that the wire from the helicopter is not allowed to become entangled with any of the ship's fittings.

CHAPTER 3 BOATWORK UNDER OARS

With a boat under oars, the coxswain stands in the after end of the boat tending the tiller and is in charge of the boat. He should give his orders in a loud clear voice and remember to always put his tiller the opposite way to that to which he wants to turn. The coxswain should also see to it that his oarsmen keep good time when pulling.

To take a boat away under oars, the coxswain will see that everyone except the bowman is seated. Rowers on the thwarts facing aft, he will then give the following orders as they may be required.

"SHIP CRUTCHES". Each oarsman drops the chain of the crutch he is going to use, through the hole in the crutch plate on the gunwale, followed by the shank of the crutch.

"TOSS OARS." Each oarsman takes an oar and placing the grip of the oar on the bottom boards and the loom between his legs, stands the oar upright with the blade in the air. At the same time the coxswain ships the tiller.

"SLIP PAINTER." The bowman repeats his order and takes the painter into the boat, after it has been let-go on board or alternatively, pulls out the toggle, letting the painter go. He then stands on the bowsheets with the boathook ready to bear off. (The bowman when standing-by should always stand on the bow sheets facing forward with the boathook held vertically in front of him, hook upwards, when he is not actually using the boathook.)

"SPRING OFF." The bowman repeats the order and then pulls the toggle out of the painter, passes the painter aft, to be hauled upon from aft. The bowman then takes up the boathook and uses it to push the bow of the boat away, while the coxswain puts the tiller towards the ship's side.

"BEAR OFF BOWMAN." The bowman repeats the order and using the boathook, pushes the bow of the boat away. The coxswain puts the tiller towards the ship's side.

"OARS DOWN." The oarsmen lower their oars outboard, resting the looms in the crutches, holding the grips and with the oars level with the blades flat above the water.

"PORT OARS DOWN" or "STARBOARD OARS DOWN" if only one side is required to lower their oars.

"BOW OARS DOWN." and "BOW OARS GIVE WAY" sometimes given to assist in getting the boat away from a ship or quay and at the same time avoiding damage to the rudder, while still too close to put all the oars down.

The bowman will, when the last oar is down, lay his boat hook on the sidebench, hook forward, taking care not to let it hang over the gunwale, and will then stand in the well of the bow, facing forward, and keep a look out, unless he is required to pull a bow oar.

"GIVE WAY TOGETHER." The oarsmen pull on their oars taking their time from the stroke oarsman, who is the aftermost oar on the starboard side of the boat.

"PORT GIVE WAY" or "STARBOARD GIVE WAY" if only one side is required to pull.

"OARS." The oarsmen take one stroke and stop pulling on their oars, taking the oars out of the water, holding them level and with the blades flat.

"HOLD WATER." The oarsmen dip the blades of their oars vertically into the water, keeping the oars still and at right angles to the boat. This is done to take way off the boat.

"BACK WATER." The oarsmen push on their oars and every stroke of the oars is a pushing stroke.

By putting the tiller to port and giving the orders "PORT GIVE WAY", "STARBOARD BACK WATER", the coxswain would turn the boat smartly round to starboard and vice-versa.

The order "OARS" must always be given by the coxswain whenever he wishes to change the action of the oarsmen, in order that they may act without confusion.

"TRAIL OARS." The oars are allowed to trail aft and the oarsmen to rest.

"LAY ON YOUR OARS." The oars are held horizontally out of the water, with the blades flat, the crew being allowed to rest on the oars.

"WAY ENOUGH." The oarsmen take one stroke with their oars, stop pulling and await the order to "TOSS OARS." The bowman, if pulling, will now boat his oar and stand-by. Coxswain sees that passengers hands and elbows are off the gunwale.

"STAND–BY BOWMAN." The bowman repeats the order, takes up the boathook and holding it vertically in front of him, with the hook up, stands on the bowsheets facing forward.

"FEND OFF BOWMAN." The bowman repeats the order, then using the boathook, prevents the bow hitting the ship or quay.

"CATCH–ON BOWMAN." The bowman repeats the order and catches-on with the boathook.

The order "TOSS OARS" must be given before the boat comes alongside.

"BOAT OARS." The oarsmen lay their oars on the side benches, blades forward. Bow oarsmen in particular, should be careful when boating their oars not to hit the bowman. At the same time the coxswain will unship the tiller

When using the boathook, the bowman must always place the shaft of the boathook forward of himself, otherwise he is apt to injure an oarsman or passenger with the other end of the boathook shaft.

"UNSHIP CRUTCHES." Oarsmen take out the crutches and let them hang by the chains. Never leave crutches shipped when alongside.

"MAKE FAST PAINTER." The bowman repeats the order, lays down his boathook taking care not to let it hang over the gunwale and makes the painter fast.

"STAND BY TO BEAR OFF WITH THE LOOMS." When this order is given, the boat will be alongside and the oarsmen on the outboard side of the boat will have their oars down. They will unship their oars from the crutches and sliding their oars inboard, will rest the looms on both gunwales.

"BEAR OFF WITH THE LOOMS." The outboard oarsmen push the boat away from the quay or ship using the looms of their oars and with the grips of the oars against the ship's side. When this has been done, they bring their oars back inboard, ship them in the crutches and await orders to pull. At the same time the bowman will stand down. This method of getting a boat away can be very useful when getting away from the lee side of a ship or when getting away from a ship with a broad rubbing strake.

To pull an oar, sit upright on the thwart with your feet on the stretcher, holding the loom and grip of the oar and with your hands the width of your chest apart. Lean forward, commence the stroke with the blade vertically in the water. Push with your feet and lean back, with the arms straight, to pull on the oar. Now take the oar out of the water, giving the loom a half turn by dropping the wrists to turn the blade flat, lean forward bending the arms, to make the next stroke. Give half a turn to the loom, using the wrists, so that the blade goes into the water vertically. The turning of the wrists is called "Feathering the oar" and takes practice. If you do not turn the oar correctly, it will hit the water at the wrong angle and be forced out of your control, in other words, you will "Catch a crab."

The oarsman on the starboard side aft is the stroke oar and gives the timing. Each oarsman will take his time from the man in front of him, except the after port oarsman, who takes his time from the stroke oar. Should the coxswain decide to give the time, he should watch the stroke oar while doing so.

The most important thing when pulling a boat, is to keep time together, otherwise, not only does it look bad but the oar will dig into the back of the oarsman in front who naturally enough is liable to resent it. Watch the man in front of you, do not look at the blade of your oar.

If an oarsman catches a crab or is out of time, the coxswain should give the order "Oars" and when all is ready, start again with "Give way together".

The coxswain will see to it that no one is allowed to sit on the gunwale at any time, neither will he sit on it himself.

To come alongside, always come alongside head to wind or tide as the case may be. Under power with a right-handed screw, if the engine is put astern, it will cant the stern to port and the bow to starboard.

CHAPTER 4 BOATWORK UNDER SAIL

BENDS AND HITCHES

In order to sail a small boat, knowledge of a few simple bends and hitches is essential. The reader is best advised if he asks a knowledgeable friend to give him a little tuition on the subject. Fifteen or twenty minutes is all the time that is required. Once learnt they will never be forgotten.

OVERHAND KNOT	Placed at intervals along a rope, they provide useful handholds.
FIGURE OF EIGHT KNOT	A useful knot to place at the end of a rope to prevent it running through a block or thimble.
REEF KNOT	Used to join two pieces of rope of equal thickness. (Unsuitable for man-made fibres).
SINGLE SHEET BEND	Used to join two pieces of rope of unequal thickness.
DOUBLE SHEET BEND	A stronger version of the single sheet bend.
CLOVE HITCH	Used to secure the end of a rope to a spar.
ROUND TURN AND TWO HALF-HITCHES	Used to make the end of a rope fast to a ringbolt. If the rope is liable to get wet, the half-hitches should be spaced and the end of the rope seized, to prevent the hitch jamming.
ROLLING HITCH	A hitch that when made fast to a spar, will not slip along the spar in the direction of the strain.
SHEEPSHANK	Used to shorten a length of rope without cutting it. When made, two large loops are formed and they should be seized onto the standing part of the rope, to ensure that the hitch does not come adrift.
BOWLINE	Probably the hardest knot to learn. Simply a loop made on the end of a rope, the size of the loop remains constant and will not slip.

Overhand knot

Reef knot

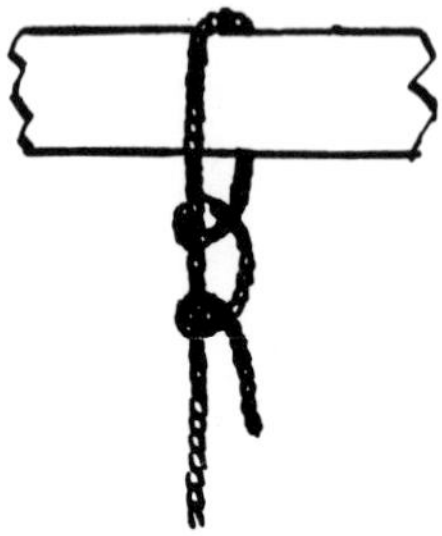

Two half hitches

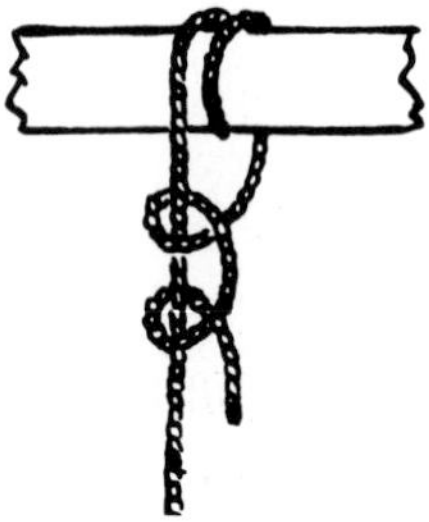

Round turn &
two half hitches

Figure of 8 knot

Bowline

Single sheet bend

Rolling hitch

Sheepshank

Knots and hitches must always be hove tight. Those above have been left loose for illustrating reasons.

THE LIFEBOAT COMPASS

A small magnetic compass filled with liquid is supplied to the lifeboat as a direction finding instrument.

The compass itself consists of a circular card 4 inches (10.16cm) in diameter balanced on a pivot and housed in a bowl with a ¼ inch (6mm) clearance between the edge of the card and the bowl. Magnetic needles are attached to the underside of the card in line with the North and South markings. Magnetic attraction of the needles by the magnetic pole gives the needles a constant direction.

The bowl is filled with an antifreeze liquid (distilled water and industrial methylated spirit) to dampen the effect of the motion of the boat on the card and has a vertical black line or arrow marked on the inside, where you can see it. This is the "Lubber line".

The bowl is slung in a ring called a gimbal which is suspended by screws through a protective case known as a binnacle, so that the bowl is free to find its own level against the motion of the boat. The screws through the binnacle casing, retaining the gimbal (gimbal pins) are required to be in line with the lubber line. It has been usual for the binnacle to house a colza oil lamp which should be kept trimmed and will burn for at least 10 hours (this lamp will not be accepted on vessels built after 1st July 1986) and for the compass card and lubber line to be luminous. The regulations require the compass to be luminous or be provided with suitable means of illumination.

The compass should be stowed in the sternsheets. To set it up, stretch a line tightly between the lifting hooks of the boat, this will give the centre line. Place the compass on the after thwart, with the gimbal pins in line with the centre line and lash it in position.

When the compass has been set up exactly on the fore and aft line of the boat with the lubber line forward. The lubber line will indicate on the compass card the direction in which the boat is heading.

Partially enclosed lifeboats and totally enclosed boats also rescue boats are required to have the compass permanently sited in position.

A compass card is normally marked like any other circle, into 360 degrees and also into 32 points (there being 11¼ degrees to every point). Lifeboat compass cards however are not normally marked in degrees but in points and half points. For it is virtually impossible to sail a lifeboat any closer to a set course, than half-a-point.

The four cardinal points North, South, East and West. The four half-cardinal points NE, SE, SW and NW and the eight three letter points NNE, ENE, ESE, SSE, SSW, WSW, WNW, and NNW are all marked by name on the card. The long un-named lines on the card are the BY-points, the short lines are the Half-points. There are 16 By-points in all. Study the card and learn how to read it.

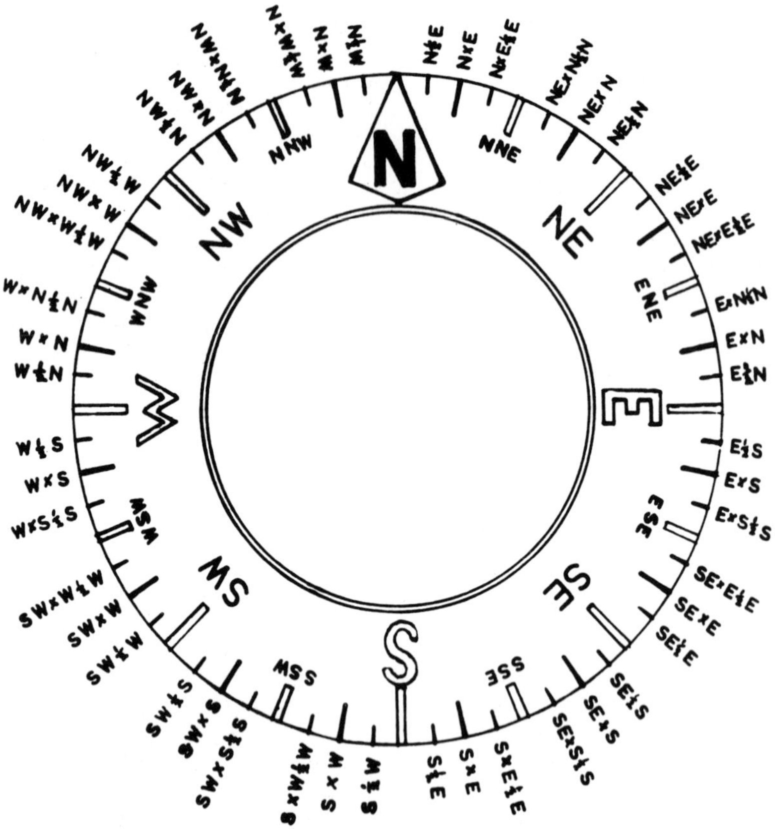

THE MARINERS COMPASS

SAILING

On ships built after 1st July 1986 all lifeboats must be motor boats and will not, under IMO Regulations, be required to carry sails.

A Lifeboat rig consists of a Standing-lug and jib.

The lug sail is a four sided sail, set fore and aft and with a yard (spar) attached to the head. Set as a "standing lug" very little of it lies on the fore side of the mast, therefore it makes no difference if it is set on either the lee or weather side of the mast, for the boat will sail equally well. At the same time a triangular jib is set forward.

Set by itself as a "dipping lug", the tack must be taken forward and about a third of the sail is set forward of the mast, therefore it must always be set on the lee side of the mast. The dipping lug derives its name from the fact that the yard must always be lowered and dipped around the mast, to keep it on the lee side, every time the boat changes its tack.

Never furl wet sails, spread and dry them first. Furled sails should be opened up and aired at reasonable intervals to ensure that they are not damaged by damp.

The mast, which is approximately two thirds the length of the boat must be of wood and is fitted with an iron band (mast head band) to take the halyard blocks and stays, the mast is also fitted with a traveller.

PARTS OF THE SAILS

LUFF	The forward edge of the sail
LEACH	The after edge of the sail
FOOT	The lower or bottom edge of the sail
HEAD	The top edge of the lug-sail The top corner of the jib
TACK	The bottom forward corner of the sail
CLEW	The bottom after corner of the sail
THROAT	The top forward corner of the lug-sail
PEAK	The top after corner of the lug-sail
HALYARD	A rope that is used to hoist a sail
TACK LANYARD	A rope that is secured to the tack of a sail
SHEET	The rope that is secured to the clew of a sail Two sheets are attached to the clew of the jib, to allow for ease of handling A SHEET MUST NEVER BE MADE FAST
BOLT ROPE	Tarred indian hemp rope. Sewn to the luff, head and corners of a sail to strengthen them. Always sewn on the port side of the sail.
CRINGLES	Loops in the bolt rope containing thimbles, into which ropes are spliced.
REEF POINTS	Small ropes attached to the lug-sail, a little above and in line with the foot, on each side. So that the sail area may be reduced by bunching the foot of the sail and tying it up with the reef points.
REEF EARING	A rope spliced into a cringle on both the luff and leach of the lug-sail, at the same height above the foot as the reef points. Used to secure tack and clew cringles, when the sail is reefed.

YARD — A spar seized to the head of the lug-sail.

YARD STROP — A small endless rope or strop taken round the yard, for the purpose of hooking the yard onto the traveller.

YARD ROVINGS — A series of lashings or lacing holding the yard to the head of the lug-sail.

EYELETS — Small brass eyelets let into the head of the lug-sail, to allow the yard rovings to pass through.

TRAVELLER — A ring of metal around the mast, free to travel up and down, a hook and eye being attached to the after side, vertically. The lug halyard is spliced into the eye and the yard strop engages on the hook, when the lug-sail is hoisted.

The lug-sail in a lifeboat may be referred to as the main sail, because the main sail is the principal sail on a fore and aft rigged vessel, similarly the jib may be referred to as the fore-sail.

A proper setting of the sails can only be obtained by constant practice. The force of the wind should be able to spill out of the sail when its work is done. The space between the jib and the lug-sail should form a funnel and create a vacuum on the lee side of the lug-sail.

See pages 110 & 111 for diagram of sails.

HOISTING SAIL

In all probability the lifeboat will be lying "Hove-to" when the decision to set sail is made. Provided the boat is not too crowded, bring everyone aft of the mast-thwart and seat them as low as possible. Let-go the lashings on the mast and sails, sort out the stays and halyards on the mast, making sure they are all clear and secure them to the mast. Set the heel of the mast in the shoe on the keelson and set the mast upright against the thwart, clamping it in position with the mast clamp. Look to see if the hook on the traveller is facing aft, if it is not, lift the mast a little, turn it and drop it back in place, then set-up the mast stays to the chain plates, with the stay lanyards, tightly. Take the jib forward, be sure the bolt rope sewn along the luff is forward and on the port side. Make fast the jib tack lanyard to the ringbolt on the stem, giving sufficient drift to allow the foot of the sail to be at least 6 ins. (15cm.) above the gunwale when the sail is hoisted. Make the jib halyard fast to the head of the jib, take a turn with the lee sheet around the lee cleat and standing on the weather side of the boat, hoist the jib. Take the halyard down the after side of the mast-thwart, as near to the side-bench as possible on the lee side, haul tight and take a turn around the thwart, make fast with two half-hitches.

Next, lay the yard on the lee side bench and the foot of the lug sail along the centre line of the boat. Be sure that the bolt rope is on the forward edge of the sail, for if it is not, you are about to hoist the sail the wrong way round. Make the tack lanyard fast on the lee side of the mast thwart by taking it under the after side of the thwart and up the forward side to the cringle, as close to the mast as possible. Allow sufficient slack for the foot of the sail to be at least 6 ins. (15cm) above the gunwale when the sail is hoisted, secure with a half hitch. Lift the traveller out of the way, then take the lanyard round the mast and back through the cringle, pull the tack close to the mast and secure with two half hitches. Be sure that the yard and tack are on the same side of the mast and take a turn with the main sheet around the after cleat on the lee side.

Satisfy yourself that the yard strop is properly placed, about a quarter of the way along the yard from the throat of the sail and hook the yard strop onto the traveller. Place a man to tend the fore end of the yard and keep it fore and aft as the sail is being hoisted. Stand on the weather side and haul on the main halyard to raise the sail. Take the halyard to the after side of the mast-thwart, as close to the side-bench as possible and on the opposite side of the boat to the jib halyard. Haul tight and make fast with a round turn and two half-hitches.

If after a sail has been hoisted two blocks (that is as high as it can go) there is slack in the luff of the sail, sweat the tack lanyard down until it is tight. Then trip the sea-anchor and haul it in, ease out the sheets and set a course.

The jib must always be hoisted first to act as a forestay for the mast.

A gap of 6 ins. (15cms) between the foot of each sail and the gunwale is required by the coxswain, who should be seated on the weather side of the sternsheets, so that he can see to leeward.

The function of the traveller is to keep the yard close to the mast.

The tack of a sail must always be made fast first in order to ensure that the tack does not fly when the sail is hoisted.

Halyards must always be taken down the after side of the mast-thwart to ensure that they clear the jib, as far outboard as possible, one on each side to act as preventer stays.

The yard is to be tended at its forward end whenever the lug-sail is hoisted or lowered, to keep it fore and aft, steady, and to ensure that no one is hit by it.

If a halyard should fly or become unrove or anything else require attention at the top of the mast, the mast must be taken down for the trouble to be put right. On no account is anyone to stand up waggling a boathook or try to climb the mast.

Sheets are never to be made fast. A single turn is taken around the cleat and the sheet held in the hand. Thus in the event of a squall approaching the sheet can be quickly thrown off the cleat and allowed to fly. In a strong breeze, the coxswain may be unable to prevent the main sheet rendering (slacking out). He should, if this occurs, pass the end of the sheet back through the clew cringle and bring it back again to the cleat, so making a "purchase" which will enable him to hold the sheet with the required single turn and equally, let the sheets fly. The main sheet is required to be long enough to make a purchase in this manner.

Always take down the sails and strike the mast before coming alongside a ship or quay. There is always the danger that the mast may hit an obstruction and be forced through the bottom of the boat or the boat be capsized. Come alongside under oars or power.

N.B. The weather side of a boat is the side onto which the wind is blowing. The lee side being the side from which the wind is blowing away.

TRIM

One of the most important aspects of sailing a boat, is the distribution of the occupants. Too much weight forward will make the boat lurch into the wind and too much weight aft will make it fall away from the wind. Generally speaking, the occupants should keep aft of the mast and centrally placed, as low as possible with as much of the weight as is needed on the weather side to keep the lee gunwale out of the water. A well trimmed boat would hold a little weather helm and have the lee gunwale just nicely out of the water.

When the tiller has to be held up to the weather side of the boat, in order to make a good course, the boat is carrying "weather helm", when tiller has to be held down to the the lee side "lee helm".

RUNNING FREE

If you are sailing with the wind anywhere on or abaft the beam, you are "running free". With the wind on your port quarter you would be "free to port" and with the wind on the starboard quarter "free to starboard" and with the wind right aft "running before the wind".

When running free, the sheets should be let out until the sail draws best and the boat is carrying a little weather helm.

Running before the wind is not recommended, it is better to have the wind a little free on one side or the other. If it is necessary to run before the wind, there are two methods. One is to set the sails "goose winged", having the jib on one side and holding the clew of the lug sail out on the other side with a boathook; the boat will tend to yaw badly and there is every danger of a gybe, either from a shift of wind or an extra bad yaw.

It is better to take the jib down and re-set the lug sail as a dipping lug, by taking it down and moving the yard strop a little further aft along the yard, until it is about a third of the way along. Let-go the tack and make it fast to the ringbolt on the stem, re-hoist the sail on the fore side of the mast and use the boathook to hold out the clew.

CLOSE HAULED

It may be necessary to try and sail the boat towards the wind, perhaps to try and keep off a rocky lee shore.

A boat sailing with the wind before the beam, that is to say somewhat into the wind, is "close hauled". If the wind is on the starboard side, she is close hauled on the "starboard tack". If the wind is on the port side, she is "close hauled" on the port tack.

A lifeboat cannot normally be sailed any closer to the wind than 6 points (67½ degrees) from it. That is to say that if the wind were blowing from North, a lifeboat could sail ENE or WNW, but no nearer to North than either of these two points.

Even so, it is very doubtful if the boat would in fact make any way at all to windward, for as a lifeboat has no deep keel, there is always a lot of leeway and you may well find that the boat still drifts to leeward.

TACKING (GOING ABOUT)

In theory, in order to sail a boat into the wind, it would be necessary to make a zig-zag course, which is known as "tacking".

To tack or go-about

1. Up helm and ease the sheets to gather away.
2. When the boat has way. Down helm, let fly jib and haul aft the main sheet.
3. When the boat is head to wind, back the jib.
4. When the boat is through, trim sheets on the lee side and set course.

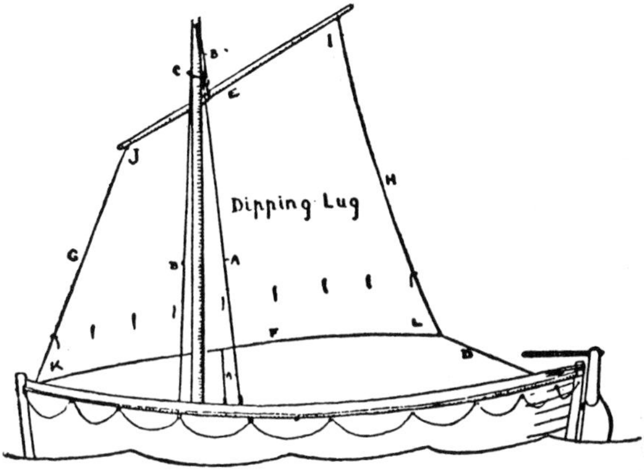

Showing a dipping lug correctly rigged.

A. Shroud
B. Halyard made fast on the weather side
C. Traveller
D. Main sheet
E. Head
F. Foot
G. Luff
H. Leach
I. Peak
J. Throat
K. Tack
L. Clew

Suitable for use when running before the wind without a fore-sail, or at any other time in the absence of a fore-sail.

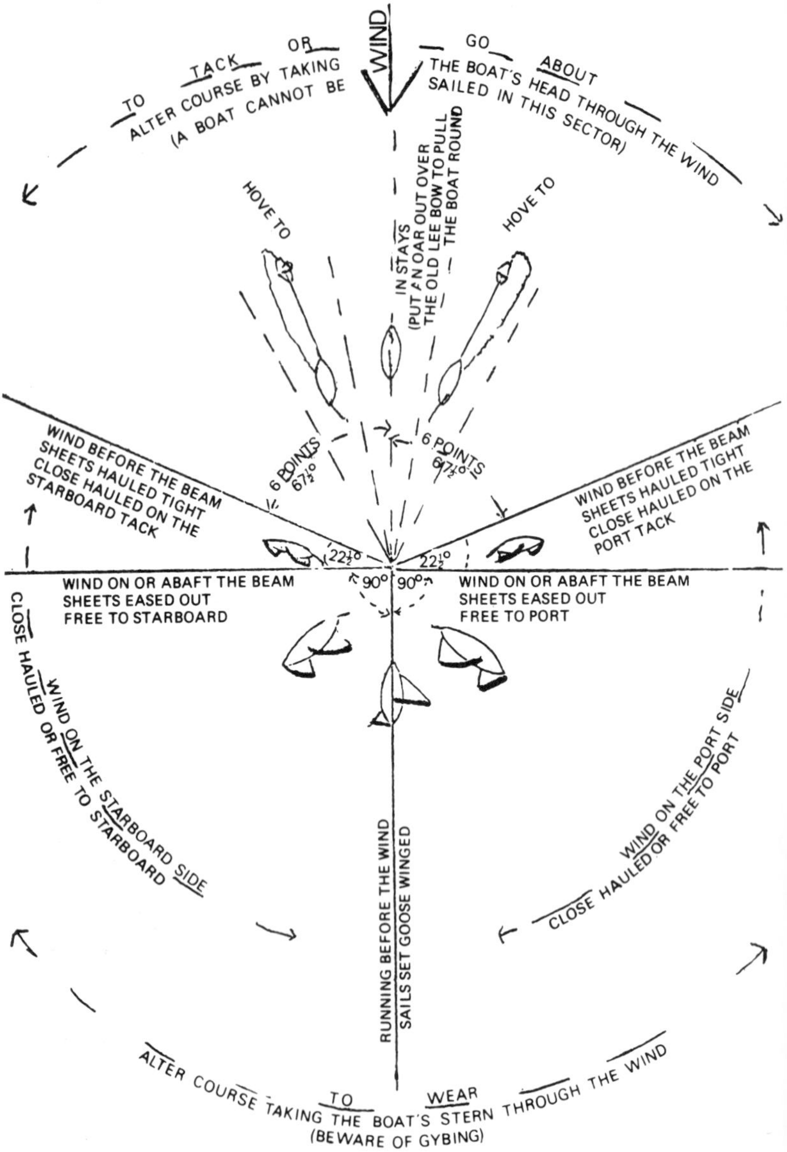

GYBE – THE WIND GETS ON THE WRONG SIDE OF THE MAINSAIL AND THROWS IT ACROSS THE BOAT

SAILING TERMS

TACKING

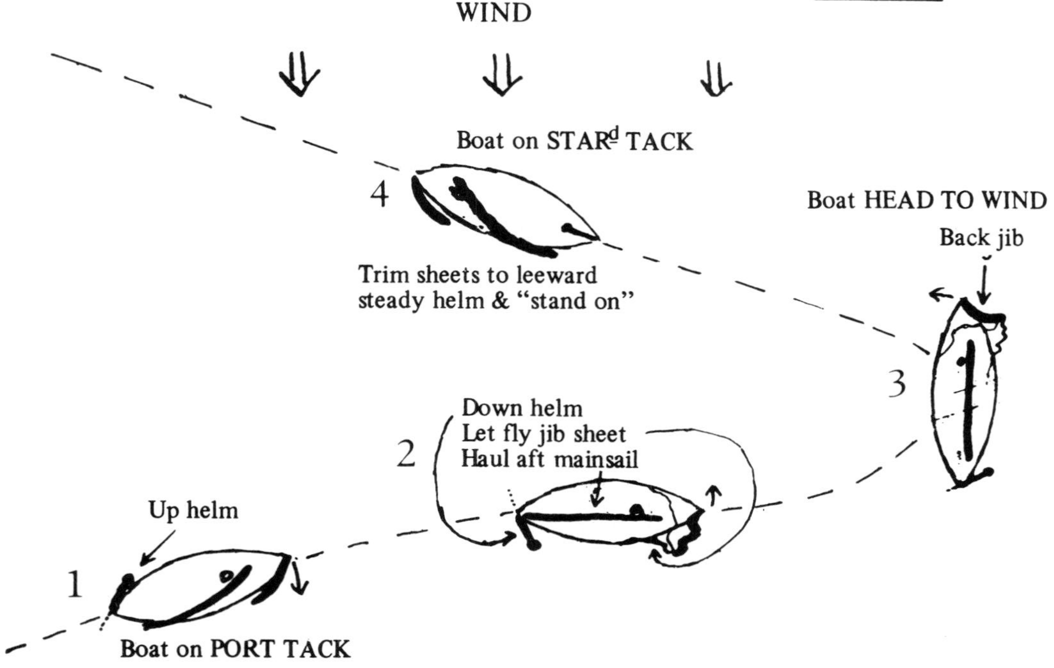
WIND
Boat on STAR^d TACK
4
Trim sheets to leeward
steady helm & "stand on"
Boat HEAD TO WIND
Back jib
3
Down helm
Let fly jib sheet
Haul aft mainsail
2
Up helm
1
Boat on PORT TACK

WEARING

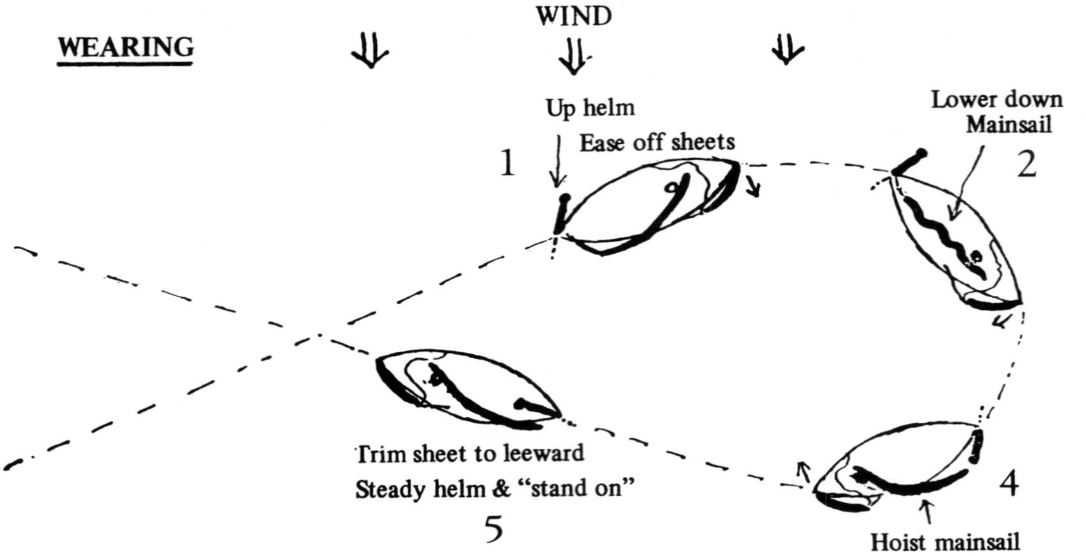
WIND
Up helm
Ease off sheets
1
Lower down
Mainsail
2
4
Hoist mainsail
Trim sheet to leeward
Steady helm & "stand on"
5

A boat under sail can be likened to a weather vane, the arrow of which always points into the wind. If a boat under sail is left to its own devices, it will always come head to wind. However, if you wish to tack a boat, that is turn the boat's head right through the wind and sail off on a new course with the wind on the other side of the boat. You will need speed to enable the boat to pass through the wind.

As a boat sailing close hauled is actually sailing against the motive power of the wind, she cannot go very fast and in order to tack, it will be necessary to gain some speed first. This can only be done by allowing the boat to "pay off", that is to say by turning the boat's head away from the wind and so taking advantage of its motion. To do this "up helm", in other words, pull the tiller to the high side, which is the weather side of the boat and "pay out" or ease the sheets a little. The boat will turn away from the wind and start to gather way. When the boat has sufficient way, "down helm" by putting the tiller to the low or lee side of the boat, at the same time, letting fly the jib sheets, to spill the wind out of the jib and haul aft as tightly as possible on the main sheet, so that the main sail is fore and aft. The wind will catch the main sail and push the stern of the boat to leeward. Aided by the rudder, the boat will come head to wind quickly and without loss of speed. As the boat comes head to wind, the luff of the main sail will shake and the wind will be split out of it. Now is the time to transfer some of the weight of the passengers from the old weather side to the new weather side and as soon as the boat is head to wind (but not before, or you will be back where you started), back the jib by holding it out on the old lee side. The wind on the jib will help to force the bow of the boat round. As soon as the luff of the main sail stops shaking, let out a little on the main sheet, so that the sail can draw. Let the jib come over to the lee side and set a course as close to the wind as you can with both sails drawing nicely.

If you attempt to tack or go-about before the boat has gathered sufficient way, the boat will come head to wind and stay there. She is then said to have "missed stays" and will remain "in stays" (caught head to wind). To get a boat out of "stays", put an oar out over the old lee bow and pull the bow round with the oar until the sails can draw.

WEARING

Sometimes, either because the wind is too strong and the seas beat you back or because the wind is so light you are unable to gather sufficient speed, it becomes impossible to go-about. When this happens it will be necessary to take the boat's stern through the wind to change your tack. This is called "wearing".

To wear

1. Up helm and ease the sheets.
2. When the wind is on the quarter, lower the main sail.
3. Bring her round on the jib.
4. When the wind is on the other quarter, raise the main sail.
5. Trim the sheets and set a course.

Pull the tiller up to the weather side and pay out on the sheets, this will bring the boat's stern towards the wind. With the wind on the quarter, the main sail must be lowered in order to avoid a gybe, continue to turn the boat with the help of the jib and as the boat passes through the wind, change over the jib sheets and transfer some of the weight of the passengers from the old weather side to the new weather side, then with the wind well on the other quarter, hoist the main sail, trim the sheets and set a new course.

Wearing causes the boat to lose a lot of ground, so that you would only wear if you found it impossible to go about by tacking or if you were running free with the wind on one quarter and wished to alter course and run free with the wind on the other quarter, when of course, it would be the best method of changing direction.

GYBE

A gybe occurs when the wind gets on the wrong side of the main sail and throws the sail across the boat. It may be caused by either a sudden shift of wind or bad steering or both and the coxswain of a lifeboat under sail must be on constant guard to see that this does not occur. No matter how light the wind, the sail will be thrown across the boat with considerable force, possibly injuring some of the persons in the boat as the foot of the main sail strikes their heads and as the sail flies out on the new lee side, a terrific strain will be placed on the mast which may well be sufficient to break it or alternatively as the lifeboat has no deep keel, capsize the boat or both. It is essential that the main sail in a lifeboat is lowered before any attempt is made to alter course by passing the stern of the boat through the wind.

REEFING

Should the wind freshen, it may become dangerous trying to sail the boat with too large an area of canvas, while being anxious to try and make land as quickly as possible and to take advantage of a favourable wind, you feel that the weather is not so bad as to make it necessary to heave-to.

It is possible to reduce the sail area by taking in a reef.

To Reef

1. Let fly sheets. Down helm and bring the boat head to wind.
2. Stream the sea-anchor.
3. Send down the main sail, unhook and lay the yard on a side-bench.
4. Make the luff earring fast to the tack cringle.
5. Make the leach earring fast to the clew cringle.
6. Bunch the foot of the sail and tie the reef points.
7. Hook the yard back on the traveller and hoist the sail
8. Retrieve the sea-anchor and set a course.

Bring the boat head to wind by putting the tiller down to the low or lee side and letting fly both sheets. Stream the boat's sea-anchor, to keep her head to wind and lower the main sail. Always reef a sail from forward to aft, commencing by seizing the luff earring to the tack cringle and then seizing the leach earring to the clew cringle. When this has been done, the reef points are to be tied with reef knots. Do not let-go the main tack lanyard, because you will want the foot of the sail to remain where it is; and so lower the height of the sail to reduce the capsizing movement of the wind. In the event of the sail not having any reef earrings attached, use a part of the tack lanyard and main sheet to seize the cringles. Do not rol the foot of the sail, this would cause it to hold water, bunch or crumple it. When all is ready, hoist the sail, trip the sea-anchor, haul it in and set a course.

SQUALL

A squall is a strong gust of wind, which could, if no precautions are taken, capsize the boat.

In the event of a squall.

1. Let fly sheets and down helm to bring the boat head to wind.
2. When the squall is over, carry-on. If it persists, heave-to.

On no account is a sheet ever to be made fast, for in a sudden and unexpected squall there would be no time to let it go and the boat would most certainly be capsized.

HEAVING-TO

If the weather deteriorates and it is no longer safe to sail the boat, it will be necessary to heave-to.

To Heave-to

1. Let fly sheets and down helm to bring the boat head to wind.
2. Stream the sea-anchor.
3. Send down the sails and strike the mast.
4. Unship the tiller and rudder. Ship the steering oar.
5. Ship the cover and trim by the stern.

If the weather is deteriorating the decision to heave-to should be made in plenty of time and before the weather is too bad. Get the sea-anchor out and make everything ready by having both the hawser and tripping line clear. Make fast the end of the hawser to the mast thwart, on the same side of the boat as the fairlead and lead the hawser to the fairlead. At the point along the hawser where it will enter the fairlead, wrap it well with canvas or cloth (part of the old boat cover will do) to protect the hawser from chafe, after the sea-anchor has been streamed. Now make the end of the tripping line fast to the mast thwart on the same side of the boat. When all is ready, wait for a lull in the wind, then down helm and let fly sheets, so that the boat comes head to wind. Put the sea-anchor out over the bow and pay out the hawser slowly as the sea-anchor takes it, at the same time paying out the tripping line also. Send down the sails and strike the mast, placing them all on the side-benches. Unship the rudder and tiller and ship the steering oar by passing it out astern and bringing the grip back through the grommet. Ship the exposure cover and trim by the stern, that is to say, keep everyone out of the bow and towards the stern. Bail out if necessary. Lie with the wind two or three points (27 to 37½ degrees) on the same bow as the fairlead and hold the boat there with the help of the steering oar. Do not try to keep the boat head to wind and again, try not to let her fall more than two or three points from the wind or you will find the boat is in the trough and in danger of capsizing. Check the parcelling on the hawser frequently to make sure that the hawser is not being chafed by the fairlead.

If you have been running free and waited too long before deciding to heave-to (so that there is a danger of broaching-to when turning the boat head to wind). Do not risk it. Just stream the sea-anchor over the stern instead of the bow and lie stern to the wind, using the grommet as a fairlead and the end of a heaving line to make a grommet for the steering oar at the bow.

A small mizzen (after) sail, if one can be rigged, can be very useful in helping to steer a lifeboat into a high wind. An oar lashed and stayed with heaving lines might be useful as a mast.

(IMO Regulations do not require partially enclosed and totally enclosed lifeboats to be provided with a steering oar and steering oar grommet.)

OIL BAG

If the weather continues to deteriorate and the boat labours badly, take the jib halyard block off the mast head band (there is usually a spike on the jacknife that will enable you to undo the shackle) and bend the two ends of the halyard together to make an endless fall. Haul in about two fathoms (3.7m) of the sea-anchor hawser and bend the jib halyard block onto the hawser, then pay the hawser out again. Put a little oil in the oil bag, cork the bag and make it fast to the jib halyard. Haul the bag a little way out towards the sea-anchor. The oil will seep out of the bag and help prevent the seas from breaking and the boat will be able to lie better to the sea-anchor. The best position for the oil bag will have to be found by experiment. If the weather is very cold and the oil to thick to seep through the bag, prick the bag a few times with a sail needle or something similar.

Although of great help in the open sea, the oil bag is useless when beaching in surf.

(IMO Regulations do not require partially enclosed and totally enclosed lifeboats to be provided with oil and an oil bag).

JURY SEA ANCHOR

If you lose your sea-anchor, make a quick temporary one by tying each each of a heaving line to the handle of a bucket and throw a bucket out over each bow while keeping the bight of the heaving line fast in the boat. Another alternative, is to make the end of the permanent painter fast round the middle of the old boat cover and pay it out.

To prepare a jury sea-anchor, furl the lug-sail and lash it together with the yard to the mast. Try and leave some pockets in the sail that will catch the water. Double a heaving line and tie the ends to each end of the mast with a rolling hitch. Bend the end of the permanent painter onto the centre of the bight of the heaving line with a double sheet bend. Haul in the temporary sea-anchor, pass the mast and sail out over the bow and pay out the painter as required until the boat is laid to the jury sea-anchor. Remember to parcel the painter and lead it through the fairlead. If the boat is not equipped with a mast and sails, the oars may be used in lieu but they will not be as efficient.

MAN OVERBOARD

Should a person jump or fall overboard.

1. Let fly sheets, down helm and bring the boat head to wind.
2. Send down the sails and man the oars.
3. Row back and take the man aboard over the weather bow or stern
   Have a man in the bows to watch the man in the water and report.
   Have the ladder over to assist the man in getting aboard.

JURY SEA ANCHOR

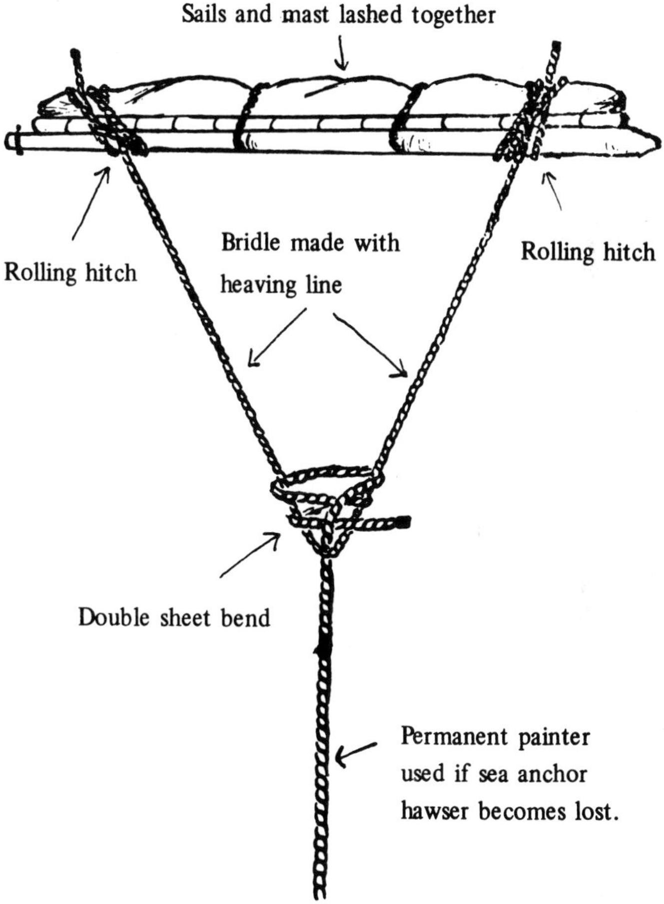
Sails and mast lashed together
Rolling hitch
Bridle made with
heaving line
Rolling hitch
Double sheet bend
Permanent painter
used if sea anchor
hawser becomes lost.

BEACHING

To beach a boat

1. Well outside the line of surf, let fly sheets, down helm and bring the boat head to wind.
2. Stream the sea-anchor. Dredge a grapnel if one is available.
3. Send down the sails and strike the mast.
4. Unship the tiller and rudder. Ship the steering oar.
5. Man the oars.
6. Let the boat drift slowly in; keeping her head to sea.
7. Beach the boat stern first. Everyone should disembark over the stern (women and children first), as quickly as possible.

If you make a sandy beach with surf running; you will have to beach the boat. Never attempt to do this in the dark, stand-off until daylight. Although the surf may not look much from seaward, it will be terrifying when the boat is in it. The important thing is to keep the boat's head to sea all the time. Whenever a large breaker is bearing down towards the boat, give way together and row into it, letting the boat ride in on the back of each sea and keeping the sea-anchor out all the time, to help you keep head to sea. If as you are coming in, the wind should drop and become an off-shore breeze, it will not have any effect on the sea and no heed need be paid to it. When the boat is beached, no one is to be allowed to leave the boat over the bow or amidships, because there is always a strong undertow which could carry them out to sea. Take the painter ashore with you. Do not lose the boat.

You will often see boats running before the surf and beaching bow first. This method is not advised for an inexperienced crew landing on a strange beach. Do remember to keep everyone seated as low as possible and so far as you can towards the bow but out of the ends of the boat.

Attention is drawn to the landing signals on pages 313 & 314.

LEAKS

If the boat is making water in a particular place, the only repairs you can make must be on the outboard side. Use the boat cover you placed in the boat before launching or if there was no boat cover, the jib. Making the cover or sail well fast at both ends, pass it under the keel from forward. Work the canvas into place over the leak and draw it as tight as you can. The more thicknesses of canvas and the tighter they can be held against the outside of the boat, the better will be the control of the leak. A leak stopper is unlikely to be included in the boat's equipment but if one is available use it.

JURY KEEL

Finding that the boat is being blown onto a rocky lee shore and that tack as you may, you cannot get away from it, it might just be possible to hold the boat off by constructing a jury keel.

To do this:– Lay two oars blade forward, one on each of the side-benches. Tie the two blades together at the tip. Pass the oars forward, taking the blades beyond the bow. Lash the loom of each oar to the gunwale, but not too tightly. Take a clove hitch round the grip of each oar in the bight of a heaving line. Force the tied blades of the oars downwards under the keel by hauling forward on the lines attached to the grips. When the oars are up and down, stay them fore and aft with the heaving lines made fast to the grips. This will help give the boat a better grip of the water and so reduce the amount of leeway the boat makes.

COMPASS COURSES

When attempting to make a compass course, no metal should be allowed within about 5 feet (1.5m) of the compass.

In northern latitudes a very good approximation of the error of the compass, which is due to variation, can be gained from the North Star.

Except in high latitudes, a course laid either due East or due West (according to the prevalent winds) will bring you to land eventually.

RULE OF THE ROAD

When two sailing vessels are approaching one another so as to involve risk of collision, one of them shall keep out of the way of the other as follows:–

(i) When each has the wind on a different side, the vessel which has the wind on the port side shall keep out of the way of the other.

(ii) When both have the wind on the same side, the vessel which is to windward shall keep out of the way of the vessel which is to leeward.

(iii) When two power vessels are meeting head-on so as to involve risk of collision, each shall alter her course to starboard.

(iv) When two power vessels are crossing so as to involve risk of collision, the vessel which has the other on her own starboard side shall keep out of the way of the other.

(v) A power driven vessel shall keep out of the way of a sailing vessel.

(vi) A vessel which is overtaking shall keep out of the way of the overtaken vessel.

At night:– Lifeboats under way, shall have ready to hand an electric torch or oil lantern showing a white light. Lifeboats under power carry a white all-round light. Green and red side lights in a combined lantern should be shown if possible.

In the U.S. territorial waters, lifeboats are exempt from carrying lights.

PART TWO. THE LIFEBOAT AND LIFERAFT
CHAPTER ONE. I.M.O. RECOMMENDATIONS FOR LIFE-SAVING APPLIANCES ON ALL SHIPS AS FROM 1st JULY 1986

COMMUNICATIONS

1. **Paragraphs 2(d) and 2(e) apply to all ships.** With respect to ships constructed before 1st. July 1986, paragraphs 2(d) and 2(e) shall apply not later than 1st. July, 1991.

2(a) **Radio life-saving appliances**

With effect from 1st July, 1986 an emergency position-indicating radio beacon shall be provided and so positioned on board so as to be capable of floating, breaking free and being automatically activated in the event of the ship sinking. (This to be a satellite E P I R B when satellite transmissions become available).

(b) **Portable radio apparatus for survival craft**

(i) A portable radio apparatus for survival craft shall be provided. The portable radio apparatus shall be stowed in a protected and easily accessible position ready to be moved to any survival craft in an emergency, except that in the case of a ship with lifeboats stowed in widely separated positions fore and aft, the portable radio apparatus shall be stowed in the vicinity of the lifeboats which are furthest away from the ship's main transmitter.

(ii) The requirements of paragraph (i) need not be complied with if a radio installation, is fitted in a lifeboat on each side of the ship or in the stern launched lifeboat capable of being launched by free-fall on a cargo ship.

(iii) On ships engaged on voyages of such duration that in the opinion of the Administration portable radio apparatus for survival craft is unnecessary, the Administration may allow such equipment to be dispensed with.

(c) **Radio telegraph installation for lifeboats**

On passenger ships engaged on voyages which are not short international voyages:

(i) Where the total number of persons on board is more than 199 but less than 1,500, a radiotelegraph installation shall be fitted in at least one of the lifeboats.

(ii) Where the total number of persons on board is 1,500 or more, at least one lifeboat on each side shall be so fitted.

(d) **Survival craft emergency position-indicating radio beacons.**

One manually activated emergency position-indicating radio beacon shall be carried on each side of the ship, they shall be so stowed that they can be rapidly placed in any survival craft except the liferaft stowed in the bow and/or stern of a cargo ship which are more than 100m. (325 ft.) from the survival craft.

(e) **Two-way radiotelephone apparatus**

(i) Two-way radiotelephone apparatus shall be provided for communication between survival craft and ship and between ship and rescue boat. An apparatus need not be provided for every survival craft; however, at least three apparatus shall be provided on each ship. This requirement may be complied with by other apparatus used on board provided such apparatus is not incompatible with the appropriate requirements of the Regulations.

(ii) For ships constructed before 1st July, 1986 such apparatus shall have precautions taken to prevent the inadvertent selection of VHF Channel 16.

3. **Distress Flares**

Not less than twelve rocket parachute flares shall be carried and be stowed on or near the navigating bridge.

4. **On-board communication and alarm systems**

(a) An emergency means comprised of either fixed or portable equipment or both shall be provided for two-way communications between emergency control stations, muster and embarkation stations and strategic positions on board.

(b) A general emergency alarm system shall be provided and shall be used for summoning passengers and crew to muster stations and to initiate the actions included in the muster list. The system shall be supplemented by either a public address system or other suitable means of communication.

PERSONAL LIFE SAVING APPLIANCES

1. **Lifebuoys**

(a) Lifebuoys shall be;

(i) so distributed as to be readily available on both sides of the ship and as far as practicable on all open decks extending to the ship's side; at least one shall be placed in the vicinity of the stern;

(ii) so stowed as to be capable of being rapidly cast loose, and not permanently secured in any way.

(b) At least one lifebuoy on each side of the ship shall be fitted with a buoyant lifeline equal in length to not less than twice the height at which it is stowed above the waterline in the lightest sea-going condition, or 30m. (97.5ft.), which ever is the greater.

(c) Not less than one half of the total number of lifebuoys shall be provided with self-igniting lights; not less than two of these shall also be provided with self-activating smoke signals and be capable of quick release from the navigating bridge; lifebuoys with lights and those with lights and smoke signals shall be equally distributed on both sides of the ship and shall not be the lifebuoys provided with lifelines.

(d) Each lifebuoy shall be marked in block capitals of the Roman alphabet with the name and port of registry of the ship on which it is carried.

2. **Lifejackets**

(a) A lifejacket shall be provided for every person on board the ship, and in addition:

(i) a number of lifejackets suitable for children equal to at least 10% of the number of passengers on board shall be provided or such greater number as may be required to provide a lifejacket for each child;

(ii) a sufficient number of lifejackets shall be carried for persons on watch and for use at remotely located survival craft stations.

(b) Lifejackets shall be so placed as to be readily accessible and their location shall be plainly indicated. Where due to the particular arrangements of the ship, the lifejackets provided may become inaccessible, alternative provisions shall be made to the satisfaction of the Administration which may include an increase in the number of lifejackets to be carried.

3. **Immersion suits**
An immersion suit, of an appropriate size, shall be provided for every person assigned to crew the rescue boat.

Operating Instructions

1. **This Regulation applies to all ships.**
2. Posters or signs shall be provided on or in the vicinity of survival craft and their launching controls and shall:
(a) illustrate the purpose of controls and the procedures for operating the appliance and give relevant instructions or warnings;
(b) be easily seen under the emergency lighting conditions;
(c) use symbols in accordance with the Recommendations of the Organisation.

Manning of survival craft and supervision.

1. **This Regulation applies to all ships.**
2. There shall be a sufficient number of trained persons on board for mustering and assisting untrained persons.
3. There shall be a sufficient number of crew members who may be deck officers or certificated persons on board for operating the survival craft and launching arrangements required for abandonment by the total number of persons on board.
4. A deck officer or certificated person shall be placed in charge of each survival craft to be used. However, the Administration having due regard to the nature of the voyage, the number of persons on board and the characteristics of the ship, may permit persons practised in the handling and operating of liferafts to be placed in charge of liferafts in lieu of persons qualified as above. A second-in-command shall also be nominated in the case of lifeboats.
5. The person in charge of the survival craft shall have a list of the survival craft crew and shall see that the crew under his command are acquainted with their duties. In lifeboats the second-in-command shall also have a list of the lifeboat crew.
6. Every lifeboat required to carry a radiotelegraph installation shall have a person assigned who is capable of operating the equipment.
7. Every motorized survival craft shall have a person assigned who is capable of operating the engine and carrying out minor adjustments.
8. The master shall ensure the equitable distribution of persons referred to in paragraphs 2, 3 and 4 among the ship's survival craft.

Survival craft muster and embarkation arrangements.

1. Lifeboats and liferafts for which approved launching appliances are required shall be stowed as close to accommodation and service spaces as possible.
2. Muster stations shall be provided close to the embarkation stations. Each muster station shall have sufficient space to accommodate all

persons assigned to muster at that station.

3. Muster and embarkation stations shall be readily accessible from accommodation and work areas.

4. Muster and embarkation stations shall be adequately illuminated by lighting supplied from the emergency source of electrical power.

5. Alleyways, stairways and exits giving access to the muster and embarkation stations shall be lighted. Such lighting shall be capable of being supplied by the emergency source of electric power.

6. Davit launched survival craft muster and embarkation stations shall be so arranged as to enable stretcher cases to be placed in survival craft.

7. An embarkation ladder extending, in a single length, from the deck to the waterline in the lightest sea-going condition under unfavourable conditions of trim and with the ship listed not less than 15 degrees either way shall be provided at each launching station or at every two adjacent launching stations. However, the Administration may permit such ladders to be replaced by approved devices to afford access to the survival craft when waterborne, provided that there shall be at least one embarkation ladder on each side of the ship. Other means may be permitted for the liferafts stowed in the bow and/or stern of a cargo ship and which are more than 100m. (325ft.) from the survival craft. *(A knotted rope is sometimes provided for these liferafts).*

8. Where necessary, means shall be provided for bringing the davit launched survival craft against the ship's side and holding them alongside so that persons can be safely embarked. *(Bowsing-lines and tackles).*

Launching stations

Launching stations shall be in such positions as to ensure safe launching having particular regard to clearance from the propellor and steeply overhanging portions of the hull and so that, as far as possible, survival craft, specially designed for free-fall launching, can be launched down the straight side of the ship. If positioned forward, they shall be located abaft the collision bulkhead in a sheltered position and, in this respect, the Administration shall give special consideration to the strength of the launching appliance.

Stowage of survival craft.

1. Each survival craft shall be stowed

(a) so that neither the survival craft nor its stowage arrangements shall interfere with the operation of any other survival craft or rescue boat at any other launching station;

(b) as near the water surface as is safe and practicable and, in the case of a survival craft other than a liferaft intended for throw-overboard launching, in such a position that the survival craft in the embarkation position is not less than 2m. (6.5ft.) above the waterline with the ship in the fully loaded condition under unfavourable conditions of trim and listed up to 20 degrees, either way, or to the angle at which the ship's weather deck edge becomes submerged, whichever is the less:

(c) in a state of continuous readiness so that two crew members can carry out preparation for embarkation and launching in less than 5 minutes;

(d) fully equipped as required;

(e) as far as practicable, in a secure and sheltered position and protected from damage by fire and explosion.

2. Lifeboats for lowering down the ship's side shall be stowed as far forward of the propellor as practicable. On cargo ships of 80m. (260ft.) in length and upwards but less than 120m (390ft) in length, each lifeboat shall be so stowed that the after end of the lifeboat is not less than the length of the lifeboat forward of the propellor. On cargo ships of 120m. (390ft.) in length and upwards, each lifeboat shall be so stowed that the after end of the lifeboat is not less than one-and-a half times the length of the lifeboat forward of the propellor. Where appropriate the ship shall be so arranged that lifeboats, in their stowed positions, are protected from damage by heavy seas.

3. Lifeboats shall be stowed attached to launching appliances. In addition liferafts shall be so stowed as to permit manual release from their securing arrangements.

4. Davit launched liferafts shall be stowed within reach of the lifting hooks, unless some means of transfer is provided which is not rendered inoperable within the limits of trim and list prescribed in paragraph 1 (b) or by ship motion or power failure.

5. Liferafts intended for throw-overboard launching shall be so stowed as to be readily transferable for launching on either side of the ship unless liferafts of the aggregate capacity required to be capable of being launched on either side are stowed on each side of the ship.

Stowage of rescue boats

Rescue boats shall be so stowed as to be:

(a) in a state of continuous readiness for launching in not more than 5 minutes;

(b) in a position suitable for launching and recovery;

(c) so placed that neither the rescue boat nor its stowage arrangements will interfere with the operation of any survival craft at any other launching station;

(d) if it is also a lifeboat, then in compliance with the requirements for the stowage of survival craft.

Survival craft launching and recovery arrangements

1. Launching appliances shall be provided for all survival craft except:

(a) survival craft which are boarded from a position on deck which is less than 4.5m (14.4ft.) above the waterline in the lightest seagoing conditions and which either:

(i) have a mass of not more than 185kg. (416lbs.) or;

(ii) are stowed for launching directly from the stowed position under unfavourable conditions of trim of up to 10 degrees and with the ship listed not less than 20 degrees either way;

(b) survival craft having a mass of not more than 185kg. (416lbs.) and which are carried in excess of the survival craft for 200% of the total number of persons on board the ship.

2. Each lifeboat shall be provided with an appliance which is capable of launching and recovering the lifeboat.

3. Launching and recovery arrangements shall be such that the appliance operator on the ship is able to observe the survival craft at all times during the launching and for lifeboats during recovery.

4. Only one type of release mechanism shall be used for similar survival craft carried on board the ship.
5. Preparation and handling of survival craft at any one launching station shall not interfere with the prompt preparation and handling of any other survival craft or rescue boats at any other station.
6. Falls, where used, shall be long enough for the survival craft to reach the water with the ship in its lightest sea-going condition, under unfavourable conditions of trim and with the ship listed not less than 20 degrees, either way.
7. During preparation and launching, the survival craft, its launching appliance, and the area of water into which it is to be launched shall be adequately illuminated by lighting supplied from the emergency source of electrical power.
8. Means shall be available to prevent any discharge of water on to survival craft during abandonment.
9. If there is a danger of the survival craft being damaged by the ship's stabilizer wings, means shall be available, powerd by an emergency source of energy, to bring the stabilizer wings inboard; indicators operated by an emergency source of energy shall be available on the navigating bridge to show the position of the stabilizer wings.
10. If open, partially enclosed or self-righting partially enclosed lifeboats are carried, a davit span shall be provided, fitted with not less than two lifelines of sufficient length to reach the water with the ship in its lightest sea-going condition, under unfavourable conditions of trim and with the ship listed not less than 20 degrees either way.

Rescue boats embarkation, launching and recovery arrangements

1. The rescue boat embarkation and launching arrangements shall be such that the rescue boat can be boarded and launched in the shortest possible time.
2. If the rescue boat is one of the ship's survival craft, the embarkation arrangements and launching station shall comply with the requirements for survival craft muster and embarkation arrangements and launching stations.
3. Launching arrangements shall comply with survival craft launching and recovery arrangements. However, all rescue boats shall be capable of being launched, where necessary utilizing painters, with the ship making headway at speeds up to 5 knots in calm water.
4. Rapid recovery of the rescue boat shall be possible when loaded with its full complement of persons and equipment. If the rescue boat is also a lifeboat, rapid recovery shall be possible when loaded with its lifeboat equipment and the approved rescue boat complement of at least six persons.

Line throwing appliances

A line throwing appliance shall be provided.

Although not an I.M.O. requirement, some Administrations insist that all lifeboats including totally enclosed lifeboats, carried on vessels sailing under their flag, shall be provided with a mast and sails and the means of rigging same.

If available, a grapnel with a length of strong line attached may be kept in each rigid lifeboat and rescue boat. In an emergency off a lee shore, a grapnel can be a very useful piece of equipment.

I.M.O. ADDITIONAL REQUIREMENTS FOR PASSENGER SHIPS BUILT AFTER 1st JULY 1986

SURVIVAL CRAFT AND RESCUE BOATS

1. **Survival Craft**

(a) Passenger ships engaged on international voyages which are not short international voyages shall carry:

(i) Partially enclosed, self-righting partially enclosed or totally enclosed lifeboats on each side of such aggregate capacity as will accommodate not less than 50% of the total number of persons on board. The Administration may permit the substitution of lifeboats by liferafts of equivalent total capacity provided that there shall never be less than sufficient lifeboats on each side of the ship to accommodate 37.5% of the total number of persons on board. These liferafts may be either rigid or inflatable and shall be served by launching appliances equally distributed on each side of the ship; and

(ii) in addition, liferafts of such aggregate capacity as will accommodate at least 25% of the total number of persons on board. These liferafts shall be served by at least one launching appliance on each side which may be those provided in compliance with paragraph (1) above or equivalent approved appliances capable of being used on both sides. However, stowage of these liferafts need not be within reach of the lifting hooks.

(b) Passenger ships engaged on short international voyages and complying with the special standards of subdivision shall carry :

(i) Partially enclosed, self-righting partially enclosed or totally enclosed lifeboats equally distributed, as far as practicable, on each side of the ship and of such aggregate capacity as will accommodate at least 30% of the total number of persons on board and rigid or inflatable liferafts of such aggregate capacity that, together with the lifeboat capacity, the survival craft will accommodate the total number of persons on board. The liferafts shall be served by launching appliances equally distributed on each side of the ship; and

(ii) in addition, liferafts of such aggregate capacity as will accommodate at least 25% of the total number of persons on board. These liferafts shall be served by at least one launching appliance on each side which may be those provided by paragraph (1) above or equivalent approved appliances capable of being used on both sides. However, stowage of these liferafts need not be within reach of the lifting hooks.

(c) Passenger ships engaged on short international voyages and not complying with the special standard of subdivision, shall carry survival craft complying with the requirements of paragraph (a) (i) above.

(d) All survival craft required to provide abandonment by the total number of persons on board shall be capable of being launched with their full complement of persons and equipment within a period of 30 minutes from the time the abandon ship signal is given.

(e) In lieu of meeting the requirements of paragraphs (a), (b) and (c) above. Passenger ships of less than 500 tons gross tonnage where the total number of persons on board is less than 200 may, comply with the following:

- (i) They shall carry on each side of the ship rigid or inflatable liferafts of such aggregate capacity as will accommodate the total number of persons on board.
- (ii) Unless the liferafts required by paragraph (i) above can be readily transferred for launching on either side of the ship, additional liferafts shall be provided so that the total capacity on each side will accommodate 150% of the total number of persons on board.
- (iii) If the rescue boat is also a partially enclosed, self-righting partially enclosed or totally enclosed lifeboat, it may be included in the aggregate capacity required by paragraph (e), (i) above, provided that the total capacity available on either side of the ship is at least 150% of the total number of persons on board.
- (iv) In the event of any one survival craft being lost or rendered unserviceable, there shall be sufficient survival craft available for use on each side to accommodate the total number of persons on board.

2. **Rescue Boats**

(a) Passenger ships of 500 tons gross tonnage and over shall carry at least one rescue boat on each side of the ship.

(b) Passenger ships of less than 500 tons gross tonnage shall carry at least one rescue boat.

(c) A lifeboat may be accepted as a rescue boat provided it also complies with the requirements for a rescue boat.

3. **Marshalling of Liferafts**

(a) The number of lifeboats and rescue boats that are carried on passenger ships shall be sufficient to ensure that in providing for abandonment by the total number of persons on board not more than six liferafts need be marshalled by each lifeboat or rescue boat.

(b) The number of lifeboats and rescue boats that are carried on passenger ships engaged on short international voyages and complying with the special standards of subdivision shall be sufficient to ensure that in providing for abandonment by the total number of persons on board not more than nine liferafts need be marshalled by each lifeboat or rescue boat.

Personal Life-saving Appliances.

1. **Lifebuoys.**

(a) A passenger ship shall carry not less than the number of lifebuoys prescribed in the following table:

Length of ship in metres	Length of ship in feet	Minimum number of lifebuoys
Under 60	Under 197	8
60 and under 120	197 and under 394	12
120 and under 180	394 and under 590	18
180 and under 240	590 and under 787	24
240 and over	787 and over	30

(b) Passenger ships of under 60m (197ft.) in length shall carry not less than six lifebuoys provided with self-igniting lights.

2. **Lifejackets**

In addition to a lifejacket for every person on board and a number of additional lifejackets for persons on watch, every passenger ship shall carry lifejackets for not less than 5% of the total number of persons on board. These lifejackets shall be stowed in conspicuous places on deck or at muster stations.

3. **Lifejacket lights**

(a) **This paragraph applies to all passenger ships.** With respect to passenger ships constructed before 1st July. 1986, the requirement of this paragraph shall apply not later than 1st July, 1991.

(b) On passenger ships engaged on international voyages which are not short international voyages each lifejacket shall be fitted with a light.

4. **Immersion suits and thermal protective aids**

(a) **This paragraph applies to all passenger ships.** With respect to passenger ships constructed before 1st July, 1986, the requirements of this paragraph shall apply not later than 1st July, 1991.

(b) Passenger ships shall carry for each lifeboat on the ship at least three immersion suits and, in addition, a thermal protective aid for every person to be accommodated in the lifeboat and not provided with an immersion suit. These immersion suits and thermal protective aids need not be carried:

(i) for persons to be accommodated in totally or partially enclosed lifeboats; or

(ii) if the ship is constantly engaged on voyages in warm climates where in the opinion of the Administration, thermal protective aids are unecessary.

(c) The provisions of paragraph 4 (b) (i) apply also to totally or partially enclosed lifeboats not complying with the 1981 (as amended) I.M.O. requirements, provided that they are carried on ships constructed before 1st July, 1986.

Survival craft and rescue boat embarkation arrangements

1. On passenger ships, survival craft embarkation arrangements shall be designed for:

(a) All lifeboats to be boarded and launched either directly from the stowed position or from an embarkation deck but not both;
(b) davit launched liferafts to be boarded and launched from a position immediately adjacent to the stowed position or from a position to which, the liferaft is transferred prior to launching.

2. Rescue boat arrangements shall be such that the rescue boat can be boarded and launched directly from the stowed position. Notwithstanding the requirements of paragraph 1 (a) above, if the rescue boat is also a lifeboat and the other lifeboats are boarded and launched from an embarkation deck, the arrangements shall be such that the rescue boat can also be boarded and launched from the embarkation deck. With the number of persons assigned to crew the rescue boat on board.

Stowage of additional liferafts

On passenger ships, every liferaft shall be stowed with its painter permanently attached to the ship and with a float free arrangement so that as far as practicable the liferaft floats free and, if inflatable, inflates automatically when the ship sinks.

Muster stations

Every passenger ship shall have passenger muster stations which shall:
(a) be in the vicinity of, and permit ready access for the passengers to, the embarkation stations unless in the same location;
(b) have ample room for marshalling and instruction of the passengers.

Drills

1. **This regulation applies to all passenger ships after 1st July, 1986.**
2. On passenger ships an abandon ship drill and fire drill shall take place weekly.

I.M.O. ADDITIONAL REQUIREMENTS FOR CARGO SHIPS BUILT AFTER 1st JULY, 1986

Survival craft and rescue boats

1. **Survival Craft**

(a) Cargo ships shall carry:

(i) Totally enclosed lifeboats of such aggregate capacity on each side of the ship as will accommodate the total number of persons on board. The Administration may, however, permit cargo ships, except oil tankers, chemical tankers and gas carriers, operating under favourable climatic conditions and in suitable areas, to carry self-righting partially enclosed lifeboats, provided the limits of the trade area are specified in the Cargo Ship Safety Equipment Certificate; and

(ii) in addition, a rigid or inflatable liferaft or liferafts, capable of being launched on either side of the ship and of such aggregate capacity as will accommodate the total number of persons on board. If the liferaft or liferafts cannot be readily transferred for launching on either side of the ship, the total capacity available on each side shall be sufficient to accommodate the total number of persons on board.

(b) In lieu of meeting the requirements of paragraph (a) above, cargo ships may carry:

(i) one or more totally enclosed lifeboats, capable of being free-fall launched over the stern of the ship of such aggregate capacity as will accommodate the total number of persons on board; and

(ii) in addition one or more rigid or inflatable liferafts on each side of the ship, of such aggregate capacity as will accommodate the total number of persons on board. The life-rafts on at least one side of the ship shall be served by launching appliances.

(c) In lieu of meeting the requirements of paragraphs (a) and (b) above, cargo ships of less than 85m. (276ft.) in length other than oil tankers, chemical tankers and gas carriers, may comply with the following:

(i) They shall carry on each side of the ship, one or more liferafts of such aggregate capacity as will accommodate the total number of persons on board.

(ii) Unless the liferafts required by paragraph (i) above can be readily transferred for launching on either side of the ship, additional liferafts shall be provided so that the total capacity available on each side will accommodate 150% of the total number of persons on board.

(iii) If the rescue boat required by paragraph 2 is also a life-boat complying with the Regulations, it may be included in the aggregate capacity required by paragraph (c) (i) above, provided that the total capacity available on either side of the ship is at least 150% of the total number of persons on board.

(iv) In the event of any one survival craft being lost or

rendered unserviceable, there shall be sufficient survival craft available for use on each side to accommodate the total number of persons on board.

(d) Cargo ships where the survival craft are stowed in a position which is more than 100m. (325ft.) from the stem or stern shall carry, in addition to the liferafts required by paragraphs (a) (ii) and (b) (ii) above. a liferaft stowed, as far forward or aft, or one as far forward and another as far aft, as is reasonable and practicable. Such liferaft or liferafts need not be fitted with a float free arrangement but may be securely fastened so as to permit manual release and need not be of the type which can be launched from an approved launching device.

(e) With the exception of the survival craft which are boarded from a position on deck which is less than 4.5m (14.4ft.) above the water-line in the lightest sea-going condition, all survival craft required to provide for abandonment by the total number of persons on board, shall be capable of being launched with their full complement of persons and equipment within a period of 10 minutes from the time the abandon ship signal is given.

(f) Chemical tankers and gas carriers carrying cargoes emitting toxic vapours or gases shall carry, totally enclosed lifeboats with a self-contained air support system.

(g) Oil tankers, chemical tankers and gas carriers carrying cargoes having a flashpoint not exceeding 60 degrees C (Closed cup test) shall carry, totally enclosed lifeboats with a self-contained air support system and which are also fire-protected.

2. **Rescue Boats**

Cargo ships shall carry at least one rescue boat. A lifeboat may be accepted as a rescue boat, provided that it also complies with the requirements for a rescue boat.

3. In addition to their lifeboats, cargo ships constructed before 1st July, 1986 shall carry not later than 1st July, 1991.

(a) One or more rigid or inflatable liferafts of such aggregate capacity as will accommodate the total number of persons on board. The liferaft or liferafts shall be equipped with a lashing or an equivalent means of securing the liferaft which will automatically release it from a sinking ship;

(b) where the survival craft are stowed in a position which is more than 100m. (325ft.) from the stem or stern, in addition to the liferafts required by paragraph (a) above, a liferaft stowed as far forward or aft, or one as far forward and another as far aft as is reasonable and practicable. Such liferaft or liferafts need not be fitted with a float free arrangement but may be securely fastened so as to permit manual release.

Personal Life-saving Appliances

1. **Lifebuoys**

(a) Cargo ships shall carry not less than the number of lifebuoys prescribed in the following table.

Length of ship in metres	Length of ship in feet	Minimum number of lifebuoys
Under 100	Under 325	8
100 and under 150	325 and under 487.5	10
150 and under 200	487.5 and under 650	12
200 and over	650 and over	14

(b) Self-igniting lights for lifebuoys on tankers shall be of an electric battery type.

2. **Lifejacket lights**

(a) **This paragraph applies to all cargo ships.** With respect to cargo ships constructed before 1st July 1986. This paragraph shall apply not later than 1st July 1991.

(b) On cargo ships each lifejacket shall be fitted with a light.

3. **Immersion suits and thermal protective aids**

(a) **This paragraph applies to all cargo ships.** With respect to cargo ships constructed before 1st July 1986. This paragraph shall apply not later than 1st July 1991.

(b) Cargo ships shall carry for each lifeboat on the ship at least three immersion suits or, if the Administration considers it necessary and practicable, one immersion suit for every person on board the ship; however, the ship shall carry, in addition to the thermal protective aids required to be carried for 10% of the number of persons every liferaft, lifeboat and rescue boat is permitted to accommodate or two, whichever is the greater. Thermal protective aids for persons on board not provided with immersion suits. These immersion suits and thermal protective aids need not be required if the ship:

(i) has totally enclosed lifeboats on each side of the ship of such aggregate capacity as will accommodate the total number of persons on board; or

(ii) has totally enclosed lifeboats capable of being launched by free-fall over the stern of the ship of such aggregate capacity as will accommodate the total number of persons on board and which are embarked and launched directly from the stowed position, together with liferafts on each side of the ship of such aggregate capacity as will accommodate the total number of persons on board; or

(iii) is constantly engaged on voyages in warm climates where in the opinion of the Administration, immersion suits are necessary.

(c) Cargo ships of less than 85m (276ft.) other than oil tankers, chemical tankers and gas carriers shall carry immersion suits for every person on board unless the ship:

(a) has davit launched liferafts; or

(b) has liferafts served by equivalent approved appliances capable of being used on both sides of the ship and which do not require entry into the water to board the liferaft; or

(c) is constantly engaged in voyages in warm climates where in the opinion of the Administration immersion suits are unnecessary.

(d) the immersion suits required by this Regulation may be used by the crews of rescue boats.

(e) the totally enclosed lifeboats referred to in paragraph 3. (b), (i) and 3. (b), (ii) carried on cargo ships constructed before 1st July, 1986 need not comply with the Regulations for totally enclosed lifeboats.

Survival craft embarkation arrangements

1. Cargo ship survival craft embarkation arrangements shall be so designed that lifeboats can be boarded and launched directly from the stowed position and davit launched liferafts can be boarded and launched from a position immediately adjacent to the stowed position or from a position to which the liferaft is transferred prior to launching.
2. On cargo ships of 20,000 tons gross tonnage and upwards, lifeboats shall be capable of being launched, where necessary utilizing painters, with the ship making headway at speeds up to 5 knots in calm water.

Stowage of liferafts

On cargo ships, every liferaft, other than those required to be stowed either right forward or right aft on ships where the survival craft are more than 100m.(325ft.) from the stem or stern, shall be stowed with its painter permanently attached to the ship and with a float-free arrangement so that the liferafts floats free and, if inflatable, inflates automatically when the ship sinks.

Department of Transport Merchant Shipping Notice
No. **M.1173**

INFLATABLE LIFERAFTS STOWAGE, AND FLOAT FREE ARRANGEMENTS

Notice to Owners, Masters, Skippers and Seamen of Merchant Ships and Fishing Vessels

This notice supersedes M. 789, 936 and 1085

PART I—STOWAGE

(a) Stowage positions

Liferafts should be stowed in positions which will ensure their serviceability when they are needed and from which they may be readily launched. They should not be stowed in positions where there is risk of damage through cargo, stores etc. being handled on deck. When a hydrostatic release is fitted the stowage position should be such that the liferaft will float free if the ship sinks before the liferaft can be launched manually. Whenever practicable rafts should be stowed well clear of the propeller.

While it is considered that the present practice of stowing boats and liferafts alongside the accommodation is generally most suitable, the concentration of these appliances in a small area is undesirable. As ships vary in layout, it is impracticable to lay down precise instructions as to where liferafts should be sited. On small ships, however, the liferafts should not be placed adjacent to the boat. Where more than one liferaft is provided they should be distributed on each side of the ship and so sited, fore and aft, that an incident (fire or collision) is unlikely to make all liferafts inaccessible.

(b) Protection of stowed rafts

Stowage should give the maximum possible protection from smoke, funnel deposits and sparks, oil, heat, flooding, weather etc. If icing-up is likely, as in the case of trawlers operating in Arctic waters, and a proportion of the rafts are fitted with hydrostatic release gear, the remainder should be stowed in protected positions,eg adjacent to casings, so as to ensure that they can still be readily launched in an emergency. Rafts should not be stowed in positions where they may lie in trapped water nor should they be allowed to come into contact with any materials containing copper or copper compounds. Liferafts should be stowed in accordance with the manufacturer's recommendations. Liferafts should not be stowed inside machinery casings since they might be inaccessible in case of an engine room fire and because high temperatures should be avoided as they may cause deterioration of the rubber and accelerate other forms of degradation.

(c) Deck illumination in way of stowages

All ships must be provided with a safety lighting system for illuminating the decks on which the liferafts are stowed. The lights should be arranged to illuminate sufficiently the stowage positions so that the raft may be readily prepared for launching. In the case of the 6 man liferaft stowed forward on certain ships the emergency illumination may be provided by a safety lamp or a gas-light torch, a hand safety lamp is preferable as it can usually be hung or clipped at some convenient position to illuminate the liferaft stowage.

(d) Securing painters

In general the end of the operating cord of every raft should be fastened to any eye-plate on the deck or other suitable strong point so that on being launched the raft is held to the ship. On very large ships where the raft is stowed on the centre line for transference to either side, eg on a purpose built trolley, it may not be practicable to secure the end of the painter to a strong point at the stowage position. In this case the master should carefully instruct the crew that the painter should be secured to a strong point near the ship's side before being launched. It should be impressed on all members of a ship's crew that, if for any reason the operating cord of a raft has to be untied from the strong point before the raft is launched, then the cord should be made fast again to the rail or other suitable strong point before launching takes place. Lives may well be lost through failure to take this elementary precaution.

(e) Height of stowage above waterline

In general rafts should not be stowed more than 60 feet above the lightest sea-going water-line. Owners or builders of ships where this condition cannot be met should be instructed to check that rafts are of a type which are certificated by the Department as having been satisfactorily drop tested from a sufficient height.

(f) Hand launching arrangements

Where the regulations require that the liferafts carried shall be readily transferable to the water on either side of the ship and this is not practicable, a raft or rafts of sufficient aggregate capacity must be stowed on each side of the ship.

To avoid the necessity of manhandling liferafts over guard-rails or bulwarks, portable rails or hinged openings may be necessary. Suitable protection should be provided at liferaft stowage positions to prevent the possibility of a person falling over the side.

(g) Display of illustrated instruction posters

Posters illustrating launching and boarding of liferafts and any necessary precautions should be displayed in waterproof covers adjacent to embarkation positions of rafts which are launched by means of a launching appliance, and adjacent to stowage positions of other rafts.

(h) Removal of transport fastenings on delivery to the ship

Any additional lashing for transport purposes should be removed in accordance with the manufacturers' instructions before the raft is stowed on board the ship.

Attention is drawn to M.S. No. M.1042 on pages 246 & 247.

(i) Embarking and disembarking liferafts

Liferafts should be handled carefully when being embarked and stowed in their operating position or disembarked for servicing. They should not be dropped, for instance, from a gunwhale down on to the deck, and they should not be rolled.

(j) Stowage of rafts in soft valises

With the requirement to fit float free arrangements being introduced the number of liferafts contained in soft valises which are carried in ships has reduced considerably. Where however liferafts in soft valises are provided they may be stowed on wooden platforms or in boxes or on raised gratings. Such platforms and boxes should have drainage holes provided, and construction should as far as possible be rat-proof.

A most suitable type of box is that made with a loose drop-on lid, and the sides and ends hinged so that when the lid is removed the walls collapse. A normal box fitted with fixed sides and ends and a hinged lid is not suitable because the raft in its valise may settle and jam against the sides. Each stowage should be a reasonably close fit to prevent movement of the liferaft which might result in fabric chafe. Under no circumstances should any other item be stowed in the box.

(k) Stowage of rafts in rigid containers

A liferaft packed in an accepted rigid container may be stowed in a cradle without any protection beyond that afforded by its own container. In some older type liferafts where the fastenings of the container are intended to be parted by the inflation of the liferaft, these are not to be removed or in any way weakened. Liferafts should not be stowed in an inverted position, and the drain holes in the bottom of the raft container should be kept clear of all obstructions, eg deck cradle structural members. Liferafts should not be stowed in a vertical position.

(l) Securing of stowed rafts

Bottle-screws should not be used in the holding-down arrangements. It is of the utmost importance that the raft may be released easily for putting over the ship's side or floating free.

(m) Hosing down

Care should be taken when hosing down decks etc not to wet the raft or its containment any more than is unavoidable. On no account should a hose be played on a raft container.

(n) Stowage of rafts adjacent to ships compasses

In deciding on the stowage position of the liferaft, particularly on small ships, consideration should be given to the possible effect on the ship's compass and direction finder of the gas cylinder or any steel in the stowage arrangements. Under these conditions the liferaft and stowage, if necessary should be regarded as a fixed magnetic material for the purpose of paragraph 11 of Notice No. M.616.

(o) Ramp stowage

On passenger ships, especially in Classes II and II(A), fitted with a large number of inflatable liferafts, the rafts may be stowed clear of lifeboats along the ship's side on especially constructed inclined ramps. The arrangements should be such as to control the release of the liferafts one at a time.

(p) Clearance during lowering of liferafts by launching appliances

With the ship upright, any liferafts whilst being lowered by a launching apppliance should clear any projection eg belting by at least 152 mm (6 in.). Nevertheless, fairings should be fitted above and below the beltings in way of the launching positions.

Where it is necessary to lower liferafts past openings between bulwarks or rails and the deck above, or past overhanging decks, satisfactory arrangements should be made by the provision of fending bars, or other equally effective means, to prevent the liferafts from lodging on the rail or being damaged or the passengers from being injured due to the liferaft swinging under the overhang when the vessel is listing.

(q) Shipside ladders on passenger ships

In passenger ships a 3-string shipside ladder should be provided at each liferaft embarkation position extending to the light waterline. There should be satisfactory access to and means of mounting these ladders which should be kept clear of the boarding position on davit launched liferafts.

(r) Single raft stowed forward

In ships having a 6-man raft stowed forward, means of embarking into it should be provided. A suitably knotted lifeline of sufficiently large size and adequate length is considered satisfactory for this purpose.

PART II—FLOAT FREE ARRANGEMENTS

A. On Merchant Ships and Pleasure Craft (13·7 m and over)

1. The Merchant Shipping (Life-saving Appliances) Regulations 1980 as amended, contain certain requirments for the carriage of

liferafts on the above ships. In particular, paragraph (1)(s) of Part 1 of Schedule 9 states that

(i) The liferaft shall be so stowed as to be readily available in case of emergency. It shall be stowed in such a manner as to permit it to float free from its stowage, inflate and break free from the vessel in the event of sinking.

(ii) If used, lashings shall be fitted with an automatic release system of a hydrostatic or equivalent nature approved by the Secretary of State.

(iii) The liferaft required by paragraph (9) of regulation 11 of these Regulations may be securely fastened.

2. The Department accepts the fitting of certain hydrostatic release units as effective compliance with the requirements of the above Regulation. With regard to the liferaft referred to in paragraph (1)(s)(iii) of Part 1 of Schedule 9, a hydrostatic release device need not be fitted.

A weak link should be incorporated in the hydrostatic release system to ensure that a raft which has been released hydrostatically is not dragged under by the sinking ship before it has time to inflate to its full dimensions. The weak link system should be of sufficient strength to pull the painter from the raft container and operate its inflation mechanism, but should break at a force of 227 Kg (500 lb) ± 45 Kg (100 lb).

The arrangement of painter attachment, weak link and manual release should be such that when the raft is hand released for jettisoning it is attached to the ship by the full strength of the painter system.

A Senhouse slip and lashing should be provided between the hydrostatic release and the liferaft holding-down straps for manual release of the raft.

3. Hydrostatic releases should be installed strictly in accordance with the manufacturer's instructions and if of a type which requires servicing should be serviced annually by an approved servicing station who should record the date of servicing on the small tally plate attached to the unit. To enable the latter requirement to be met hydrostatic releases should not be secured permanently to the deck.

4. Only those Hydrostatic Release Units which are of a type accepted by the department and to which a Certificate of Inspection and Test has been issued should be used. Details of Hydrostatic Release Units which have been accepted may be obtained from any of the Department's Marine Offices or Fishing Vessel Survey Offices.

B. Fishing Vessels

Although there is no mandatory requirement in the Fishing Vessel (Safety Provisions) Rules 1975 for liferafts to be fitted with a float free facility the Department strongly recommends that hydrostatic release units should be fitted to all liferafts carried on these vessels.

Department of Transport
Marine Directorate
London WC1V 6LP
April 1985

Attention is drawn to Merchant Shipping Notices Nos. M.1062, M.1063, M.1064 and M.1065 which deal with the worldwide list of service stations for inflatable liferafts.

U.K. REQUIREMENTS UNDER THE 1974 CONVENTION FOR SHIP CONSTRUCTED BEFORE 1st JULY, 1986.

On all such ships, provided that they have not already been implemente the following requirements are to be implemented not later than 1st Jul 1991. An asterisk denotes requirements already implemented by the U.K Administration, either partly or in full.

*The provision of survival craft emergency position-indicating radi beacons.
*The provision of two-way telephone apparatus.
*The provision of operating instructions in the vicinity of survival craft.
*The provision of trained personnel for manning and supervision o survival craft.
*The provision of a training manual.
*The provision of emergency lists, muster lists, emergency instruction and muster stations.
*The provision of an emergency alarm signal.
*The provision of lifejackets.
*The provision of lifejacket lights.
*The provision of a line throwing apparatus.
*The provision of distress rockets.
*The provision of lifebuoys.
*The provision of immersion suits and thermal protective aids.
*The provision of liferafts.
*The provision of emergency lighting.
*The provision of launching appliances.
*The provision of side ladders.
*The provision of means to prevent the discharge of water into surviva craft.
*The provision of a portable radio transmitter/receiver for survival craft.
*The practice of regular abandon ship and fire drills and training in same
*The practice of maintaining operational readiness, maintenance an inspections.
*The fitting of retro-reflective tape to all survival equipment.
*The periodic servicing of all inflatable equipment and where carrie E.P.I.R.B's.
*The periodic servicing of hydrostatic release equipment.

The following 1974 SOLAS requirements remain unaltered, on vesse constructed prior to 1st July, 1986 as at December, 1983

Every lifeboat must be attached to a separate set of gravity davits except that luffing davits may be fitted where the lifeboats weigh not mor than 2¼ tons (2,250kg) fully loaded with equipment and persons. Luffin davits may not be fitted to tankers of 1,600 gross tons and over or to th emergency lifeboats on passenger ships.

Lifeboats, liferafts and buoyant apparatus must be so stowed that the can be put safely into the water on either side of the ship with 15 degree list either way.

On every ship, the lifeboats, Class "C" boats, D.o.T. inflatable boat liferafts and buoyant apparatus shall be so stowed that they can all b safely launched in the shortest possible time. The overall launching perio in passenger ships which carry liferafts under launching appliances shall not exceed 30 minutes.

Open lifeboats, provided that they are fitted with portable exposure covers are acceptable as lifeboats, mechanically propelled lifeboats, and motor lifeboats on all ships including tankers and as emergency boats on passenger ships.

In the U.K. all the inflatable liferafts shall be supplied with BMT sea anchors and have their ballast pockets modified before 1st January 1985.

FOREIGN GOING PASSENGER SHIPS

Emergency Boats

All passenger ships shall carry two boats attached to davits - one on each side of the ship - for use in an emergency. These boats shall be not more than 8.5 metres (28ft) in length. They may be counted as lifeboats, provided they comply fully with the requirements for lifeboats and may be the motor lifeboats. However, these boats shall not be required to be fitted with skates. Where appropriate, they should be equipped with a radio and searchlight.

Lifeboats, liferafts, buoyant apparatus and motor lifeboats radio.

Passenger ships engaged on voyages which are not short international voyages shall carry:

(i) Lifeboats on each side of such aggregate capacity as will accommodate half the total number of persons on board. Provided that the Administration may permit the substitution of lifeboats by rigid or inflatable liferafts of the same total capacity so however that there shall never be less than sufficient lifeboats on each side of the ship to accommodate 37½% of all on board and

(ii) rigid or inflatable liferafts of sufficient aggregate capacity to accommodate 25% of the total number of persons on board, together with buoyant apparatus for 3% of that number. Provided that ships which have a factor of subdivision of 0.33 or less shall be permitted to carry, in lieu of liferafts for 25% of all on board and buoyant apparatus for 3% of all on board, buoyant apparatus for 25% of that number.

(iii) Where the total number of persons on board a passenger ship engaged on voyages which are not short international voyages, is more than 199 but less than 1,500, a radio telegraph apparatus capable of transmission and reception shall be fitted in at least one of the motor lifeboats carried on the ship.

(iv) Where the total number of persons on board such a ship is 1,500 or more, such a radiotelegraph apparatus shall be fitted in a motor lifeboat on each side of the ship.

(v) The radio apparatus shall be installed in a cabin large enough to accommodate both the equipment and the person using it.

(vi) The arrangements shall be such that the efficient operation of the transmitter and receiver shall not be interfered with by the engine while it is running, whether a battery is on charge or not.

(vii) The radio battery shall not be used to supply power to any engine starting motor or ignition system.

(viii) The motor lifeboat engine shall be fitted with a dynamo for recharging the radio battery and for other services.

(ix) A searchlight shall be fitted in each such motor lifeboat.

A practice muster of the crew must take place before the ship leaves its home port and every seven days, when practicable. A practice muster of the passengers must be held within 24 hours of leaving port.

On passenger ships engaged on short international voyages:

If the lifeboats provided are not sufficient to accommodate all on board rigid or inflatable liferafts shall be provided so that the accommodation provided in the lifeboats and liferafts on the ship shall be sufficient for all on board provided that:—

(i) the number of lifeboats shall, in the case of ships of 58 metres (190ft) in length and over, never be less than four, two of which shall be carried on each side of the ship, and in the case of ships of less than 58 metres (190ft) in length, shall never be less than two, one of which shall be carried on each side of the ship.

(ii) in addition to the liferafts and lifeboats required to be carried to provide accommodation sufficient for all on board. Liferafts sufficient to accommodate 10% of the total number of persons for whom there is accommodation in the lifeboats carried on that ship and buoyant apparatus for at least 5% of the total number of persons on board.

On every passenger ship:

(i) one lifeboat carried on each side of the ship shall be a motor life boat capable of proceeding at 6 knots in calm water with sufficient fuel for 24 hours continuous operation. The engine and accessories shall be suitably enclosed to ensure operation under adverse weather conditions and the engine casing shall be fire-resisting. Provision shall be made for going astern. Provided that in passenger ships in which the total number of persons which the ship is certified to carry, together with the crew, does not exceed 30, only one such motor lifeboat shall be required.

For every person the ship is certified to carry, a lifejacket and in addition, lifejackets for at least 5% over and above that number.

CARGO SHIPS

Lifejackets

On all cargo ships, a lifejacket for every person on board and in addition, lifejackets for 25% of the persons on board. These additional lifejackets should be kept in some such position as the bridge, to cater for situations in which it is not possible for all the crew to collect their lifejackets from their accommodation. The number of additional lifejackets should be related to the number of persons that the vessel is certified to carry, as follows:— More than 16 persons, additional lifejackets for not less than 25%. Between 4 and 16 persons, not less than 4 additional lifejackets. Less than 4 persons, 2 additional lifejackets. A suitable dry stowage position, unlocked and clearly marked, should be provided.

Every lifejacket shall be provided with a light.

A cargo ship of under 500 gross tons shall carry:

(a) On each side of the ship, lifeboats of sufficient capacity to accommodate all persons on board, also sufficient rigid or inflatable liferafts to accommodate all persons on board. Where there are 16 or more persons on board, the ship shall carry at least two liferafts;

or

(b) A lifeboat, Class C boat or D.o.T. inflatable boat which shall be capable of being launched on one side of the ship, and at least two liferafts, each being of sufficient capacity to accommodate all persons on board.

A minimum of four lifebuoys.

A cargo ship of 500 gross tons or over but under 1,600 gross tons shall carry:

(a) on each side of the ship, lifeboats sufficient to accommodate all persons on board and sufficient rigid or inflatable liferafts to accommodate all persons on board. These liferafts shall be equipped with lashings or equivalent means of securing the liferaft which will automatically release them from a sinking ship. Where there are 16 or more persons on board, the ship shall carry at least two liferafts;

or

(b) A motor lifeboat or Class C boat or D.o.T. inflatable boat fitted with a motor and capable of being launched on one side of the ship and sufficient liferafts on each side of the ship, equipped with lashings which will automatically release them from a sinking ship, to accommodate all persons on board.

Where the distance from the embarkation position to the water in the ship's lightest sea-going condition exceeds 4.5m (14ft 9in) these liferafts shall be designed for use with a launching appliance. At least one launching appliance shall be supplied on each side of the ship and no appliance shall be allocated more than two liferafts.

A minimum of 8 lifebuoys.

A cargo ship of 1,600 gross tons or over shall carry:

On each side of the ship sufficient lifeboats to accommodate all the persons on board, the lifeboats shall not be less than 7.3m (24ft) in length and one of them must be a motor lifeboat capable of proceeding at 4 knots for 24 hours when loaded and in a flat calm.

Rigid or inflatable liferafts which can be readily transferred to the water on either side of the ship to accommodate at least half the total number of persons on board. These liferafts shall be equipped with a lashing or equivalent means of securing the liferaft which will automatically release them from a sinking ship.

Where the distance from the embarkation position to the water in the ship's lightest sea-going condition exceeds 4.5m (14ft. 9in.) these liferafts shall be designed for use with a launching appliance. At least one launching appliance shall be supplied on each side of the ship and no appliance shall be allocated more than two liferafts.

A minimum of 8 lifebuoys.

On tankers of 1,600 gross tons and over:

One lifeboat on each side is to be a motor lifeboat capable of proceeding at six knots for 24 hours when fully loaded and in a flat calm.

On tankers of 3,000 gross tons and over:

On each side of the ship two lifeboats of sufficient total capacity to accommodate all the persons on board. Two lifeboats are to be carried aft and two amidships, except that tankers with no amidships superstructure will carry all four boats aft. On such tankers where it is not possible to carry four boats aft, two motor lifeboats only, need be carried. No lifeboat shall then exceed 26 feet (7.9m) in length.

Cargo ships with no amidships superstructure of 150 metres (492ft) and over shall carry in addition, a six person liferaft together with a knotted lifeline or jacob's ladder and a gas-tight torch or hand safety lamp, stowed as far forward or aft as is reasonable and practicable for the use of members of the crew who might be trapped forward or aft.

On all cargo ships, lifeboats, liferafts and buoyant apparatus shall be capable of being put into the water safely and rapidly even under unfavourable conditions of trim and of 15 degrees of list either way.

Passenger ship, Lifeboat/Passenger launches:

On passenger ships, lifeboat/passenger launches may be provided with buoyant seat cushions sufficient for all persons on board, in lieu of life jackets, lifebuoys and buoyant apparatus. Such a cushion must be capable of supporting for each person for which it is provided 16lbs (7.25kg) of iron in fresh water for 24 hours and is required to have Department of Transport approval.

Equipment on ships constructed prior to 1st July, 1986:

Every lifeboat attached to a set of davits, other than a lifeboat which is carried as an alternative to a Class C boat, inflatable boat or other boat shall be so arranged that even under unfavourable conditions of trim and of up to 15 degrees of list either way it can be put into the water when loaded with its full compliment of persons and equipment.

Every Class C boat, inflatable boat or other boat, including a lifeboat carried as an alternative, attached to a single arm davit shall be so arranged that when loaded with its equipment and a launching crew of two persons it can be put into the water on one side of the ship when the ship is upright or is listed 15 degrees towards that side.

When a lifeboat is attached to any set of davits not of sufficient strength that the lifeboat can be safely lowered into the water when loaded with its full complement of persons and equipment. Each davit shall be conspicuously marked with a red band 150mm (6in) wide painted on a white background.

Every lifeboat shall be constructed with rigid sides.

In any lifeboat fitted with a rigid shelter, the shelter shall be capable of being readily opened from both inside and outside and shall not impede rapid embarkation or disembarkation or the launching and handling of the lifeboat. Such a shelter where fitted may be accepted as a cover of highly visible colour capable of protecting the occupants against injury.

In addition to the 1986 IMO lifeboat equipment requirements (page 184) every lifeboat shall carry the following additional equipment:–

(a) A single banked complement of buoyant oars *(one oar for each thwart),* two spare buoyant oars, and a steering oar; (the steering oar blade is to be painted a distinguishing colour), one set and a half of crutches, attached to the lifeboat by lanyard or chain;

(b) Two plugs for each plug hole (except where proper automatic valves are fitted) attached to the lifeboat by lanyards or chains;

(c) A rudder attached to the lifeboat and a tiller. *(Attached with a lanyard and kept shipped at all times whenever possible. Rudder fittings must be so contrived that the rudder cannot be accidentally unshipped. The rudder must be lifted right up the rudder post and half turned to unship it. The tiller is also attached by a lanyard to the lifeboat.);*

(d) Grab lines secured from gunwale to gunwale under the keel *(2½in.*

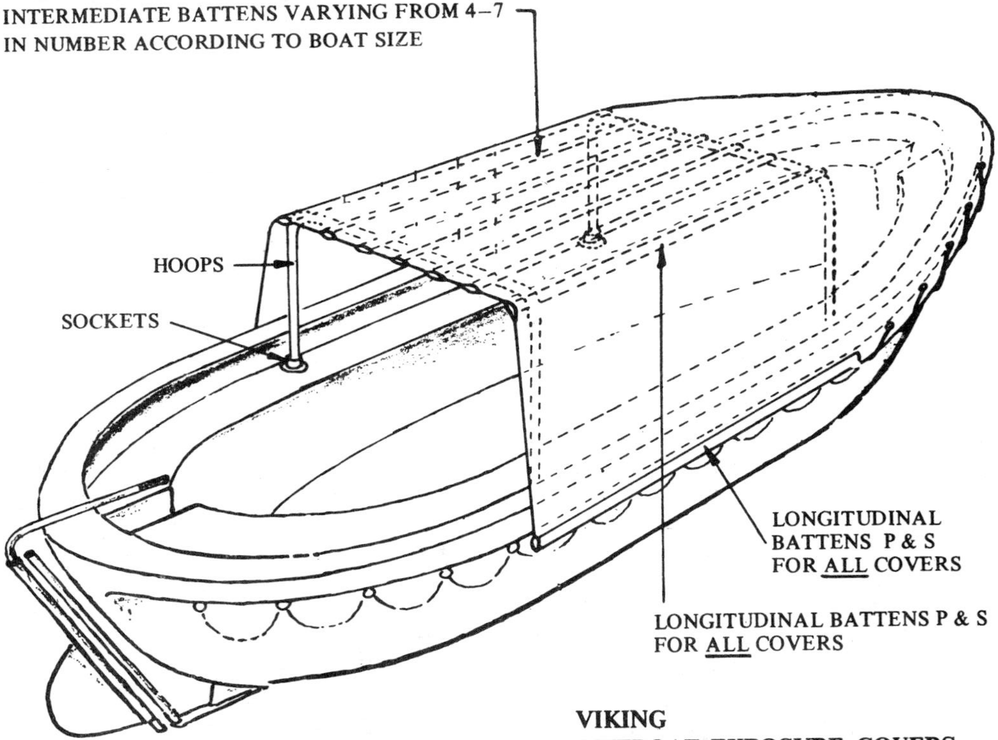

VIKING
LIFEBOAT EXPOSURE COVERS

(20mm) manila rope, knotted at intervals and with a sheepshank at the keel);

(e) An all-round globe lamp with oil sufficient for 12 hours;

(f) A watertight box containing two boxes of matches not readily extinguished by the wind. Three rustproof graduated drinking vessels (graduated to 10, 20 & 50 cc.)

(g) A mast or masts, with galvanised wire stays together with orange coloured sails which shall be marked for identification purposes with the first and last letter of the name of the ship to which the lifeboat belongs;

(h) Two painters of sufficient length and size. One shall be secured to the forward end of the lifeboat with a strop and toggle *(toggle painter)* so that it can be released and the other shall be firmly secured to the bow of the lifeboat *(permanent painter)* and be ready for use; *(stow the permanent painter in the bow sheets on the bottom boards, stow the toggle painter on top ready for use.);*

(i) A vessel containing 4.5 litres (1 gallon) of vegetable, fish or animal oil. A means shall be provided *(oil bag)* to enable oil to be easily distributed on the water, and shall be so arranged that it can be attached to the sea-anchor. *(the sea-anchor hawser to be three times the length of the lifeboat with a tripping line 3.5m (11.5ft) longer);*

(j) A jack-knife fitted with a tin opener to be kept attached to the lifeboat with a lanyard *(in lieu of a jack-knife and 3 tin openers);*

(k) Two light buoyant heaving lines *(in lieu of 2 rescue quoits);*

(l) A cover of highly visible colour capable of protecting the occupants against injury or exposure.

(m) In the first-aid outfit, the six morphine ampoule syringes, have been removed from the required contents.

A mast step, mast clamp and chain plates are to be provided for stepping and securing the mast. Ringbolts provided inboard on both the stem and stern posts for securing painters and steering oar grommet.

A steering oar grommet or crutch is to be provided on the stern post for use with the steering oar.

Lifeboats are to be marked in permanent characters with the Department of Trade stamp, the surveyor's initials, the date the boat was built, the length, depth and breadth of the boat, on one side of the stem or sheer strake. The number of persons the boat is certified to carry, is to be marked on both sides: The name of the ship, port of registry, and the number of the boat are to be painted on each bow of the lifeboat.

If dis-engaging gear is fitted, the means of letting-go both hooks must be placed aft and must be so constructed that the hooks will release together, and only when the boat is waterborne. Floating blocks are to be fitted with a suitable long link or ring that will not jam for attaching to the sling hooks if dis-engaging gear is not fitted.

A fairlead must be provided on the bow, for the purpose of taking either the sea-anchor hawser, a painter or tow rope.

Mechanically propelled boats and motor lifeboats need not carry a mast and sails and need carry only half the quantity of oars.

Every lifeboat certified to carry more than 60 persons shall be a mechanically propelled or motor lifeboat.

Every lifeboat certified to carry more than 100 persons shall be a motor lifeboat.

THWARTSHIP SECTION OF BOAT IN CHOCKS

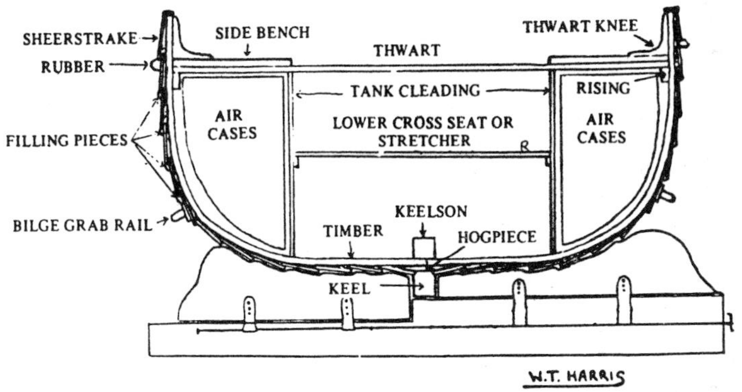

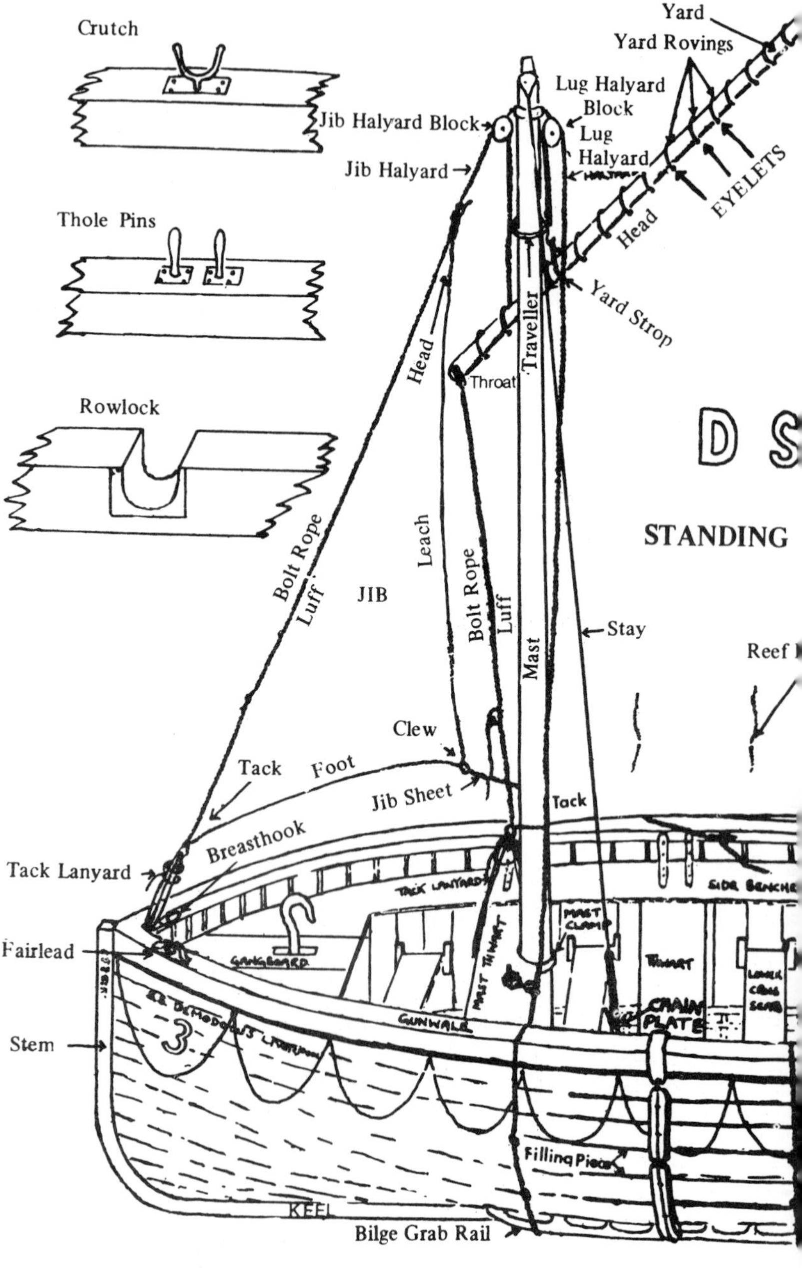
Crutch
Thole Pins
Rowlock
Yard
Yard Rovings
Lug Halyard Block
Lug Halyard
Jib Halyard Block
Jib Halyard
EYELETS
Head
Yard Strop
Traveller
Head
Throat
D S
STANDING
Bolt Rope
Luff
JIB
Leach
Bolt Rope
Luff
Mast
Stay
Clew
Tack
Foot
Jib Sheet
Tack
Breasthook
Tack Lanyard
TACK LANYARD
Fairlead
GANGBOARD
MAST THWART
MAST CLAMP
THWART
CHAIN PLATE
GUNWALE
Stem
3
Filling Piece
KEEL
Bilge Grab Rail

BOAT RIGGED FOR PORT TACK
SHEWING PARTS OF SAIL AND BOAT
Construction of Bow
Sling Hook
Breasthook
Gangboard
Stem
Apron
Breasthook
Stem Band
Deadwood
Sling
Plate
Timbers
KEELSON
Hog Piece
Keel
Forefoot
Keel Plate
Leach
Reef Cringle
Reef Earring
Clew
Main Sheet
Tiller
TIMBERS
Steering Oar
Grommet
GANGBOARD
Sternpost
Rudder
Grab Lines
Keel Grab Lines
KEEL
W.T. HARRIS

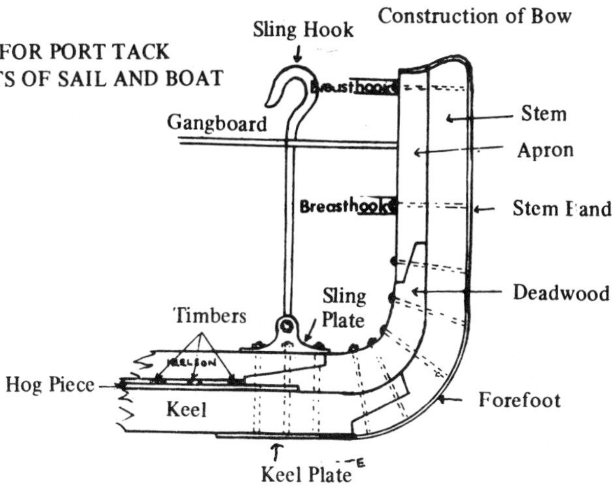

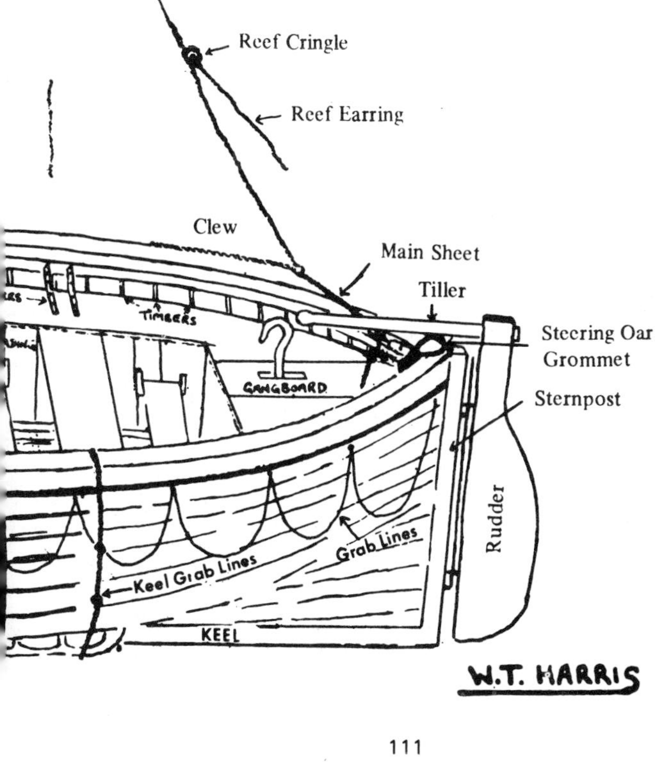

Motor lifeboats on passenger ships and tankers must be capable of proceeding at 6 knots for a period of 24 hours when loaded and in a flat calm.

Motor lifeboats on other cargo ships shall be capable of proceeding at 4 knots for a period of 24 hours when loaded and in a flat calm

A mechanically propelled lifeboat shall be capable of proceeding at a speed of 3.5 knots when fully loaded in smooth water over a distance of a quarter of a mile, and must be capable of proceeding astern.

Mechanically propelled and motor lifeboats are required to be provided with two boathooks.

Motor lifeboats shall also be provided with at least two portable fire extinguishers of 4.5 litres (1 gallon) each suitable for extinguishing oil fires, together with a receptacle containing sand and a scoop. *(Foam extinguishers are not to be exposed to frost. All fire extinguishers are to be recharged annually).* They are to be fitted with compression ignition engines protected by a fire-proof engine cover and in certain cases be watertight above crankshaft level. A cock is to be provided on each end of the fuel line together with a drip tray beneath the fuel tank. A wood lifeboat is required to have a second drip tray beneath the engine. On motor lifeboats certified to carry 100 or more persons an additional pump (which may be engine driven) is required. At least one of the pumps shall be manual.

Every set of davits shall be attached to a wire rope span so positioned that when the boat is in the lowering position the span is as near as practicable over the centre line of the boat. *(The davit head span must not be set up too tightly, in order to ensure that it does not hamper the free action of the davits. When new spans are fitted, they should be 4 to 6 inches (10 to 15cm) longer than the distance between the two eye plates).* There shall be at least two lifelines fitted to the span, long enough to reach the water with the ship at her lightest sea-going draught and listed 15 degrees either way. *(Emergency boats will normally have 4 lifelines fitted to the span).*

Floating blocks attached to lifeboat falls shall be of a non-toppling type and in the case of emergency lifeboats provision shall be made to prevent the falls from cabling. *(This can be achieved by fitting a light wire rope span between the two floating blocks.)*

Cordage rope falls shall be of manila or other suitable material and shall be durable, unkinkable, firm laid and pliable. Winding reels or flaking boxes shall be provided.

The points of attachment of lifeboats, Class C boats and inflatable boats to the falls shall be at such height above the gunwale as to ensure stability when lowering. *(The hooks of lifeboats are required to face inboard to facilitate letting-go when dis-engaging gear is not fitted.)*

No lifeboat shall be less than 4.9m (16.25ft) except when carried as an alternative to a Class C boat.

The cubic capacity of a lifeboat shall be measured in metres and shall be determined by Stirling's (Simpsons') Rules as given by the following formula:–

Cubic capacity = L/12 (4A+2B+4C), Where L denotes length and A.B.C. denotes the areas of cross sections, forward, midships and aft.

In wood lifeboats it may be determined by the formula:–

Cc = length x breadth x depth x 0.6.

In mechanically propelled and motor lifeboats an allowance shall be deducted for the machinery space.

In the pre-decimal era, ten cubic feet were allowed per person when calculating the number of persons a lifeboat would accommodate. With the metric system it is somewhat more involved. However, it shall in no case exceed the number of adult persons wearing lifejackets which can be seated without in any way interfering with the use of oars or the operation of propulsion equipment.

Class C boats

A Class C boat is an open boat with rigid sides at least 12.5ft. (3.8m) long, having a square stern. It must carry internal buoyancy of 7½% of its cubic capacity. The Department of Transport stamp, dimensions and number of persons it is certified to carry must be stamped on it in permanent characters. It must be capable of being launched on one side of the ship either when the ship is upright of listed 15 degrees towards that side.

Every Class C boat carried on cargo vessels of under 500 tons and pleasure yachts of 21 metres (69ft) in length or over shall be equipped as follows:–

(a) A single banked complement of buoyant oars *(one oar for each thwart)* and one spare buoyant oar provided that there shall never be less than three oars; one set of crutches attached to the boat by lanyards or chains; a boat hook;
(b) two plugs for each plug hole (except where proper automatic valves are fitted) attached to the boat by lanyards or chains; a bailer and a bucket;
(c) a rudder attached to the boat and a tiller;
(d) a lifeline becketed round the outside of the boat;
(e) a locker, conspicuously marked as such, suitable for the stowage of small items of equipment;
(f) a painter of sufficient length and size secured to the forward end of the boat with strop and toggle so that it can be released;
(g) means to enable persons to cling to the boat if upturned in the form of bilge keels or keel rails;
(h) a waterproof electric torch suitable for morse signalling together with one spare set of batteries and one spare bulb in a waterproof container;
(i) two light buoyant heaving lines;
(j) a hatchet;
(k) 6 distress flares;
(l) 2 buoyant smoke signals;

An optional outboard motor may be attached.

Department of Transport Inflatable Boats (D.o.T.I. boats)

Every inflatable boat shall comply with the following requirements when it is substituted for a Class C boat:–

1. The overall length of the boat shall be not less than 3.8m (12.5ft) and the boat shall be of such form and proportion as to have ample stability in a seaway when afloat in the empty, laden or swamped condition. The boat shall be suitable for the accommodation of at least 6 persons.

2. The boat shall be of sufficiently robust construction to survive when fully loaded, without such deterioration as would involve any loss of seaworthiness, for 30 days afloat under extremes of temperatures 60

degrees C. (150°F) to minus 30 degrees C (–22°F) and in weather likely to be encountered at sea anywhere in the world.

3. All materials used in the construction of the boat and its accessories shall be able to withstand world-wide sea-going climatic conditions. The boat and its accessories shall be resistant to the effects of humidity when stowed on board a ship and all fabrics, cordage, webbing and thread shall be rotproof. The boat shall be so constructed that it shall not be affected by oil or oil products.

4. The boat shall possess a sufficient margin of durability to ensure that its performance will not be affected after 24 months stowage on board a ship in a weather deck stowage with a minimum of additional protection.

5. The main buoyancy chambers forming the boundry of the boat shall on inflation provide at least 0.17 cubic metres (6 cu.ft.) of volume for each person the boat is certified to accommodate. The diameter of the main buoyancy chambers of single tube boats shall be at least 0.43m (17 in.).

6. The main buoyancy chambers shall be divided into at least two compartments along each side and one compartment in the bow, making a minimum total of five compartments.

7. In boats of more than one tube the volume of either tube shall not exceed 60% of the total volume.

8. At least one thwart shall be fitted so that the boat can be rowed satisfactorily.

9. The floor of the boat shall be waterproof and shall provide an efficient working platform.

10. A transom which shall not be inset by more than 20% of the overall length of the boat shall be provided.

11. A bow cover of highly visible colour and extending for at least 15% of the overall length of the boat shall be provided.

12. A non-return valve shall be fitted to each buoyancy chamber for manual inflation.

13. A safety relief valve designed to operate at a pressure not exceeding 125% of the designed working pressure of the buoyancy chambers shall be fitted in each buoyancy chamber. Means of deflating shall be fitted to each chamber.

14. The boat shall be provided with the following equipment:–

(a) a painter of adequate length and size;
(b) a grabline secured around the outside of the boat and a grabline fitted around the inside of the boat.
(c) a drain plug;
(d) a steering grommet attached to the transom;
(e) handholds or straps for the purpose of righting the boat, from the inverted position;
(f) a sea-anchor attached to the boat by a line of adequate strength at least 9m (30ft) in length. *(to which a swivel should be attached);*
(g) an efficient manually operated bellows or pump;
(h) two buoyant rescue quoits each attached to 18m (60ft) of light buoyant line;
(i) at least two buoyant oars and two buoyant paddles;
(j) a safety knife;
(k) bridle slinging arrangements to enable the boat to be lowered into or raised from the water;

(l) a bailer and two sponges;

(m) a repair kit in a suitable container for repairing punctures in the buoyancy compartments;

(n) one waterproof electric torch suitable for morse signalling together with one spare set of batteries and one spare bulb in a waterproof container.

15. The boat shall be marked with the number of persons which it is deemed fit to accommodate, date of manufacture, maker's name or trade mark, serial number, name of ship and port of registry.

Sea anchors are to be capable of preventing the loaded boat drifting at more than 1 knot in a wind force of 5 or 6.

Drainage arrangements shall be fitted capable of draining the boat within 2½ minutes when flooded. To the extent that the water levels inside and outside the boat are the same it shall not be possible to accidentally flood the boat through these drainage arrangements.

Arrangements must be made on board the ship for the stowage of the boat which must be kept at all times in the fully inflated condition. They are to be kept covered at all times to protect them, particularly their buoyancy tubes, from funnel deposits such as sparks and the stowage must give protection so far as possible from excessive heat, oil and weather etc. Boats must not be stowed in positions where they may lie in trapped water nor must they be allowed to come in contact with any materials containing copper or copper compounds. The cover must protect the whole of the boat and should be arranged for quick release and removal in an emergency.

Provided that the stowage meets with the above conditions the boats may be stowed either flat or on their sides or ends. The holding down arrangements must be such that they can be readily released in an emergency.

Whilst it is recognised that in an emergency (man overboard) the boat will probably be launched by hand, mechanical means for launching and recovery of the boat must be provided. It is not necessary for the launching device, in its operating position, to be capable of lifting the boat from its stowed position and then launching it, provided that the boat can be readily moved from its stowed position to the launching device.

To the average seaman who has been trained "never to sit on the gunwale" of a boat, it will be strange at first to find himself instructed to sit on the buoyancy tubes. This is however the correct procedure when on board an inflatable boat.

As an optional extra these boats may be fitted with an outboard motor. Where an outboard motor has been fitted, the boat must be equipped with a 5lb (2.25kg) dry powder fire extinguisher suitable for oil fires and a set of tools and spares.

Engines shall be permanently attached to the boats they are intended to propel unless they weigh less than 40kg (90lbs.) and the total weight of engine, fuel tank and fuel does not exceed 60kgs. (135lbs).

Portable engines weighing less than 40kg (90lbs.) will have securing lines in the bow of the boat for securing the fuel tank. There shall be a means of ensuring that fuel cannot escape when fuel pipes are disconnected. Outboard motors (which are petrol driven) must never be laid horizontally and if unshipped, must be stowed upright in a rack. Cases have occured when outboard motors have been laid horizontally, of sea water getting into the engine and causing corrosion which has put the engine completely out of service.

Outboard motors are to be test run once a week.

When beaching an inflatable boat, use the same method as you would when beaching a lifeboat. It may help to slightly deflate the buoyancy tubes as this will give the boat a better grip of the water.

If through accidental handling the boat capsizes, it can be readily righted by one man. A line is stowed in a pocket, on the side. When required, it should be pulled out and used as a bridle to right the boat. It should then be re-stowed in its housing, otherwise it could foul the propeller and bring the boat to a sudden stop.

The craft should always be operated with the buoyancy compartments fully and evenly inflated. This will ensure prime performance and reduce likelihood of chafe. Check and top up until the relief valves blow, every three days. The proofing is resistant to oil and fungus attack but a periodic wash with clean fresh water to ged rid of all sand, mud, salt, etc. especially in the crevices inside between floor and tube, will preserve effective life.

Oil can be removed with a rag moistened in white spirit or naptha prior to the washing process.

Check that the equipment is all in its rightful place and in good condition by weekly checks - don't overlook the contents of the stowage box.

Do not allow water to accumulate in the bottom of the craft - the boat cover should keep this out but remember that there are drain holes in the bottom of the transom.

When lifting the boat either into or out of the water a lifting sling is employed. Each leg of the sling is the same length, therefore it can be fastened either way round. Fit the hooks of the sling into ringbolts on the transom and "D" rings on the bow. Ensure that the ropes are not twisted or entwined.

To start a cold engine:– Pull the choke full out, set the twist grip throttle to "Start" and pull the cord firmly. When the engine fires, push the choke into the middle position and as soon as the engine is running evenly, push the choke fully in.

On no account attempt to run the engine (or even start it) on neat petrol. The excessive heat generated would quickly cause the engine to seize up. Petrol and oil as per manufacturer's instructions should be pre-mixed. *(Ratios vary from 16 : 1 up to 50 : 1 depending on make).*

FISHING VESSELS

RULES FOR LIFE-SAVING APPLIANCES

Requirements for vessels of 75 metres in length and over

76.—(1) Every vessel of 75 metres in length and over to which these Rules apply shall carry:—

(a) at least two lifeboats, one of which shall be a motor lifeboat, attached to davits so arranged that there is at least one lifeboat on each side of the vessel, the lifeboats on each side of the vessel being of sufficient aggregate capacity to accommodate half the total number of persons on board the vessel;

(b) at least two liferafts of sufficient aggregate capacity to accommmodate not less than one-and-a-half times the total number of persons on board.

(2) In every such vessel:—

(a) the lifeboat davits shall be of the gravity type except that davits which serve a lifeboat weighing not more than 2·3 tonnes in the turning out condition may be of the luffing type;

(b) the liferafts shall be so stowed that they can be readily transferred to the water on either side of the vessel.

(3) Every such vessel shall carry:—

(a) a portable radio equipment which complies with Part I of Schedule 13 to these Rules (a manual radio) or two portable radio equipments which comply with Part II of the said Schedule (radio telephone) the batteries of which shall be renewed annually. Such radio equipment shall be kept in a suitable place ready to be moved into a lifeboat or liferaft in case of emergency and in vessels where the disposition of superstructures or deckhouses is such that the main transmitter and lifeboats or liferafts are a substantial distance apart such equipment shall be kept in the vicinity of those lifeboats or liferafts which are furthest away from the main transmitter;

(b) for every person on board weighing 32 kilogrammes or more a lifejacket which shall comply with these Rules and for every person on board weighing less than 32 kilogrammes a lifejacket which shall comply with these Rules;

(c) a line throwing appliance;

(d) at least four lifebuoys which shall comply with the following provisions:—

(i) half the lifebuoys carried shall have self-igniting lights attached. Two of the lifebuoys having such lights attached shall be provided with self-activating smoke signals capable of producing smoke of a highly visible colour for at least 15 minutes;

(ii) one lifebuoy on each side of the vessel shall have attached to it a buoyant line of at least 27 metres in length but any such lifebuoy having a line attached shall not have a self-igniting light;

(iii) the two lifebuoys equipped with self-igniting lights and self-activating smoke signals shall be carried one on each side of the navigating bridge and, if reasonably practicable, so fitted as to be capable of quick release;

(e) not less than 12 parachute distress rocket signals which comply with the requirements of these Rules.

Requirements for vessels of 45 metres in length and over but less than 75 metres in length.

77.—(1) Every vessel of 45 metres in length and over but less than 75 metres in length to which these Rules apply shall carry either:—

(a)(i) at least two lifeboats attached to davits so arranged that there is at least one lifeboat on each side of the vessel, the lifeboats on each side of the vessel being of sufficient aggregate capacity to accommodate half the total number of persons on board the vessel; and

(ii) at least two liferafts of sufficient aggregate capacity to accommodate not less than one-and-a-half times the total number of persons on board; or.

(b)(i) on each side of the vessel one or more liferafts of sufficient aggregate capacity to accommodate the total number of persons on board. Each liferaft shall be of approximately the same capacity; and

(ii) a liferaft of sufficient capacity to accommmodate at least half the total number of persons on board and which can be readily placed in the water on either side of the vessel provided that this additional liferaft shall not be required where the liferafts specified in sub-paragraph (b)(i) above can be readily placed in the water on the opposite side of the vessel to that on which they are stowed; and

(iii) a lifeboat, Class C boat or inflatable boat capable of being launched on one side of the vessel with its equipment and a launching crew of two persons when the vessel is upright or listed to 15 degrees in either direction provided that any lifeboat, Class C boat or inflatable boat carried in compliance with this sub-paragraph of this Rule shall be fitted with a suitable engine.

(2) In every such vessel lifeboat davits provided for the lifeboats carried under the provision of paragraph (1)(a)(i) above shall be of the gravity type except that davits which serve a lifeboat weighing not more than 2·3 tonnes in the turning out condition may be of the luffing type.

(3) In every such vessel liferafts carried under the provisions of paragraph (1)(a)(ii) above shall be so stowed that they can be readily transferred to the water on either side of the vessel.

(4) In every such vessel where the distance from the embarkation deck to the waterline in the lightest sea-going condition exceeds 4·6 metres the liferafts carried in compliance with paragraph (1)(b)(i) above shall be of the davit launched type. At least one launching

appliance shall be provided on each side of the vessel for every two rafts carried.

(5) Every such vessel shall carry:—

(a) a portable radio equipment which complies with part I of Schedule 13 to these Rules (a manual radio) or two portable radio equipments which comply with Part II of the said Schedule (radio telephones) the batteries of which shall be renewed annually. Such radio equipment shall be kept in a suitable place ready to be moved into a lifeboat or liferaft in case of emergency and in vessels where the disposition of superstructures or deckhouses is such that the main transmitter and lifeboats or liferafts are a substantial distance apart such equipment shall be kept in the vicinity of those lifeboats or liferafts which are furthest away from the main transmitter;

(b) for every person on board weighing 32 kilogrammes or more a lifejacket which shall comply with these Rules and for every person weighing less than 32 kilogrammes a lifejacket which shall comply with these Rules;

(c) a line throwing appliance;

(d) at least four lifebuoys which shall comply with the following provisions:—

(i) half the lifebuoys carried shall have self-igniting lights attached. Two of the lifebuoys having such lights attached shall be provided with self-activating smoke signals capable of producing smoke of a highly visible colour for at least 15 minutes;

(ii) one lifebuoy on each side of the vessel shall have attached to it a buoyant line of at least 27 metres in length but any such lifebuoy having a line attached shall not have a self-igniting light;

(iii) the two lifebuoys equipped with self-igniting lights and self-activating smoke signals shall be carried one on each side of the navigating bridge and, if reasonably practicable, so fitted as to be capable of quick release;

(e) not less than 12 parachute distress rocket signals which comply with the requirements of these Rules.

Requirements for vessels of 24·4 metres in length and over but less than 45 metres in length

78.—(1) Every vessel of 24·4 metres in length and over but less than 45 metres in length to which these Rules apply shall carry either:—

(a) (i) a lifeboat attached to a mechanically controlled single arm davit of sufficient capacity to accommodate the total number of persons on board the vessel; and

(ii) liferafts on the following scale:—
vessels with 16 or more persons on board—at least two liferafts of sufficient aggregate capacity to accommodate the total number of persons on board;
vessels with fewer than 16 persons on board—at least one liferaft of sufficient capacity to accommmodate the total number of persons on board; or

(b) (i) a lifeboat, Class C boat or suitable inflatable boat which shall be capable of being launched on one side of the vessel; and

(ii) at least two liferafts of sufficient aggregate capacity to accommodate twice the total number of persons on board;

(2) In every such vessel liferafts carried in compliance with this Rule shall be so stowed that they can be readily transferred to the water on either side of the vessel.

(3) Every such vessel shall carry:—

(a) a portable radio equipment which shall comply with Schedule 13 to these Rules provided that the batteries of any radio equipments which comply with Part II of the said Schedule shall be renewed annually. Such radio equipment shall be kept in a suitable place ready to be moved into a lifeboat or liferaft in case of emergency and in vessels where the dispositions of superstructures or deckhouses is such that the main transmitter and lifeboats or liferafts are a substantial distance apart such equipment shall be kept in the vicinity of those lifeboats or liferafts which are furthest away from the main transmitter;

(b) for every person on board weighing 32 kilogrammes or more a lifejacket which shall comply with the requirements of Part I of Schedule II to these Rules and for every person on board weighing less than 32 kilogrammes a lifejacket which shall comply with the requirements of Part II of the Schedule.

(c) a line throwing appliance;

(d) at least four lifebuoys which shall comply with the following provisions:—

(i) · half the lifebuoys carried shall have self-igniting lights attached. Two of the lifebuoys having such lights attached shall be provided with self-activating smoke signals capable of producing smoke of a highly visible colour for at least 15 minutes;

(ii) one lifebuoy on each side of the vessel shall have attached to it a buoyant line of at least 27 metres in length but any such lifebuoy shall not have a self-igniting light;

(iii) the two lifebuoys equipped with self-igniting lights

and self-activating smoke signals shall be carried one on each side of the navigating bridge and, if reasonably practicable, so fitted as to be capable of quick release;

(e) not less than 12 parachute distress rocket signals which comply with the requirements of these Rules.

Requirements for vessels of 17 metres in length and over but less than 24·4 metres in length.

79.—Every vessel of 17 metres in length and over but less than 24·4 metres in length to which these Rules apply shall carry:—

(a) at least two liferafts of sufficient aggregate capacity to accommodate twice the number of persons on board. The liferafts shall be so stowed that they can be readily transferred to the water on either side of the vessel;

(b) a portable radio equipment which complies with Schedule 13 to these Rules. The batteries of any radio equipment which comply with Part II of the said Schedule shall be renewed annually;

(c) for every person on board weighing 32 kilogrammes or more a lifejacket which shall comply with the requirements of Part I of Schedule II to these Rules and for every person on board weighing less than 32 kilogrammes a lifejacket which shall comply with the requirements of Part II of the said Schedule;

(d) a line throwing appliance;

(e) at least four lifebuoys which shall comply with the following provisions:—

(i) half the lifebuoys carried shall have self-igniting lights attached. Two of the lifebuoys having such lights attached shall be provided with self-activating smoke signals capable of producing smoke of a highly visible colour for at least 15 minutes;

(ii) one lifebuoy on each side of the vessel shall have attached to it a buoyant line of at least 27 metres in length but any such lifebuoy having a line attached shall not have a self-igniting light;

(iii) The two lifebuoys equipped with self-igniting lights and self-activating smoke signals shall be carried one on each side of the navigating bridge and, if reasonably practicable, so fitted as to be capable of quick release;

(f) not less than 12 parachute distress rocket signals which comply with the requirements of Schedule 18 to these Rules.

Requirements for vessels of 12 metres in length and over but less than 17 metres in length

80.—Every vessel of 12 metres in length and over but less than 17 metres in length to which these Rules apply shall carry:—

(a) One or more liferafts of sufficient aggregate capacity to accommodate the total number of persons on board. The liferafts shall be so stowed that they can be readily transferred to the water on either side of the vessel;

(b) for every person on board weighing 32 kilogrammes or more a lifejacket which shall comply with the requirements of Part I of Schedule II to these Rules and for every person on board weighing less than 32 kilogrammes a lifejacket which shall comply with the requirements of Part II of the Schedule.

(c) at least two lifebuoys which shall comply with the following provisions:—

(i) half the lifebuoys carried shall have self-igniting lights attached. One of the lifebuoys having such a light attached shall be provided with a self-activating smoke signal capable of producing smoke of a highly visible colour for at least 15 minutes;

(ii) one lifebuoy shall have attached to it a buoyant line of at least 18 metres in length but any such lifebuoy having a line attached shall not have a self-igniting light;

(d) a line throwing appliance;

(e) not less than 12 parachute distress rocket signals which comply with the requirements of Schedule 18 to these Rules.

Requirements for vessels less than 12 metres in length

81.—(1) Every vessel of less than 12 metres in length to which these Rules apply shall carry:—

(a) lifebuoys at least in equal number to half the total number of persons on board and in no case less than two lifebuoys;

(b) a buoyant heaving line at least 18 metres in length shall be attached to one lifebuoy;

(c) not less than six red star distress signals each of which shall be capable of emitting two or more red stars either together or separately at or to a height of not less than 45 metres; each such star shall burn with a luminosity of at least 5,000 candelas for not less than 5 seconds.

In every fishing vessel a motor lifeboat shall have a compression ignition engine and shall be capable of going ahead in smooth water when loaded with its full complement of persons and equipment at a speed of 4 knots and shall be provided with fuel for 24 hours continuous operation at that speed.

In every fishing vessel all lifeboats or Class C boats attached to davits shall be served by wire rope falls and winches except when the lifeboat or Class C boat weighs 2·3 tonnes or less in the lowering condition.

In every fishing vessel arrangements shall be made for warning the crew when the vessel is about to be abandoned.

In every vessel of 45 metres in length and over, one ladder shall be carried at each set of lifeboat davits, in every vessel which carries a Class C boat or lifeboat which cannot be lowered with its full complement of persons, suitable means shall be provided for embarking persons in the boat.

In every vessel where liferafts are the prime survival craft, ladders or other suitable means shall be provided for safe embarkation into the liferafts.

In every vessel means shall be provided for the electric lighting of the launching gear for lifeboats, inflatable boats or davit launched liferafts during the preparation for and process of launching and also for the lighting of the stowage position of liferafts.

In every vessel water in the lifeboats and Class C boats shall be contained in tanks marked "water" and the water is to be frequently changed.

It is recommended that fishing vessels should be provided with float free attachments to liferafts, lifejacket lights and immersion suits and/or thermal protective aids.

C—MUSTER AND DRILLS

Muster list

119.—(1) The skipper of every vessel of 24·4 metres in length and over to which these Rules apply shall prepare or cause to be prepared a muster list showing in respect of each member of the crew the special duties which are allotted to him and the station to which he shall go in the event of an emergency (hereinafter referred to as "emergency station").

(2) In every such vessel, the muster list shall specify definite signals to be made on the whistle or siren for calling the crew to the emergency station and shall include the emergency signal which shall consist of a succession of seven or more short blasts followed by one long blast. In every vessel of 45 metres in length and over the signals made on the whistle or siren shall be supplemented by bells or other means of warning which shall be electrically operated and which shall be capable of being operated from the bridge. The muster list shall also specify the means of indicating when the vessel is to be abandoned.

(3) In every such vessel of 24·4 metres in length and over the muster list shall show the duties assigned to the different members of the crew in connection with:—

(a) the preparation and launching of the boats and liferafts attached to davits or to other launching appliances;

(b) the preparation and launching of liferafts not attached to davits and other lifesaving appliances;

(c) the operation of fire appliances for extinction of fire.

(4) In every such vessel the muster list shall be prepared, or if a new list is not necessary revised, each time a new agreement with the crew has been signed and before the vessel proceeds to sea and shall be dated and signed by the skipper.

(5) In every such vessel if, after the muster list has been prepared, any change takes place in the crew which necessitates an alteration in the muster list, the skipper shall either revise the list or prepare a new list.

(6) In every such vessel, copies of the muster list shall be posted in the crew's quarters and at the main control station before the vessel proceeds to sea and shall be kept so posted while the vessel is at sea.

Training

120.—(1) In vessels of 24·4 metres in length and over to which these Rules apply musters of the crew shall take place at the commencement of each voyage and at intervals of not more than 14 days thereafter, and if more than 25% of the crew have been replaced at any port one of such musters shall take place within 48 hours of leaving that port to ensure that the crew understand and are drilled in the duties assigned to them in the event of an emergency.

(2) In vessels of 12 metres in length and over but less than 24·4 metres in length to which these Rules apply the skipper shall ensure that the crew are trained in the use of all lifesaving and fire appliances and equipment with which the vessel is provided and shall ensure that all members of the crew know where the equipment is stowed and such training shall be carried out at intervals of not more than one month.

(3) In vessels of 75 metres in length and over to which these Rules apply drills shall be so arranged that every lifeboat is swung out at least once per month and, if reasonable and practicable, lowered at least once every four months.

(4) In vessels of 24·4 metres in length and over but less than 75 metres in length to which these Rules apply the Class C boat or inflatable boat shall be swung out at each drill and, if equipped with an engine, the engine shall be operated.

Inspections

121.—(1) In vessels of 24·4 metres in length and over to which these Rules apply lifesaving and fire appliances and equipment shall be inspected when musters of the crew are held, and in any case at intervals of not more than one month to ensure that all equipment is in good condition and always ready for immediate use.

(2) In vessels of 12 metres in length and over but less than 24·4 metres in length to which these Rules apply inspections of the lifesaving equipment and fire appliances shall be made at intervals of not more than one month.

Survival

Never sail when the weather or weather forecast is formidable.

It is important to always inform friends, relatives and the Harbour Master of:– Your expected time of departure, destination, route and estimated time of arrival. In small boats, have a "grab bag" containing a selection of useful articles handy, to take in the survival craft in an emergency.

Ensure that you are in possession of all necessary information, that your equipment is both sufficient and efficient and has been kept in good order. Have adequate reserve supplies of fuel, fresh water and provisions on board. Do not forget to carry distress pyrotechnics and an E.P.I.R.B.

In 1981, 20% of the U.K., pleasure boat drownings were alcohol associated, many of them children who were not supervised properly by drinking parents. Recent research has shown that after three hours exposure to wind, sun-glare, noise, vibration and alcohol, a boat operator may be fatigued to the point where reaction time doubles. At night they also incur colour loss in red and green. Research also shows that fishermen (the largest sub-group) spent twice as much on beer as they did on tackle.

Being drunk in charge of a dinghy, may sound funny but it can be just as lethal as a car. Apart from increasing the danger of accident, alcohol can be a main contributory factor in hypothermia, it lowers the body temperature and also causes dehydration.

THE MERCHANT SHIPPING (LIFE-SAVING APPLIANCES) REGULATIONS 1986.

Ships of Class XII (Pleasure craft)

(1) This Regulation applies to new ships of Class XII *(Keel laid on or after 1st July 1986).*

(2) Every such ship of 21·3 metres in length and over shall carry:—

- (a) at least two liferafts so stowed that they can be readily transferred to the water on either side of the ship, of sufficient aggregate capacity to accommodate twice the total number of persons on board;
- (b) four lifebuoys, two of which shall be fitted with buoyant lifelines and two with self-igniting lights and self-activating smoke signals;
- (c) a lifejacket suitable for a person weighing 32 kg. or more for each such person on board;
- (d) a lifejacket suitable for a person weighing less than 32 kg. for each such person on board;
- (e) a lifejacket light fitted on each of the lifejackets required by sub-paragraphs (c) and (d);
- (f) either
  - (i) 6 rocket parachute flares or
  - (ii) 6 red star distress rocket signals;
- (g) a line throwing appliance;

(h) posters or signs showing operating instructions on or in the vicinity of survival craft and their launching controls;
(i) a training manual;
(j) instructions for on-board maintenance of life saving appliances; and
(k) a copy of the table "Life-Saving Signals and rescue Methods, SOLAS No. 2", published by the Department of Transport.

and any such ship of 25·9 metres in length and over shall carry in addition a rescue boat or inflated boat. A lifeboat may be accepted as a rescue boat provided that it also complies with the requirements for a rescue boat. The lifeboat, rescue boat or inflated boat shall be served by a launching appliance.

(3) Every such ship to which this Regulation applies of 13·7 metres in length or over but less than 21·3 metres in length and engaged on either a voyage to sea in the course of which it is more than 3 miles from the coast of the United Kingdom or on a voyage to sea during the months of November to March, inclusive, shall carry:—

(a) One or more liferafts complying with the requirements of Part I, II or III of Schedule 4 so stowed as to be readily transferrable to the water on either side of the ship, and of sufficient aggregate capacity to accommodate the total number of persons on board;
(b) two lifebuoys, one of which shall be fitted with a self-igniting light and self-activating smoke signal;
(c) a buoyant lifeline at least 18 metres in length;
(d) a lifejacket suitable for a person weighing 32 kg. or more for each such person on board;
(e) a lifejacket suitable for a person weighing less than 32 kg. for each such person on board;
(f) a lifejacket light fitted on each of the lifejackets required by sub-paragraphs (d) and (e);
(g) if operating in partially smooth .waters or proceeding to sea either
   (i) 6 rocket parachute flares, or
   (ii) 6 red star distress signals; and
(h) posters or signs showing operating instructions on or in the vicinity of survival craft and their launching controls;
(i) a training manual;
(j) instructions for on-board maintenance of life-saving appliances; and
(k) a copy of the table "Life-Saving Signals and Rescue Methods, SOLAS No. 2", published by the Department of Transport.

(4) Every such ship of 13·7 metres in length or over but less than 21·3 metres in length which does not proceed to sea or which onlty proceeds to sea during the months of April to October, inclusive, on voyages in the course of which is not more than 3 miles from the coast of the United Kingdom shall carry:—

(a) one lifebuoy for each two persons on board provided that at least two lifebuoys are carried; such ships which operate only in smooth waters shall not be required to carry more than two lifebuoys. One lifebuoy shall be fitted with a self-igniting light and a self-activating smoke signal;

(b) a buoyant line at least 18 metres in length;

(c) a lifejacket suitable for a person weighing 32 kg. or more for each such person on board;

(d) a lifejacket suitable for a person weighing less than 32 kg. for each such person on board;

(e) a lifejacket light fitted on each of the lifejackets required by sub-paragraphs (c) and (d) in the case of ships which proceed to sea;

(f) if operating in partially smooth waters or proceeding to sea either

(i) 6 rocket parachute flares or

(ii) 6 red star distress rocket signals;

(g) a copy of the table "Life-Saving Signals and Rescue Methods, SOLAS No. 2", published by the Department of Transport.

(5) In lieu of carrying life-jackets complying with the requirements of Part I or II of Schedule 10 every such ship may carry lifejackets complying with British Standard Specifications BS 3595: 1981 provided such lifejackets do not depend wholly upon oral inflation. Lifejackets of the partially inherently buoyant type for persons weighing 32 kg. or more shall have buoyancy in the uninflated state of not less than 89 newtons.

General requirements

(5) The following tests and inspections shall be carried out weekly:—

(a) all survival craft, rescue boats and launching appliances shall be visually inspected to ensure that they are ready for use;

(b) all engines in lifeboats and rescue boats shall be run ahead and astern for a total period of not less than 3 minutes provided the ambient temperature is not lower than that at which the engine is required to start; and

(c) the general emergency alarm shall be tested.

(9) Every inflatable liferaft, inflated and rigid inflated rescue boat, inflated boat, inflatable lifejacket and hydrostatic release unit shall be serviced at a service station approved by the Secretary of State at intervals not exceeding 12 months, provided that in any case such interval may be extended by a period not exceeding 5 months.

Survival craft muster and embarkation arrangements.

An embarkation ladder shall be provided at each launching station or at every two adjacent launching stations, extending, in a single length, from the deck to the waterline in the lightest seagoing condition under unfavourable conditions of trim and with the ship listed not less than 15 degrees either way and where such distance exceeds 1 metre such ladders may be replaced by approved devices to afford access to survival craft when waterborne, provided that there shall be at least one embarkation ladder on each side of the ship. Handholds shall be provided to assist in a safe passage from the deck to the ladder and vice-versa.

Every liferaft shall be stowed with its painter permanently attached to the ship and with a float-free arrangement so that the liferaft floats free and, if inflatable, inflates automatically when the ship sinks.

4-man capacity liferafts are only permitted in vessels where the total number of persons on board is less than 6. If more than 5 persons are carried, the minimum capacity of any liferaft is to be 6.

The following requirements for approved liferafts may be dispensed with on vessels of less than 21.3 metres (70 feet) in length:

(a) means for collecting rain shall not be required to be provided.

(b) the height of 18m from which a liferaft is required to be dropped without damage may be reduced to the height of its stowage but in no case shall it be less than 6m (19.5ft);

(c) the method of insulating the floor against cold need not be complied with;

(d) the minimum temperature at which the liferaft is required to operate may be -18 degrees C; (-8 degrees F.)

(e) towing arrangements shall not be required to be provided.

(f) **The following equipment is to be provided;**

(i) One buoyant rescue quoit, attached to at least 30m (97.5ft) of buoyant line;

(ii) for liferafts which are fit to accommodate not more than 12 persons; one safety knife and one bailer;
for liferafts which are fit to accommodate 13 or more persons; two safety knives and two bailers;

(iii) two sponges;

(iv) two paddles;

(v) one repair outfit capable of repairing punctures;

(vi) one topping-up pump or bellows (not applicable to rigid rafts)

(vii) one rust-proof drinking vessel, graduated to 10, 20 and 50 cc.;

(viii) a first-aid outfit complying with the regulations;

(ix) one waterproof electric torch suitable for morse signalling together with one spare set of batteries and one spare bulb in a waterproof container;

(x) two parachute distress rocket signals;

(xi) six anti-seasickness tablets per person the liferaft is fit to accommodate;

(xii) instructions on how to survive in a liferaft;

(xiii) one copy of the Life-Saving Signals SOLAS 2;

(xiv) one sea anchor permanently attached to the liferaft,

(xv) two safety tin openers;

(xvi) three hand-held distress flare signals;

(xvii) watertight receptacles containing ½ litre (1 pint) of fresh water per person the raft is certified to accommodate.

The container and liferafts shall be marked DOT (UK) approved.

The following additional items of equipment are recommended for all craft

Two anchors, each with a warp or chain of appropriate size.
One bilge pump.
An efficient compass and a spare.
Charts covering the intended area of operations.
Navigational drawing instruments. (parallel rulers, dividers etc.)
Daylight distress orange smoke signals.
An appropriate length of buoyant tow rope.
A fog signal.
A leak stopper.
A first-aid box.
A mini flare pack.
A radio receiver for weather forecasts.
A hand lead line.
A waterproof torch or hand lamp.
A light buoyant heaving line 30m (97.5ft) long.
A small supply of fresh water (tinned).
A radar reflector of adequate performance, mounted at least 3m (10ft) above sea level.
An E.P.I.R.B.
A sea-anchor.
A line to be used as inboard lifelines in bad weather.
A suitable engine tool kit.
A "grab bag" of useful survival kit.
Spare engine fuel and oil. A line suitable for use as a tow line.
The name, number or sail number of the boat should be painted on the vessel in letters at 22 cm (9in) high, to assist identification.

Stow liferafts and inflatable boats so that they cannot lie in trapped water or be contaminated by heat, oil, tar, wet paint, acids, copper, sparks exhaust and funnel deposits.

NEVER ABANDON UNTIL IT IS IMPERATIVE BECAUSE THE VESSEL IS ACTUALLY SINKING.

If in addition, you have a dinghy or second raft on board, take it with you as a second line of defence. There have been many occasions when days, weeks and even months have been spent aboard a liferaft. For identification purposes, all liferafts should be marked in large characters on both canopy and under side with the vessel's name or sail number.

All equipment in the vessel, particularly heavy gear, such as cookers, gas bottles, batteries etc., should be well secured (do not rely on gravity) so that they cannot be thrown out of position. Locker doors and drawer should have safety fasteners on them to prevent them opening and spilling the contents.

Inflatable liferafts and boats must be serviced annually. This is most important.

When purchasing a liferaft, ensure that it is a type in which, the sea-anchor hawser is anchored on the opposite side of the raft to the entrance and a tripping line is provided, that the painter is anchored on the same side of the raft as the entrance (so that you can cut it in a hurry), that hand-lines are provided on both the inside and outside of the liferaft, that there is a "V" reinforcement both inside and outside between upper and lower buoyancy tubes, unless the tubes have been welded together as opposed to the normal glued joint, that the sea-anchor is of the new conical type. If possible a double floor and circular entrance.

An elongated liferaft which would allow for some sleeping space is preferable to a circular one, and will not spin.

Carry the E.P.I.R.B. close to the liferaft so that you can take it with you in an emergency.

Inflatable liferafts that are supplied packed in a canvas valise must be protected from rats. Given the opportunity, rats will completely destroy a raft overnight.

Do not purchase a raft manufactured with rubber proofed multi-ply cotton fabric. These have been condemned due to possible deterioration from fungal attack and oxidation of the rubber coating.

In the U.K., E.P.I.R.B's are required to be licensed by the Home Office.

Small boat owners who limit their sailing to a few weeks annually, are advised that much of their life-saving equipment (liferafts, lifejackets, pyrotechnics and E.P.I.R.Bs. in particular) can be obtained on hire for short periods.

HM Coastguard

Yacht and Boat Safety Scheme

Mayday, Mayday, Mayday!

Another call for help
It doesn't matter whether the incident is large or small, far out to sea or inshore; it is the Coastguards who, once alerted, make sure that rescue services are quickly on the scene. Helicopters, life-boats, Coastguard cliff rescue teams, ships in the area – they are in touch with them all.

And there is another, equally important side to the Coastguard's work – giving local information and free advice on how to avoid getting into difficulties at sea. You can always 'talk safety' with a Coastguard.

Small craft safety
Every year, Coastguards are alerted to more than 1,000 small craft overdue or in distress. But often they do not have enough information (such as the appearance, equipment and sailing plans of the missing craft) to concentrate a search in the right area.

As the owner or user of a small craft you can help the Coastguard (and yourself) by joining the Yacht and Boat Safety Scheme. It costs nothing, and one day might save your life as well as the lives of your family or friends.

The aims of the Yacht and Boat Scheme are simple:

- to provide your nearest Rescue Centre with the information needed to mount a successful search and rescue operation
- to promote closer links between the Coastguards and all small-craft owners and users.

Yacht and Boat Safety Scheme

How do you join?
All you need to do is fill in a simple postcard. You can get one from Coastguard stations, marinas, yacht clubs, harbourmasters' offices and wherever you see the circular 'Issuing Authority' sign.

Describe your craft, its equipment and your normal sailing area, and then drop the card into a post-box. Remember to up-date the information whenever details change.

What information do the Coastguards need?
The following details will be helpful: the name of your club or association; type of craft or rig; name of craft, how and where displayed; colours of hull, topsides, sail; sailing/fishing number; speed and endurance under power; special identification features; life-raft and serial number; dinghy type and colour; life-jackets carried; radio – HF/MF trans/rec, VHF channels and call sign; type of distress signals carried; usual base, mooring, activity and sea areas; shore contact's name, address, telephone number; owner's name, address, telephone number; date.

Keeping in touch – VHF radio
You will sail more safely if your craft is fitted with VHF radio, so that you can pass distress, urgency and safety messages (including your latest position reports) direct to the Coastguards.

Pass distress and urgency messages on Channel 16. If you want to report your position, or change your plans, you should call up on Channel 16 and say you have 'safety traffic'. You will be asked to switch to Channel 67 (now a special safety channel between small craft and the Coastguard) and your message will be dealt with.

When using Channel 67 for safety messages, you should give the name of the Coastguard Rescue Centre which holds your Safety Scheme card.

Keeping in touch – MF radio
Yachts or boats fitted with MF radio transceivers only, or outside coastal VHF range, should contact the Post Office Coast Radio Stations periodically to give their TR (position report); then, if it becomes necessary, Coastguards can use the last known TR to establish a search datum.

Passages outside coastal waters
If you intend to sail outside UK waters you are recommended to contact your nearest Coastguard station and give details. These can then be noted, and through their contact with overseas rescue organisations the Coastguards will be able to advise you of any measures which could improve your safety.

Don't forget – Coastguards are there to help, and their services are free.

LIFEGUARD COMPACT INFLATABLE LIFEJACKETS

British Standards Institution BS 3595

COMPACT ONE – 15 MANUAL CO_2 INFLATION LG 1035

This Lifejacket is worn neatly folded against the chest or with the collar lightly inflated (the main body of the Lifejacket remaining folded). In emergency the CO_2 gas is released into the body of the Lifejacket by pulling the toggle at the bottom of the folded Lifejacket. An oral tube and non-return valve are also provided for the wearer to top up the Lifejacket. The patented 'flatfold' system conceals the CO_2 cylinder and inflator mechanism within the body of the Lifejacket so that only the inflation toggle protrudes below the Lifejacket.

COMPACT ONE – 15 AUTOMATIC CO_2 INFLATION LG 1035A

This Lifejacket is virtually the same as the Compact One – 15 Manual but has been fitted with an automatic mechanism that releases the CO_2 gas into the body of the Lifejacket when immersed in water. There is a manual override that allows the wearer to inflate the Lifejacket prior to entry into the water and an oral tube and non-return valve are available for topping up.

COMPACT FOAMFLOAT ORAL INFLATION LG 1074

This Lifejacket has an inherent buoyancy of 13.5 lb. (6.1 kg) provided by closed cell foam. When fully inflated the total buoyancy rises to 35 lb. (15.9 kg).

The foam volume is kept in line with the body shape of the wearer and the neckline is exceptionally smooth.

* Abrasion and flame resistant covers are available where Lifejackets are used in a working situation – details are available from the manufacturers.

* Servicing is recommended on a regular basis – the service interval will depend on the frequency of usage.

D.o.T. Approved.

LIFEGUARD "ONE-TO-FIVE" LIFEJACKET

You know how much your children love the water, so it makes sense to ensure that they are completely protected – for their safety and your peace of mind.

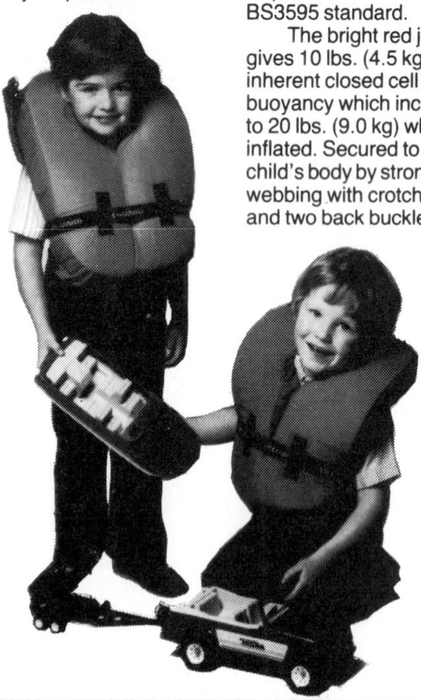

The Lifeguard 'One-to-Five' is a lightweight, air/foam lifejacket specially designed for the under-fives and to the buoyancy requirements of the BS3595 standard.

The bright red jacket gives 10 lbs. (4.5 kg) of inherent closed cell foam buoyancy which increases to 20 lbs. (9.0 kg) when inflated. Secured to your child's body by strong nylon webbing with crotch strap and two back buckles, there's no chance of it rising over your child's head once it is correctly adjusted.

The smooth neckline ensures safe and comfortable head support above the water. With the lifejacket fully inflated your child will float face upwards at the approved angle.

With a Lifeguard lifejacket you can enjoy yourself as much as the kids!

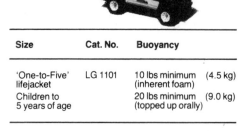

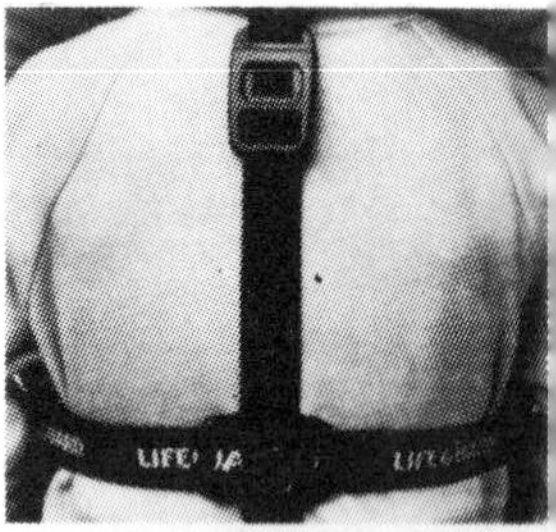

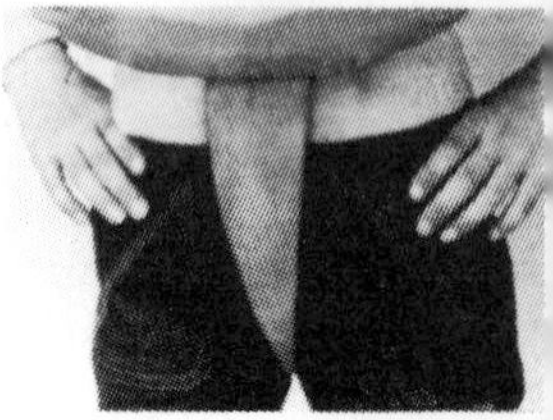

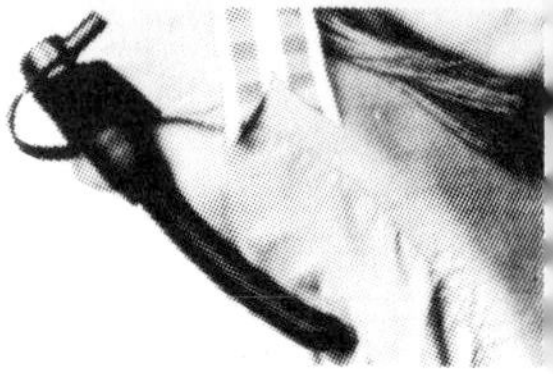

Size	Cat. No.	Buoyancy	
'One-to-Five' lifejacket	LG 1101	10 lbs minimum (inherent foam)	(4.5 kg)
Children to 5 years of age		20 lbs minimum (topped up orally)	(9.0 kg)

"LIFEGUARD COMPACT 1 CO2 INFLATABLE LIFEJACKET DONNING INSTRUCTIONS

(a) Hold the lifejacket with both hands by the collar with the webbing harness and instructions towards you.

(b) Put your head through the aperture allowing the back strap from the collar to hang behind.

(c) Pass the waistband from the left side around the waist, through the loop in the back strap to be tied off at the loop on the right hand side.

(d) The loop at the front of the lifejacket is for use in lifting the wearer out of the water.

Oral inflation

(a) The lifejacket may be topped up by mouth, through the non-return valve located at the front on the right hand side. Lift off the cap and blow into the mouthpiece. Always replace the cap after use to prevent the ingress of dirt and water.

(b) On no account should the lifejacket be inflated by mouth if there is a fully charged cylinder in the inflation mechanism. Releasing the C.O.2 into an already inflated lifejacket is liable to damage or even burst the fabric.

The manually operated inflator

Is operated by pulling sharply on the operating cord.

The automatic inflator

Is operated either automatically by water immersion, or manually by pulling sharply on the operating cord.

The Cylinders

Check the gas content before starting a voyage by weighing the cylinder. The minimum total weight is stamped on the cylinder. Fully charged it should weigh approximately 133g (4½oz.) if empty it will weigh approximately 100g (3½oz.) and should be thrown away.

Use and Maintenance

(a) The Compact lifejacket should not be worn beneath restrictive clothing nor beneath a safety harness of a type which will restrict inflation. Certain types of safety harness suitable for wear with a lifejacket are available and the owner of the lifejacket should ensure that the harness chosen does not interfere with the lifejacket's performance.

(b) It should not be stowed near heat, corrosive liquids or fumes.

(c) The gas holding performance of the lifejacket should be checked periodically by inflating it by mouth and leaving it inflated for an hour or two. If it deflates, have it serviced.

(d) If a water activated light is required, this can be supplied as an optional extra.

(e) A scheme for annual servicing and testing the lifejacket is in operation. For details contact the manufacturer's.

(f) After use at sea or if it becomes dirty it should be washed in fresh warm water. Do not use detergents.

"LIFEGUARD" COMPACT FOAMFLOAT INFLATABLE LIFEJACKET DONNING INSTRUCTIONS

(a) Check that the waistbelt is undone and passes through the loop at the end of the back strap.
(b) Holding the lifejacket at the sides, with waistbelt towards the body pass the head through the aperture.
(c) With the backstrap suspended down the back, pass the waistbelt from left to right through the sleeve in the backstrap, and around the waist.
(d) LG. 1074 (13.5lb/6kg. inherent buoyancy) Tighten and tie the waist webbing securely to the loop on the harness base.

LG. 1125 (20lb/9kg. inherent buoyancy D.o.T. approved)Pass the end buckle through the larger buckle on the harness waist and adjust the waistbelt to a comfortable secure fit.
(e) Check that the lifting loop is exposed at the front neck.

Oral inflation

The lifejacket is inflated by mouth, through the non-return oral valve. Lift off the cap and blow into the mouthpiece until the lifejacket is fully inflated. Always replace the dust cap after use to prevent the ingress of dirt and water.

Use and Maintenance

As for Compact 1 CO_2 inflatable lifejacket. Do not use as a cushion.

"LIFEGUARD" ONE-TO-FIVE LIFEJACKET DONNING INSTRUCTIONS

1. The buckles are opened by pressing the square centre clip while pulling both sections apart. Place the lifejacket over the child's head with the crotch strap to the front. Pass the crotch strap between the legs from front to back and connect the buckle at the underside of the headrest. Adjust to a comfortable fit by pulling the webbing through the adjuster bar of the buckle. Fasten and adjust the waistbelt similarly afterwards.

2. **Inflation/Deflation**

To inflate, lift the dust cap and blow by mouth into the valve. To deflate, insert the probe into the valve. Replace the cap afterwards. The valve is located and intended for use by a person other than the wearer of the lifejacket as a child of this age group might not be able to inflate the lifejacket.

3. **Performance**

The one-to-five lifejacket provided 10lbs (4.5kg) immediate buoyancy which can be increased to 20lbs (9kg) by mouth inflation. In the water the lifejacket will bring the wearer to an upward facing attitude, this is achieved in 5 seconds if the lifejacket is fully inflated.

4. **Recovery**

The wearer can be safely lifted out of the water by means of the waistbelt at the centre front.

5. **Care and Maintenance**

Do not leave the lifejacket in a compressed state for long periods of stowage nor allow it to be used as a cushion. Keep the dustcap on the valve. Maintain as for a Compact 1 CO_2 lifejacket.

The Subrella Leak Stopper

SUBRELLA FOR SAFETY

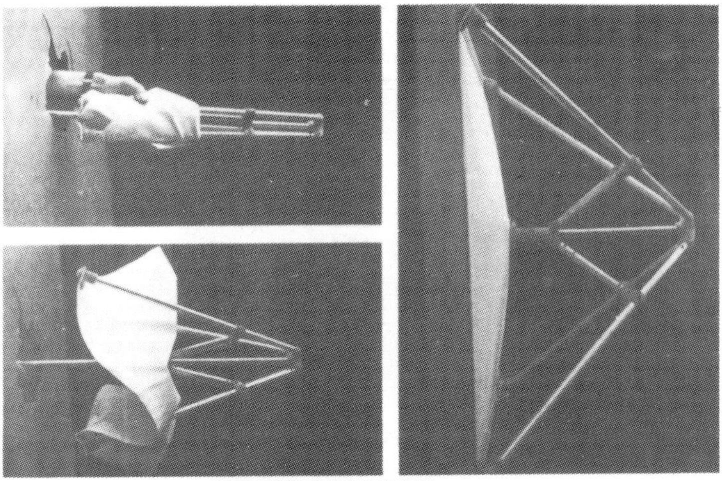

OPERATING INSTRUCTIONS :–

1. Secure the lanyard.
2. Pass the **SUBRELLA** through the damaged area.
3. Tug the **SUBRELLA** sharply back to deploy.
4. Take up lanyard slack.
5. (Note the method of folding for **re stowing**).
6. (Do not tie the lanyard when **re stowing**).

The danger of being holed is always present and very few mariners are prepared for this emergency. The **SUBRELLA** is a tried and tested device that is pushed through the damaged area and then opens on the outside. It is opened simply by pulling it sharply back towards the hull and without any springs or levers. The action of the water automatically spreads the patch over the hole and the pressure forms an effective seal in a matter of seconds. The **SUBRELLA** can also be used from the outside.

Manufactured of tough anti-corrosive materials, the **SUBRELLA** is available in three sizes.

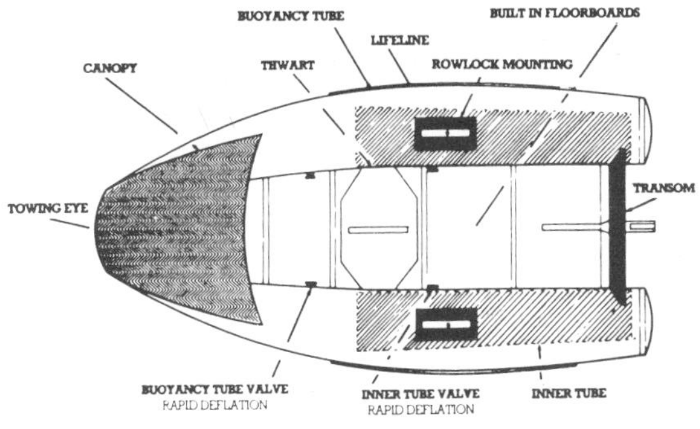

The Tinker Traveller

The J. M. Henshaw "Tinker Traveller" under sail.
Reproduced from "Yachting World."

The J. M. Henshaw "Tinker Traveller" with canopy raised.
Reproduced from "Yachting World."

Have a survival drill with all the family every time you use the boat as a tender. Remember the best position to stream the sea-anchor is from the beam. This considerably reduces the possibility of capsize. Keep the drop keel **up.** Do not let the wind balloon the canopy.

INFLATABLE DINGHYS SUITABLE TO ACT AS A SURVIVAL CRAFT ON PLEASURE YACHTS.

On yachts of less than 13.7 metres (45ft.) in length an inflatable dinghy may be substituted for an inflatable liferaft and on yachts of 25.9 metres (85 ft.) or over in length, an inflatable boat may be substituted for the life-boat or Class C boat they are required to carry.

The J. M. Henshaw (Marine) Ltd. "Tinker Tramp" (4 man) and "Tinker Traveller" (6 man) are small compact inflatable dinghys which can be collapsed, carried by one person and stowed in the boot of a car. They can be quickly rigged by one person for use as a tender either with or without an outboard motor. In an emergency they can be launched quickly by one person for use as a survival craft. However, even when fitted with CO_2 and survival canopies they are NOT liferafts.

These dinghys are hard to capsize, virtually unsinkable, easy to row, fast under power and easy to sail. For survival purposes they can be fitted with a roll up canopy and CO_2 gas bottles for rapid inflation. They can also be fitted with a mast, sails and drop keel, when they will sail comfortably to within 45 degrees of the wind direction.

Construction Details:–

A rigid floor and transom which fold for packing.

One outer air-tube running the length of the dinghy on each side, these tubes each contain an interior independent air-tube with a 4 inch (100mm) all round gap between inner and outer tubes around the outside of the boat, with either or both outer tubes damaged, buoyancy is maintained by the inner tubes. All the tubes are provided with inflation/deflation valves. A towing eye, thwart, mounted rowlocks, rudder and tiller and a bow dodger. The equipment includes 2 six foot (1.8m) oars, a pump and a repair kit. Optional extras are a mast complete with sail, together with a drop keel, a roll up canopy complete with support poles and two CO_2 bottles.

The outer tubes can be inflated either manually or by CO_2, if required, the inner tubes are inflated manually.

When intended for use as safety craft, the dinghy should be carried on deck with the inner tubes inflated and the canopy fitted but not erected and the painter made fast, with a safety knife handy for cutting the painter. All lashings used to secure the dinghy should be secured by means of senhouse slips to facilitate quick release. As with all other inflatables, the dinghy should be protected from sparks, funnel and exhaust deposits, acids, copper, bitumens, wet paint and friction. Remove sand grit by washing with a hose, remove tar and oil with a rag moistened in light spirit.

On yachts of less than 21.3m (70ft) which do not proceed more than 3 miles from the U.K. Coast or to sea during the months of November to March, an inflatable dinghy provides a suitable safety craft. Inclusion of the following equipment is advisable:– one painter/tow rope, one torch, a safety knife, a bailer, two sponges, tinned fresh water, pyrotechnic distress signals, one BMT sea-anchor attached to a hawser, one compass and a manually operated E.P.I.R.B.

Tests in the BMT testing tank show that in heavy weather the best position for the sea-anchor is to take the hawser over the air tube and secure it amidships just abaft the mast step and lie beam on. The keel should never be dropped when lying to a sea anchor. Under these conditions the tank failed to capsize the boat.

On yachts Of 25.9m (85ft) and over, they may be carried as an alternative to a lifeboat, Class C boat or D.o.T.I. boat. They are then required to be at least 3.8m (12.5ft) in length and to be provided with a launching device (although they may be launched by hand) capable of recovering the dinghy, together with the following equipment:–

(a) at least two buoyant oars and two buoyant paddles;
(b) a bailer and two sponges;
(c) a crutch or steering grommet on the transom;
(d) a grab line secured around the outside of the boat and a grabline fitted around the inside of the boat;
(e) a painter of adequate size and length;
(f) handholds or straps for the purpose of righting the boat from the inverted position;
(g) an efficient manually operated pump or bellows;
(h) a four legged bridle slinging arrangement to allow the boat to be lowered into or raised from the water;
(i) a repair kit in a suitable container for repairing punctures;
(j) a sea-anchor attached to a line at least 9m (30ft) in length;
(k) a safety knife;
(l) two buoyant rescue quoits attached to 18m (60ft) of light buoyant line
(m) a waterproof electric torch suitable for morse signalling together with one spare set of batteries and one spare bulb in a waterproof container;
(n) a pocket for loose equipment;

In addition, draining arrangements are required to be fitted.

In addition, it is advisable to carry the following extra equipment:–
A compass, tinned fresh water, pyrotechnic distress signals and a manually operated E.P.I.R.B.

IMPORTANT All inflatable Boats, Liferafts and Lifejackets are required to be serviced annually at an approved service station. Pyrotechnics and batteries are to renewed at the recommended intervals.

R.O.R.C. Recommendations for the contents of a "Grab bag."

(a) a second sea-anchor and line;
(b) two safety tin openers;
(c) a first-aid kit;
(d) one graduated rustproof drinking vessel;
(e) two 'Cyalume' sticks or throwable floating lamps;
(f) one daylight signalling mirror and one signalling whistle;
(g) two red parachute flares;
(h) three hand flares;
(i) non-thirst provoking food rations and barley sugar or equivalent;
(j) watertight receptacles containing fresh water (at least 1½ litres per person);
(k) one copy of the Illustrated Table of Life-saving Signals; SOLAS 2.
(l) nylon strings and polythene bags.
(m) a small flat piece of hardwood, for use as a cutting board.

EXTRACTS FROM THE 1979 FASTNET RACE INQUIRY

The results of the 1979 Fastnet Race inquiry held by the Royal Yachting Association and the Royal Ocean Racing Club show that of the 303 yachts which took part in the race, only 85 finished the course. 25 yachts were abandoned of which 5 were lost, believed sunk and 15 crew members lost their lives.

The storm was at its height between midnight and 0800 hours on the 14th of August when winds reached force 11 and wave heights were in the order of 40 - 44 feet (approx 13 metres). Although a storm warning had been broadcast some 6 hours beforehand, very few of the yachts were aware of the conditions they were about to experience. A brief summary of some of the important factors regarding safety which came to light through the inquiry are as follows:–

Safety Harness

There were at least 26 instances of harness failure which resulted in the loss of 6 lives. The following points were raised. Harnesses should have two lines, each with its own hook, should be simple to put on and those which combined harness/lifejackets in a single unit were to be preferred. Some harnesses had a buckle which could slip if the harness was worn inside out. At least one man was lost because he had used the guard rail as a point of attachment and the guardrail failed, another was lost when the belt of a jacket/harness pulled out, yet another harness line broke at a point where there was a knot in the line and on another a stainless steel hook straightened and released itself. One man was nearly lost when the ring on his harness broke and on another the harness buckle was reported as holding under the strain, but liable to come undone when there is no load on the line.

Lifejackets

Three reports were received of bodies being sighted or recovered floating face down in the water although a lifejacket was being worn. In one instance the wearer's head appeared to have slipped out of the collar. With regard to various types of lifejackets there was a marked lack of agreement, opinions differed on the merits of permanent buoyancy versus inflation. However, it was pointed out that lifejackets with oral/manual inflation should have a relief valve incorporated.

Liferafts

235 liferafts of various makes were carried, 15 of which were used, some merely to transfer crew from yacht to helicopter. 12 liferafts were washed overboard, 8 were stowed in the cockpit and 4 on deck. In many cases rafts stowed in cockpits were secured only by the lid of the locker; some lockers opened accidentally and rafts either fell out or were washed out, and one of the rafts stowed on deck went overboard still secured to the chocks on which it was stowed. One crew were unable to use either of two liferafts, one was washed off the cabin top, the other could not be extracted from its stowage under the cockpit sole because the floorboards were jammed. Many of the liferafts capsized on launching and a number capsized with survivors aboard. One raft painter parted with one member of the crew aboard and was lost. A number of rafts had the painter secured on the opposite side of the raft to the canopy opening; this caused much

lifficulty both in boarding the raft and cutting the painter. On some rafts the canopy tore, in particular at the opening. On one raft the bottom ring and floor broke away from the top ring and canopy. The drogues were of the old parachute type and mostly, when used, failed either by the breaking of the shrouds or the tearing away of the anchorage. Many did not use their drogues, for under the sea conditions prevailing they felt it better to allow the raft to drift with the speed of the sea. 53% of the survivors (i.e. those who spent longest in the raft) felt that cold was an important factor and that an inflatable floor would have been a considerable improvement. With regard to foil "space blanket", trials carried out some years ago indicated that they were of little use. The blankets are very efficient at preventing heat loss by radiation, but the major heat loss suffered by survivors in a raft is by conduction through the raft floor, against which a foil blanket offers little protection. On some liferafts there was a lack of handholds on the outside of the raft with the result that crews were unable to hold onto the raft, or turn it round to gain access to the canopy door. Seven lives were lost in incidents associated with rafts of which three were directly attributable to the failure of the raft, moreover, the yachts which these seven people abandoned were subsequently found afloat and were towed into harbour.

CHAPTER 2

IMO. REQUIREMENTS FOR LAUNCHING APPLIANCES AND EMBARKATION APPLIANCES.

1. **General Requirements.**

(a). Each launching appliance together with all its lowering and recovery gear shall be so arranged that the fully equipped survival craft or rescue boat it serves can be safely lowered against a trim of up to 10 degrees and a list of up to 20 degrees either way;

(i) when embarked as required with its full complement of persons;

(ii) notwithstanding the requirements of paragraph (i) above, lifeboat launching appliances for oil tankers, chemical tankers and gas carriers with a final angle of heel greater than 20 degrees calculated in accordance with the International Convention for the Prevention of Pollution from Ships, 1973, as modified by the 1978 Protocol, shall be capable of operating at the final angle of heel on the lower side of the ship;

(iii) without persons in the survival craft or rescue boat,

(b) A launching appliance shall not depend on any means other than gravity or stored mechanical power which is independent of the ship's power supplies to launch the survival craft or rescue boat it serves, in the fully loaded and equipped condition and also in the light condition.

(c) A launching mechanism shall be so arranged that it may be actuated by one person from a position on the ship's deck, and from a position within the survival craft or rescue boat; the survival craft shall be visible to the person on deck operating the launching mechanism.

(d) Each launching appliance shall be so constructed that a minimum amount of routine maintenance is necessary. All parts requiring regular maintenance by the ship's crew shall be readily accessible and easily maintained.

(e) The winch brakes of a launching appliance shall be of sufficient strength to withstand:

(i) a static test with a proof load of not less than 1.5 times the maximum working load; and

(ii) a dynamic test with a proof load of not less than 1.1 times the maximum working load at maximum lowering speed.

(f) The launching appliance and its attachments other than winch brakes shall be of sufficient strength to withstand a static proof load on test of not less than 2.2 times the maximum working load.

(g) Structural members and all blocks, falls, padeyes, links, fastenings and all other fittings used in connection with launching equipment shall be designed with not less than a minimum factor of safety on the basis of the maximum working load assigned and the ultimate strength of the material used for construction. A minimum factor of safety of four-point-five shall be applied to all davit and winch structural members and a minimum factor of six shall be applied to falls, suspension chains, links and blocks.

(h) Each launching appliance shall, as far as practicable, remain effective under conditions of icing.

(i) A lifeboat launching appliance shall be capable of recovering the lifeboat with its crew.
(j) The arrangements of the launching appliance shall be such as to enable safe boarding of the survival craft in accordance with these regulations.

2. **Launching appliances using falls and a winch.**
(a) Falls shall be of rotation resistant and corrosion resistant steel wire rope.
(b) In the case of a multiple drum winch, unless an efficient compensatory device is fitted, the falls shall be so arranged as to wind off the drums at the same rate when lowering, and to wind onto the drums evenly at the same rate when hoisting.
(c) Every rescue boat launching appliance shall be fitted with a powered winch motor of such capacity that the rescue boat can be raised from the water with its full complement of persons and equipment.
(d) An efficient hand gear shall be provided for recovery of each survival craft and rescue boat. Hand gear handles or wheels shall not be rotated by moving parts of the winch when the survival craft or rescue boat is being lowered or when it is being hoisted by power.
(e) Where davit arms are recovered by power, safety devices shall be fitted which will automatically cut off the power before the davit arms reach the stops in order to avoid overstressing the falls or davits, unless the motor is designed to prevent such overstressing.
(f) The speed at which the survival craft or rescue boat is lowered into the water shall be not less than that obtained from the formula:

$$S = 0.4 + (0.02 \times H)$$

where S = speed of lowering in m/s
and H = height in m from davit head to the waterline at the lightest sea-going condition.

(g) The maximum lowering speed shall be established by the Administration having regard to the design of the survival craft or rescue boat, the protection of its occupants from excessive forces, and the strength of the launching arrangements taking into account inertia forces during an emergency stop. Means shall be incorporated into the appliance to ensure that this speed is not exceeded. *(normally a centrifugal brake).*
(h) Every rescue boat launching appliance shall be capable of hoisting the rescue boat when loaded with its full rescue boat complement of persons and equipment at a rate of not less than 0.3 m/s.
(i) Every launching appliance shall be fitted with brakes capable of stopping the descent of the survival craft or rescue boat and holding it securely when loaded with its full complement of persons and equipment; brake pads shall, where necessary, be protected from water and oil.
(j) Manual brakes shall be so arranged that the brake is always applied unless the operator, or a mechanism activated by the operator, holds the brake control in the "OFF" position.

3. **Float free launching**

Where a survival craft requires a launching appliance and is also designed to float free, the float free release of the survival craft from its stowed position shall be automatic.

4. **Free-fall launching.**

Every free-fall launching appliance using an inclined plane shall, in addition to complying with paragraph 1 of this Regulation, also comply with the following requirements:

(a) The launching appliance shall be so arranged that excessive forces are not experienced by the occupants of the survival craft during launching.

(b) The launching appliance shall be a rigid structure with a ramp angle and length sufficient to ensure that the survival craft effectively clears the ship.

(c) The launching appliance shall be effectively protected against corrosion and be so constructed as to prevent incendive friction or impact sparking during the launching of the survival craft.

5. **Evacuation slide launching and embarkation**

Every evacuation slide launching appliance shall, in addition to complying with the applicable requirements of paragraph 1 of this Regulation, also comply with the following requirements:

(a) The evacuation slide shall be capable of being deployed by one person at the embarkation station.

(b) The evacuation slide shall be capable of being used in high winds and in a sea-way.

6. **Liferaft launching appliances**

Every liferaft launching appliance shall comply with the requirements of paragraphs 1 and 2 of this Regulation, except with regard to the use of gravity for turning out the appliance, embarkation in the stowed position, and recovery of the loaded liferaft. The launching appliances shall be so arranged as to prevent permature release during lowering and shall release the liferaft when waterborne.

7. **Embarkation ladders**

(a) Handholds shall be provided to ensure a safe passage from the deck to the head of the ladder and vice versa.

(b) The steps of the ladder shall be:

(i) made of hardwood, free from knots or other irregularities, smoothly machined and free from sharp edges and splinters or of suitable material of equivalent properties;

(ii) provided with an efficient non-slip surface either by longitudinal grooving or by the application of an approved non-slip coating;

(iii) not less than 480mm (19in.) long, 115mm (4½in.) wide and 25mm (1in.) in depth, excluding any non-slip surface or coating;

(iv) equally spaced not less than 300mm (12in.) nor more than 380mm (15¼in.) apart and be secured in such a manner that they will remain horizontal.

(c) The side ropes of the ladder shall consist of two uncovered manila ropes, not less than 65 mm (2½in) in circumference, on each side. Each rope shall be continuous with no joints below the top step. Other materials may be used provided the dimensions, breaking strain, weathering, stretching and gripping properties are at least equivalent to those of manila rope.

The Longworth Sea Descent Gear Mark II
Approved for use in lieu of embarkation ladders

The Longworth Sea Descent Gear

The Longworth Sea Descent Gear has been approved for use in lieu of embarkation ladders. It enables survivors to board floating survival craft without having to go down embarkation ladders.

A wire rope fall passes through an overhead pully at embarkation deck level. The bottom of the fall has a stirrup iron attached into which the survivor places a foot, a body harness with a quick release buckle is attached to the fall at body height. With one foot in the stirrup iron and the body harness buckled on, the survivor steps overside holding the fall and is automatically lowered at a speed of 5 feet/secs. This speed is controlled by a centrifugal brake attached to the drum holding the wire fall. There is an alternative hand brake which is controlled by an operator at embarkation deck level, the descending survivor also has control of this brake by means of a rope.

As the survivor descends, energy is stored in a double flat spring moto so that when the weight of the survivor is removed from the fall, afte descent to the survival craft, the fall is automatically returned t embarkation deck level.

One operator is required to instruct and dress survivors in the bod harness, launch them overside and control their descent. The operator i able to descend afterwards without assistance.

Sea Descent Gear is to be tested every six months by the Administration

Lifeboat disengaging gear.

The "Mills Patent Instantaneous Engaging and Disengaging Gea provides a means of releasing a lifeboat from the hooks when the lifeboat is waterborne. Simultaneous release of both ends of the lifeboat is effected by pulling on the releasing handle. If for any reason it is desired to release the falls without operating the gear, each hook can be detached by hand separately.

To hoist from the water, a swivel ring on the floating block can be instantaneously engaged with the hook by simply inserting it into the jaws of the side plates, whatever twist there may be in the falls as a swivel is attached to the base of each floating block.

The connecting chain between the hooks is carried along the starboard side of the lifeboat and is protected by a casing.

The release gear consists of two pivoted hooks, one at each end of the boat. These hooks are held in the closed position by balanced weights so making sure that when the lifeboat is stowed, the hooks are always closed and engaged on the fall link. When the boat is being lowered, the weight of the boat is taken by the hooks and it is impossible to open them against this load. As the lifeboat becomes waterborne the load on the hooks decreases and they can be opened to release the boat.

Merchant Shipping Notice No. 866. Lifeboat disengaging gears.

Notice to Shipowners, Masters and Skippers of fishing vessels

Several cases have been reported to the Department where disengaging gears have been found in such condition that with the lifeboats afloat it was not possible to release the hooks by the simultaneous release mechanism; neither was it possible in some of these cases to release the hooks individually by hand.

The serious consequences of the latter situation arising in the event of a casualty to the ship are viewed with great concern by the Department and the attention of those concerned is therefore drawn to the necessity for maintaining disengaging gears in good working order. To this end the gears should be frequently tested with the lifeboat afloat - or placed on deck or ashore. It is further pointed out, that the gears should not be tested whilst the lifeboats are hanging in the davits, except for example, where the lifeboat is carried under luffing davits and where the lifeboat would be safely supported by chocks on the deck, on both sides of the keel, when the hooks are released.

N.B. Instantaneous release gear should always be kept well greased and oiled and never painted.

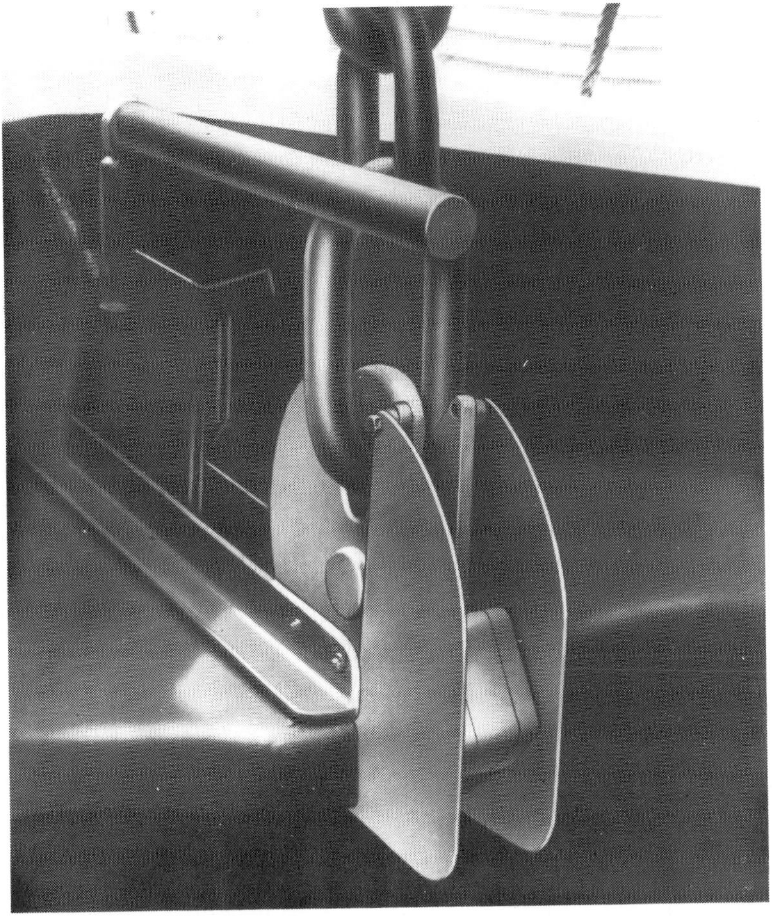

The Mills Patent Instantaneous Release Gear

Cabling

Cabling occurs when after a boat has been launched and the weight of the boat removed from the falls. The falls twist round and round and was very prevelent when manila rope falls were in use. It can occur with wire falls and will cause considerable difficulty when the boat is about to be hoisted.

There are two methods by which cabling can be overcome. One is to join the floating blocks with a light wire span and so prevent them twisting. The other is to have swivels fitted onto the bases of the floating blocks which will allow the blocks to unwind as the boat is hoisted, whenever swivels are fitted they must be kept well oiled and frequently overhauled, to prevent any possibility of the swivels freezing up.

GRAVITY DAVITS

Gravity davits are any davits which use the weight of the boat to do the work required to launch the boat overside, they may operate on pivots or have a carriage mounted on roller-track ways which are fixed either to the deck or overhead. The boat is launched by the lifting of a brake handle. The brake is required to apply itself automatically immediately the handle is released by the brake operator. The rate of the boat's descent is controlled by an independent centrifugal brake. These davits are all fitted with wire rope falls and winches. On ships constructed after 1st. July, 1986 they are required to be capable of launching a survival craft or rescue boat against an adverse list of 25 degrees. They are required to be fitted with tricing pendants to bring the boat alongside and bowsing-in tackles to replace the tricing pendants and keep the survival craft alongside. The tricing pendants must always be released and the weight of the boat transferred to the falls before persons are embarked in the boat.

Gravity davits are provided with a safety device which will prevent the davits from operating while it is in position. This device usually takes the form of a trigger to which the gripes are attached and is so adjusted that while the gripes are on, the davits cannot operate. In addition, holes are normally provided into which a bolt can be shipped to prevent the davits being inadvertantly operated in port. These bolts, known as "Harbour safety pins" are to be shipped only in port. When the falls and winches are being overhauled the lifeboat must be floated or landed before the falls are let-go for overhauling. Harbour safety pins are always to be unshipped before the vessel proceeds to sea, so that the boats are at all times ready for immediate use.

Welin overhead "Devon" gravity davits

The Welin "Devon" gravity davit has a moving embarkation platform on which the controls are mounted, incorporated in the davits which are mounted on roller tracks. Embarkation can be undertaken when the boat is stowed or at any point between the stowed and full outreach positions. Until the actual lowering of the boat into the water it is rigidly held in the davit carriage and cannot rock or sway under any circumstances, even when the ship is going at full speed. No tricing pendants or bowsing-in tackles are required and the launching can be done by a single operator either from within the lifeboat or from the platform. Welin "Devon" davits can be adapted for free-fall and are suitable for use on either oil-rigs or ships.

The "Miranda" System

"Miranda" davits employ a different approach to the launching of a boat by means of gravity.

The boat is contained in and attached to a cradle that is hoisted to the davit head by means of two single wire rope falls; there are therefore no floating blocks. The boat is attached to the cradle by means of two short wire strops placed between the cradle head and the boat's lifting hooks. The painter which is attached to the lifeboat by means of a quick release system is also attached to the cradle and not to the ship.

Embarkation can be at the stowed position or from the embarkation deck, or both. No skates or buffers are required as the cradles are fitted with tyred wheels to enable it to run down the side of an adversely listed ship and the boat remains within and attached to the cradle until it is waterborne, when the cradle will clear the boat by sinking beneath it. The coxswain can then release the strops attaching the boat to the cradle by means of the quick release gear situated at the helmsman's position and the painter is slipped by a crew member in the bow of the boat. The control wire disappears through the small hole in the deckhead as the boat speeds away.

The davits are static and contain no moving parts except the sheaves. The gripes clear automatically as the boat is lowered by means of the control wire, which is also situated at the helmsman's position. No bowsing-in tackles or tricing pendants are required.

With this system, a lifeboat or rescue boat can be launched when the ship has an adverse list of 30 degrees and 15 degrees of trim either way. The system can also be modified to allow the boat to float free, if required. However, it is unsuitable for free-fall.

Recovery at sea is a simple matter. Two crewmen, one forward and one aft, take the recovery pendants and fit the rings in the boat's lifting hooks. The helmsman now brings the boat alongside, matches the ship's speed and stations the boat parallel to the falls of the sunken cradle. The two crew-men then snap the hooks on the recovery pendants around the fall wires. As the cradle is lifted, the hooks on the recovery pendants slide down the falls until they reach the cradle head. The cradle which now contains the boat, is hoisted. After hoisting to the davit head, the boat is hung off, the recovery pendants removed from the boat's hooks, unhooked from the falls and restowed in the boat, ready for the next time. The lifting strops are now placed in the boat's hooks and the boat is hove home and griped in position. The control wire is rewound onto its spindle and the end rethreaded through the small hole in the top of the canopy.

Luffing Davits

Luffing davits, obsolete on vessels built after 1st. July, 1986, require the boat to be taken from inboard to outboard by the manual turning of a worm screw or telescopic screw. These davits are required to be capable of launching a boat against an adverse list of 15 degrees and like gravity davits, are fitted in pairs. The boat normally rests in chocks at deck level and is firmly held down by means of deck gripes.

Normally, wire rope falls and winches are fitted but under certain circumstances manila rope falls may be used. Manila rope falls attached to lifeboats 24ft. (7.3m) in length and over are required to be rove in a three-fold purchase, the hauling and standing parts of the fall being rove through the centre sheaves, in order to balance the weight. Suitable bollards for making fast shall be provided.

Nylon recovery pendant system for emergency and rescue boats.

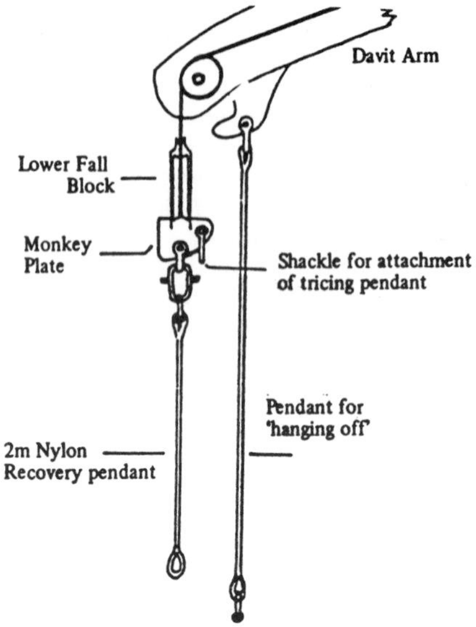

Reproduced from November 1983 *Seaways*.

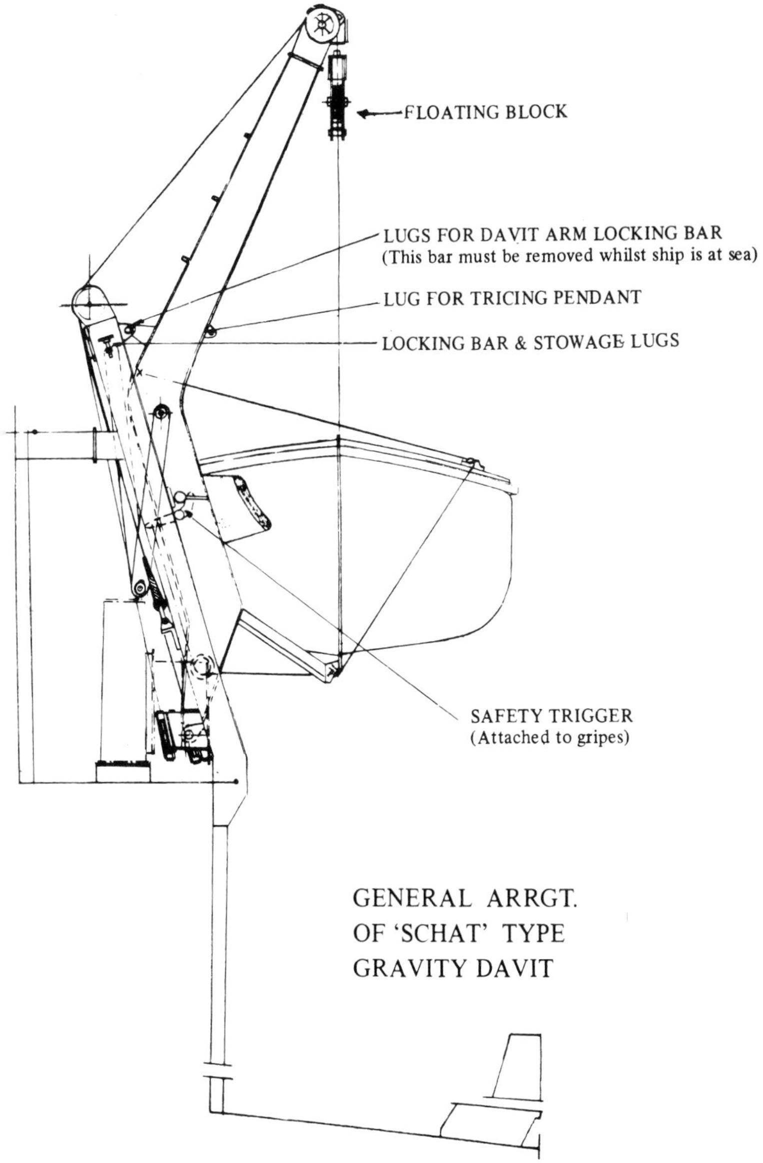

GENERAL ARRGT.
OF 'SCHAT' TYPE
GRAVITY DAVIT

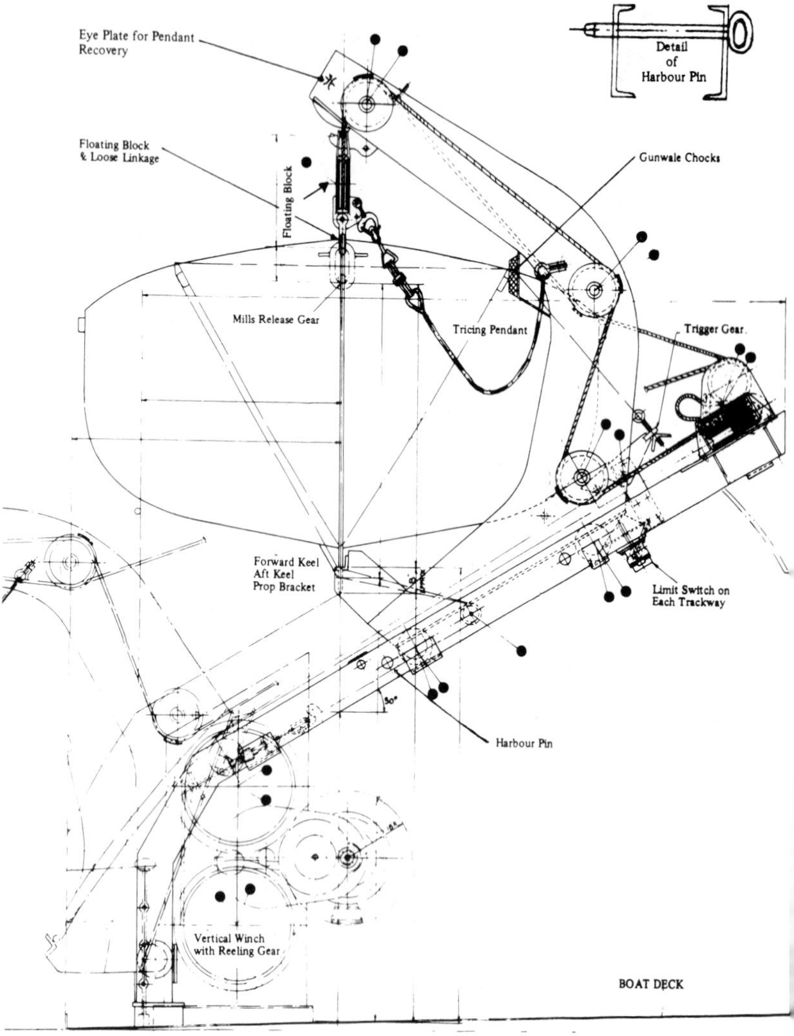

WELIN TYPE OVERHEAD GRAVITY
ROLLER TRACKWAY DAVIT

S.S. "Ribera" Machlan Roller Track gravity davits fitted with " latch-on" gripes with "Viking" G.R.P. boat attached and showing application of the gripe trigger.

WELIN 'DEVON' OVERHEAD GRAVITY DAVITS

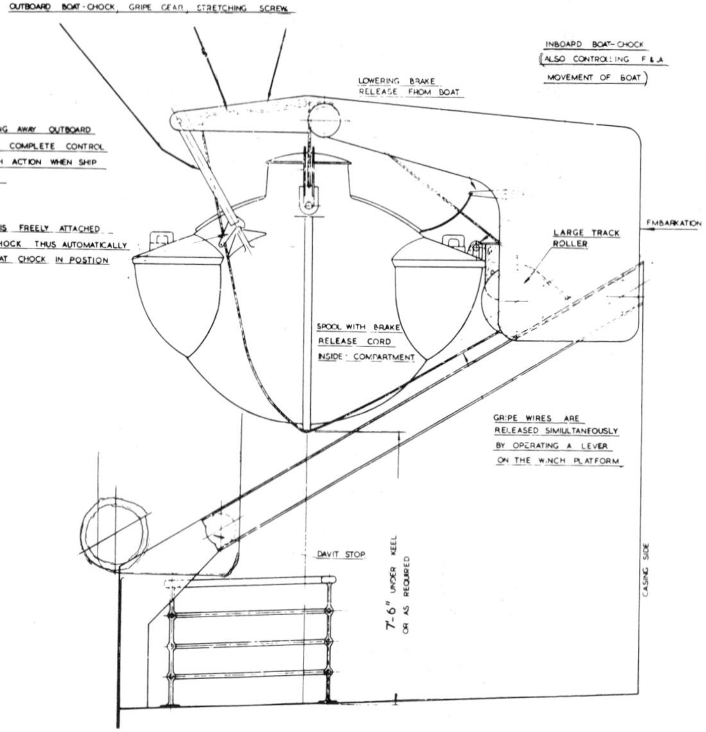

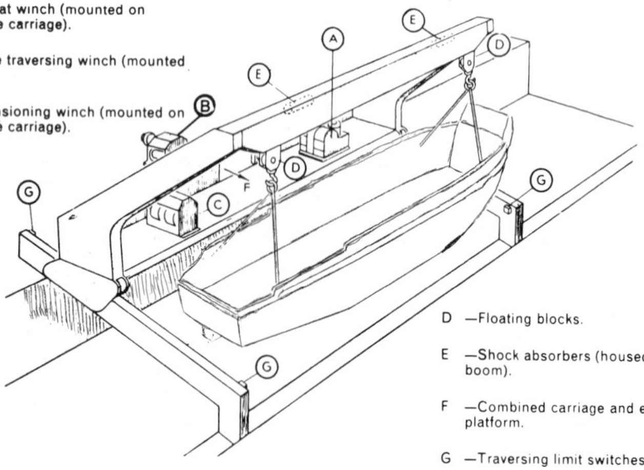

A —Main boat winch (mounted on retractable carriage).

B —Carriage traversing winch (mounted on deck).

C —Falls tensioning winch (mounted on retractable carriage).

D —Floating blocks.

E —Shock absorbers (housed in davit boom).

F —Combined carriage and embarkation platform.

G —Traversing limit switches.

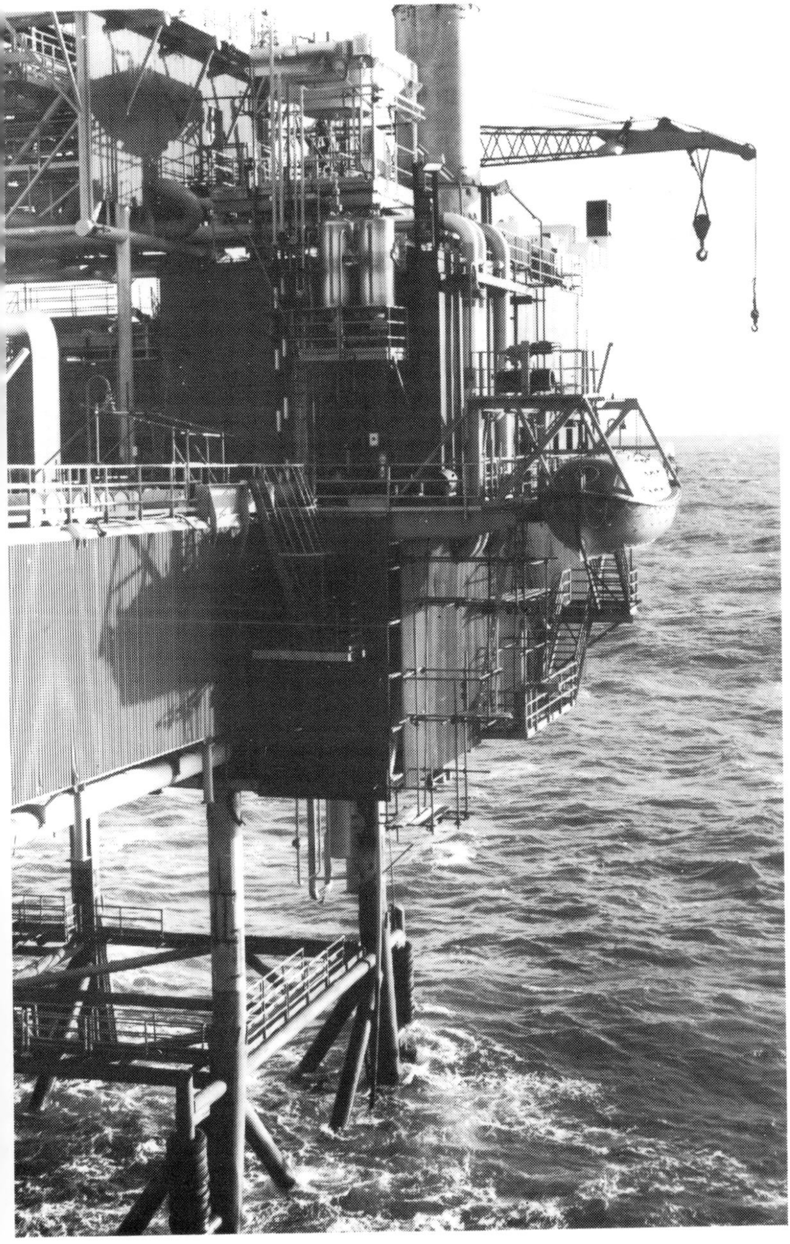

Oil rig equipped with "SCHAT" davits and a totally enclosed fire-proof lifeboat, suitable for launching by lowering or free fall.

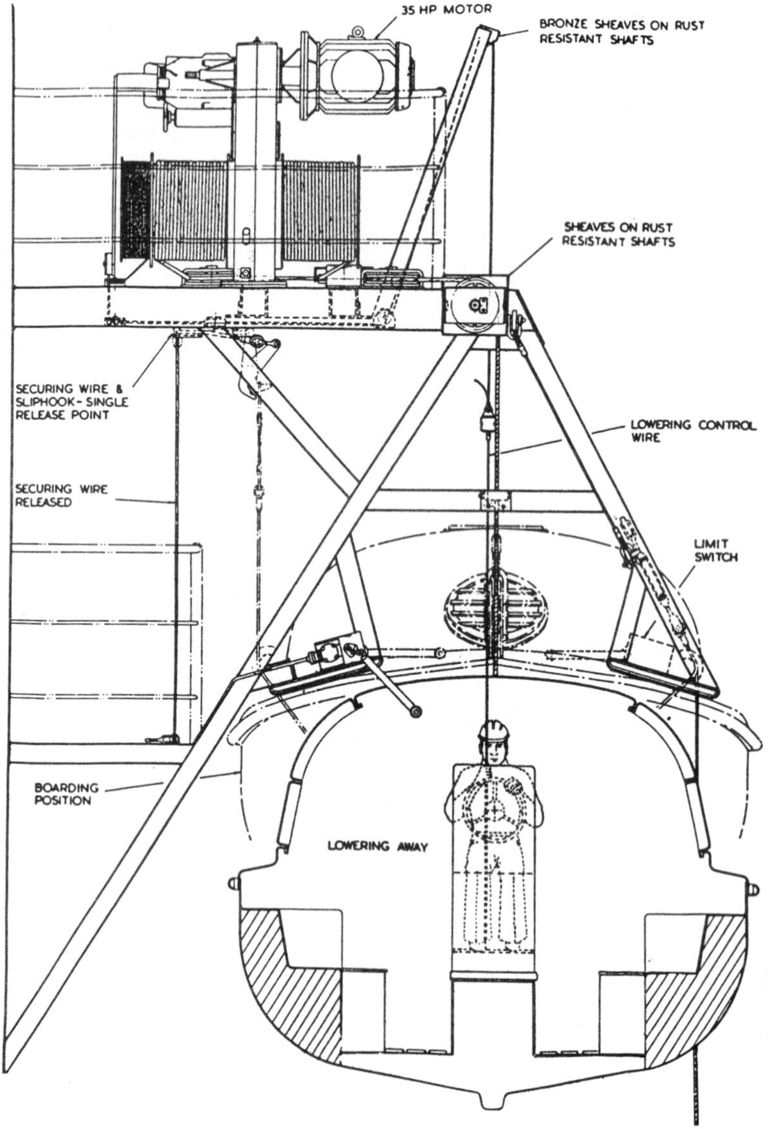

Lowering a totally enclosed lifeboat boarded in the stowed position and served by "SCHAT" ORD/DHM davits attached to an oil rig.

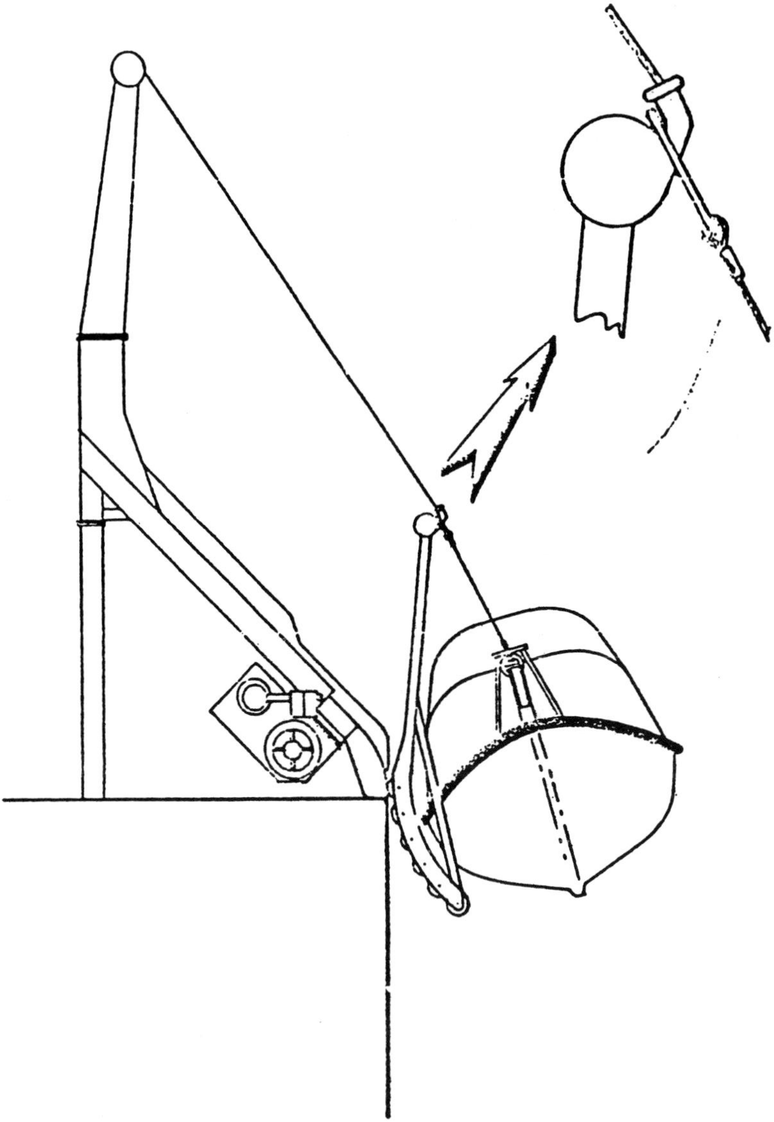

"Miranda Davits"

Showing cradle and lifeboat attachment to cradle.
The control wire, which lifts the brake, is attached to a spindle on the davits and unwinds as the boat is lowered.
When the boat leaves the ship, the end of the control wire is withdrawn from the boat.

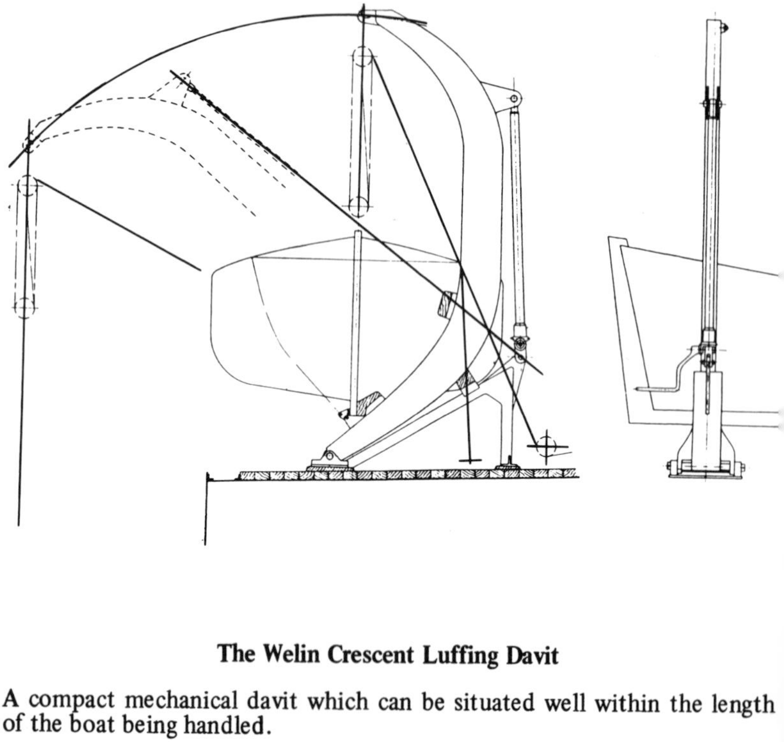

The Welin Crescent Luffing Davit

A compact mechanical davit which can be situated well within the length of the boat being handled.

This inexpensive type of luffing davit is in use in many vessels, it is simple to operate, easy to maintain and neat in appearance. Can be used with manila falls or wire falls and winches.

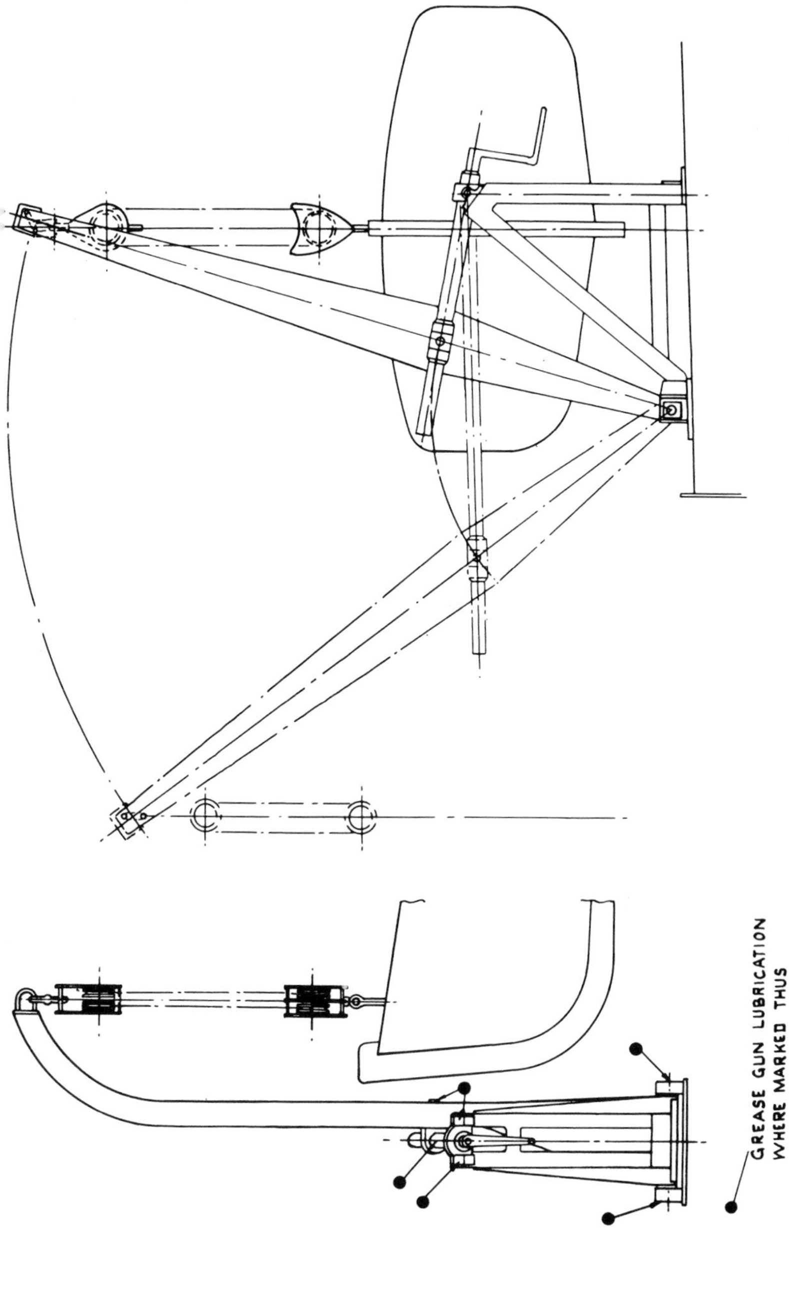

WELIN COLUMBUS LUFFING DAVITS

DAVIT CHOCKS AND GRIPE WIRES

Watercraft have devised a fitting, four of which are secured to the gunwales of all Watercraft boats and which serve as a choking position to the davit arm on the inboard side, and a gripe saddle on the outboard side.

The words "Fit davit chock here" are stencilled on the fitting, and should be covered by the davit chock itself on the inboard side.

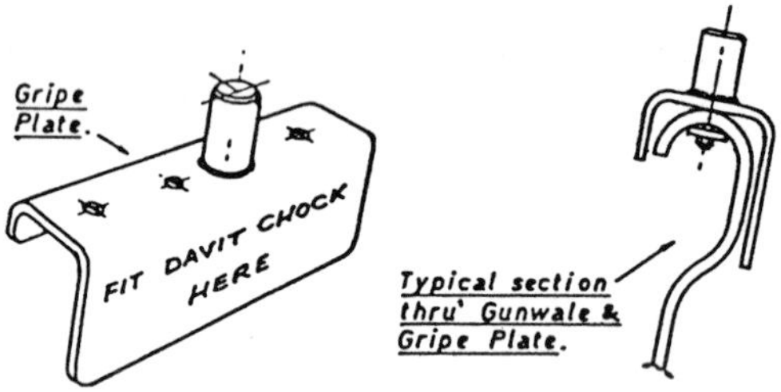

Outboard, the peg sticking up in the centre of the fitting serves as a preventer to keep the gripe in position, and the fitting itself leads the gripe over the gunwale and prevents chafe.

The inboard fitting, although the same as that outboard, is not intended for griping and should not be used for this purpose.

The chafe of the gripe on the inboard gunwale should be prevented by fitting a metal strip of suitable size and shape for the arrangement of each individual outfit.

These fittings have been designed to prevent damage to the boat by gripe wires, and chock, and the boats are reinforced in way of them. Chocks and wires fitted in other places may cause severe damage.

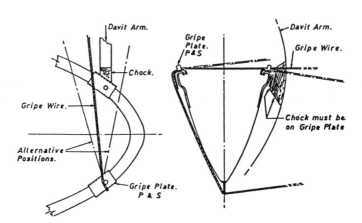

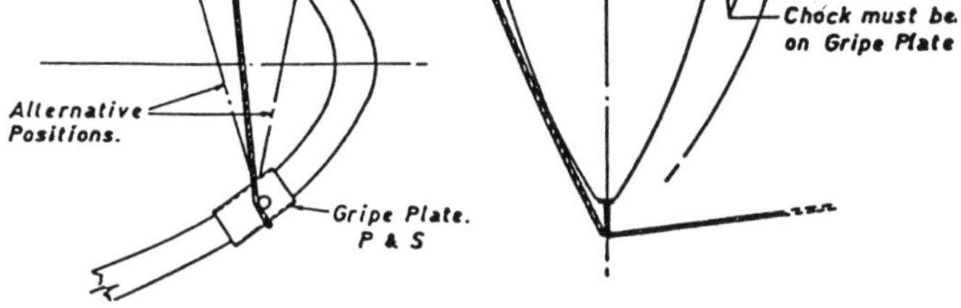

TYPICAL PLAN–SECTION AT DAVIT ARM.

DECK CHOCKS AND GRIPE WIRES

When the boat is stowed in deck chocks the same fittings are used.

In this case the chocks should be displaced fore or aft about 18 inches from the centre of the gunwale fittings which now are all used as gripe saddles.

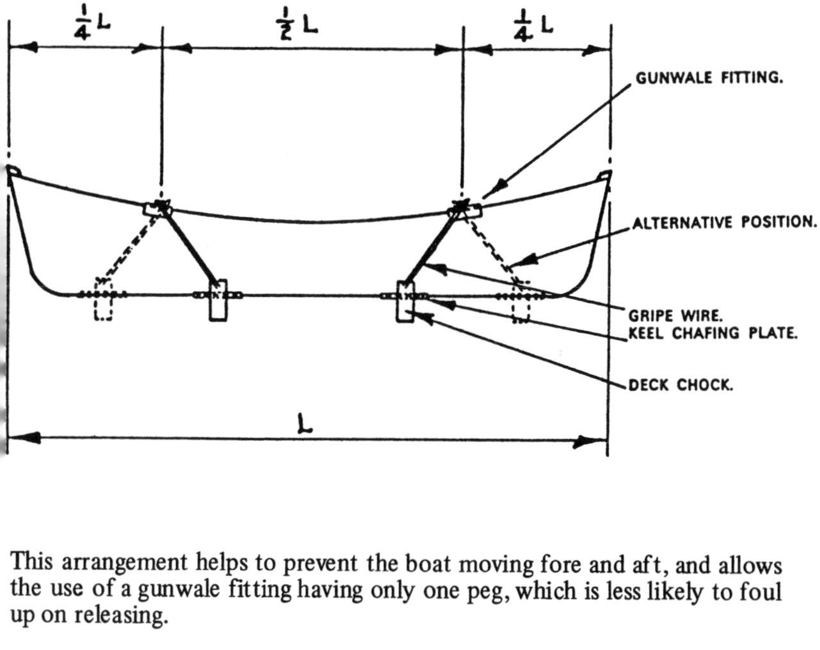

This arrangement helps to prevent the boat moving fore and aft, and allows the use of a gunwale fitting having only one peg, which is less likely to foul up on releasing.

A three-fold lifeboat fall is to be rove with the hauling and standing parts rove through the centre sheaves, as shown:–
Note that the lower or floating block lies at right angles to the upper block

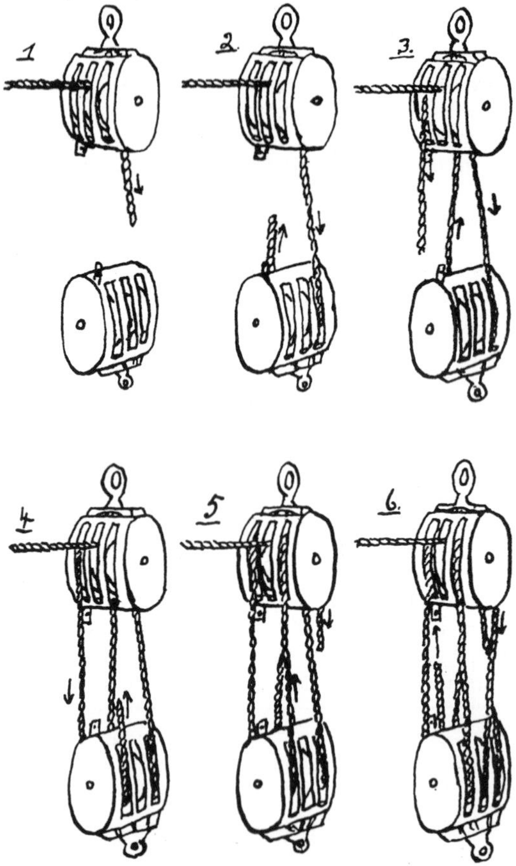

Reproduced from "The Efficient Deckhand."

When natural fibre cordage such as manila rope is used, the falls are to be left slack, to allow for shrinkage in wet weather. They are to be stowed on reels or in flaking boxes. These falls must always be set tight before swinging out the boat. There is no need to lift the boat, the outboard chock is cut away to allow the boat to swing clear when the davits are turned outboard.
Floating blocks attached to all lifeboat and rescue boat falls are required to be of a non-toppling variety.

DEPARTMENT OF TRANSPORT MERCHANT SHIPPING NOTICE
No. **M.1232**

·OLYPROPYLENE CORDAGE FOR LIFE-SAVING APPLIANCES

Notice to Owners, Masters, Officers and Seamen of Merchant Ships and Yachts; and to Owners, Skippers and Crews of Fishing Vessels

This notice supersedes Notice M. 698

(1) Polypropylene ropes intended for use with lifesaving appli-.nces should be of a type which has been accepted by the Department. ·uch ropes exceeding 12 mm diameter will incorporate a tape hroughout their length bearing at regular intervals:

(a) the initials or trade mark of the manufacturing company;
(b) the rope's brand name;
(c) the words "DOT accepted for LSA".

(2) Before accepting a rope the Department will have required vidence from manufacturers that it is sufficiently protected against he actinic degradation to withstand two years' exposure on a ship ·perating in or through the tropics.

(3) Manufacturers of ropes, samples of which have been accepted ·y the Department, will be issued with a certificate of acceptance by he Department.

(4) Responsibility for replacing worn, weathered or damaged ordage at all times lies with the master of the vessel.

(5) The following table "Cordage for LSA Purposes" indicates the haracteristics required of ropes for various lifesaving appliance uses nd the sizes considered appropriate. Unlike natural cordage the grip rovided by different types of polypropylene cordage ranges between a rip comparable with manila or sisal to little grip at all. The type of olypropylene cordage must therefore be carefully chosen to meet iffering grip requirements.

(6) Guidance on inspection and care of polypropylene cordage in se is included in BS 4928: 1973 part 1—Specification for Polypro-ylene Rope.

)epartment of Transport
Aarine Directorate
,ondon WC1V 6LP
uly 1986

Cordage for Life-Saving Appliances

Note: M denotes Manila, complying with BS 2052: 1977—Superi Special Quality or grade 1 quality.
S denotes Sisal rope, complying with BS 2052: 1977—Stan ard Quality.
P denotes Polypropylene rope of a type accepted by t Department of Transport.

Application	Type	Grip	Minimum Size of Cordage Diameter in millimetres	
			M or S	P
Lifeboat and gunwale grab-lines	M S P	Not critical	16 mm	16 mm
Buoyant apparatus grablines	M S P	Not critical	14 mm	14 mm
Lifeboat keel grablines (should be knotted)	M S P	Not critical	16 mm	16 mm
Buoyant heaving lines	Cotton or P	As Manila	8 mm	8 mm
Lifeboat boarding ladders	M S P	As Manila	16 mm	16 mm
Overside ladders for lifeboat or liferaft embarkation	M P	As Manila	20 mm	20 mm
Lifebuoy grablines	M S or P Unkinkable	Not critical	9·5 mm	9·5 mm
Lifebuoy lines	Buoyant Cotton or P	As Manila	8 mm	8 mm
Lifelines from davit spans	M P	As Manila	20 mm	24 mm
Boats' painters	M S or P boat under 8 m	As Manila	20 mm	24 mm
	8 m and under 9 m		24 mm	28 mm
	9 m boat and over		28 mm	32 mm
Buoyant apparatus painters	M S or P	As Manila		
Mass less than 140 Kg			16 mm	16 mm
Mass 140 Kg and over			20 mm	20 mm
Sea anchor for open lifeboats of 6 metres and under	M S or P	As Manila	Hawser: 20 mm Tripping Line: 12 mm	20 mm 12 mm
Open lifeboats over 6 metres			Hawser: 24 mm Tripping Line: 16 mm	24 mm 16 mm

Application	Type	Grip	Minimum Size of Cordage Diameter in millimetres	
			M or S	P
ails: alyards, sheets and roping	M S P	As Manila	As current practice	As for natural fibre cordage
owsing tackle	Should be of Manila or Polypropylene of a type accepted by the Department of Transport and having a grip comparable with Manila; the type and size of tackle and size cordage should be as follows:			
ully laden mass of boat nder 8 tonnes	Purchase Gun (3 parts)			20 mm
tonnes and under 10 tonnes	Gun (3 parts)			24 mm
0 tonnes and under 12 onnes	Luff (one double and single block)			18 mm
2 tonnes and under 15 onnes	Two fold (two double blocks)			20 mm
5 tonnes and under 20 onnes	Three/two fold (one treble and one double block)			20 mm
oats' falls	Manila, durable, unkinkable, firm laid and pliable. Breaking load to be at least 6 × maximum load when hoisting and lowering. To be not less than 20 mm. To be able to pass freely through a hole 10 mm larger than the nominal diameter of the rope. Man-made fibre cordage is not generally accepted.			

Tricing Pendants

Tricing pendants are attached to the floating blocks of the falls o gravity davits and to the shoulders of the davits themselves. Having bee made a pre-determined length, they will bring the boat alongside when it i lowered to the embarkation deck.

Made of wire, they incorporate a strong rope lashing between the end c the wire pendant and a senhouse slip, to allow it to be cut in an emergenc The shackle attaching the senhouse slip to the floating block is required t be elongated, so that it is impossible for the tongue of the senhouse slip c the link holding the tongue, to be jammed against the floating block. Th link of the senhouse slip is secured in position by a wood safety pir should this pin become wet and swell it is easily broken. Steel pins are no to be used for this purpose because they are liable to rust in place.

Care must be taken when lowering the boat, not to let it over-run and s place an undue load on the tricing pendants. Bowsing-in tackles are to b made fast to the floating blocks and the ship's side, hauled tight and mad fast and the tricing pendants released before persons are allowed to embark

A true senhouse slip is embodied in the tricing pendants to facilitate th operation of letting them go while there may be a certain amount of weigh on them.

Gripes

The function of the gripes is to hold the boat firmly down in the chock or in the case of gravity davits, firmly against the shoulder chocks of th davits.

Gripes are required to be fitted so that they can be let-go from inboard The normal method of fitting is to have the gripe wires taken over fairlead on the gunwale and fastened on the outboard side to the deck or dav frame, a senhouse slip is attached to the inboard end of the gripes fo letting-go. A strong rope lashing is incorporated next to the senhouse sli to allow it to be cut in an emergency. Care must be taken as the boat turned out, that the thimble on the inboard end of the gripes, which has t pass over the boat, does not foul anything. It is the responsibility of th two men in the boat to clear the gripes. On gravity davits it is th responsibility of the men who let-go the gripes to ensure that when trigger is fitted, it does in fact fall.

"Latch-on gripes" are an alternative method sometimes used with gravit davits. Latch-on gripes instead of passing over the boat, are led to sto bobbins on the stem and stern posts. All that is necessary is to let-go th senhouse slips and then throw the gripe wires off the bobbins.

Skates

Every lifeboat attached to davits, except emergency lifeboats, rescu boats and those attached to "Miranda" Davits, is fitted with two skates o the inboard side, for the purpose of assisting the passage of the boat dow the side of a ship with an adverse list. That is to say, the skates are there t act as skids and help slide or skate the boat down the side of the ship.

When the boat is in the water, the skates cease to have any value and wi greatly hamper the movement of the boat. Therefore, as soon as it may b convenient, unship them and tow them (normally there is sufficient woo in the skates to keep them afloat).

So very often, a ship is abandoned prematurely. Later, with the ship sti afloat, the survivors return to her; for the ship, no matter how badly she nay be damaged, will be far more comfortable than an open boat. If urvivors return to the ship, the boats must be recovered in case they should be needed again. This being so, the skates should be replaced in case they oo are needed again. Only when the ship actually sinks should the skates be discarded, for then, having served a useful purpose, they become an mbarrassment.

When lifeboats fitted to gravity davits are intended for embarkation in he stowed position, tricing pendants and bowsing-in tackles will not be itted.

When lifeboats attached to gravity davits are intended for launching by ree-fall, particularly in the case of oil-rigs, tricing pendants, bowsing-in ines and skates will not be fitted.

When lifeboats or Class C boats are attached to single arm davits, for aunching on one side of the ship only and are not required to launch the boat against an adverse list. Skates will not be fitted.

Single arm davits.

Single arm davits are mechanically controlled and are required to be itted with wire falls and a winch. They may be sited on the stern of small vessels attached to a lifeboat, Class C boat, Inflatable boat or a rescue boat. Rigid boats will be secured in chocks and griped down to the deck. Inflated boats will be secured at an approved position by approved fastenings. Single arm davits attached to boats are normally required to be able to launch the boat on one side of the ship only and are not required to launch the boat against an adverse list. Two men only are to be in the boat while it is being aunched. Survivors join the boat when it is afloat.

Single arm davits may also be placed amidships for the launching and ecovery of inflatable boats and for the launching of liferafts. When ntended for use with liferafts, the fall is required to have a tricing line attached for the purpose of recovering the hook after a liferaft has been aunched, without turning the davit inboard. They shall also be fitted with a safety hook, which when the safety catch is released, will automatically elease the liferaft as soon as it is waterborne. In lieu of a winch, some ingle arm davits intended for launching liferafts will be fitted with a spring notor for automatic recovery of the fall. Single arm davits intended for use with liferafts shall be capable of launching the liferaft when the ship is isted 20 degrees either way on vessels constructed after 1st July, 1986, or 15 degrees on other vessels.

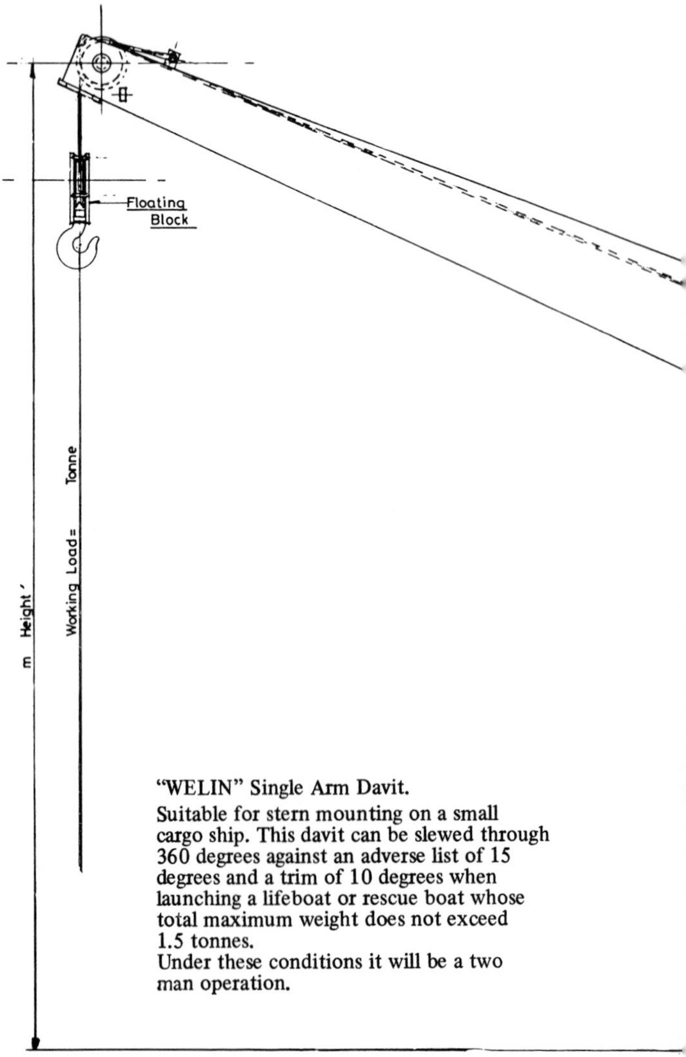

"WELIN" Single Arm Davit.

Suitable for stern mounting on a small cargo ship. This davit can be slewed through 360 degrees against an adverse list of 15 degrees and a trim of 10 degrees when launching a lifeboat or rescue boat whose total maximum weight does not exceed 1.5 tonnes.
Under these conditions it will be a two man operation.

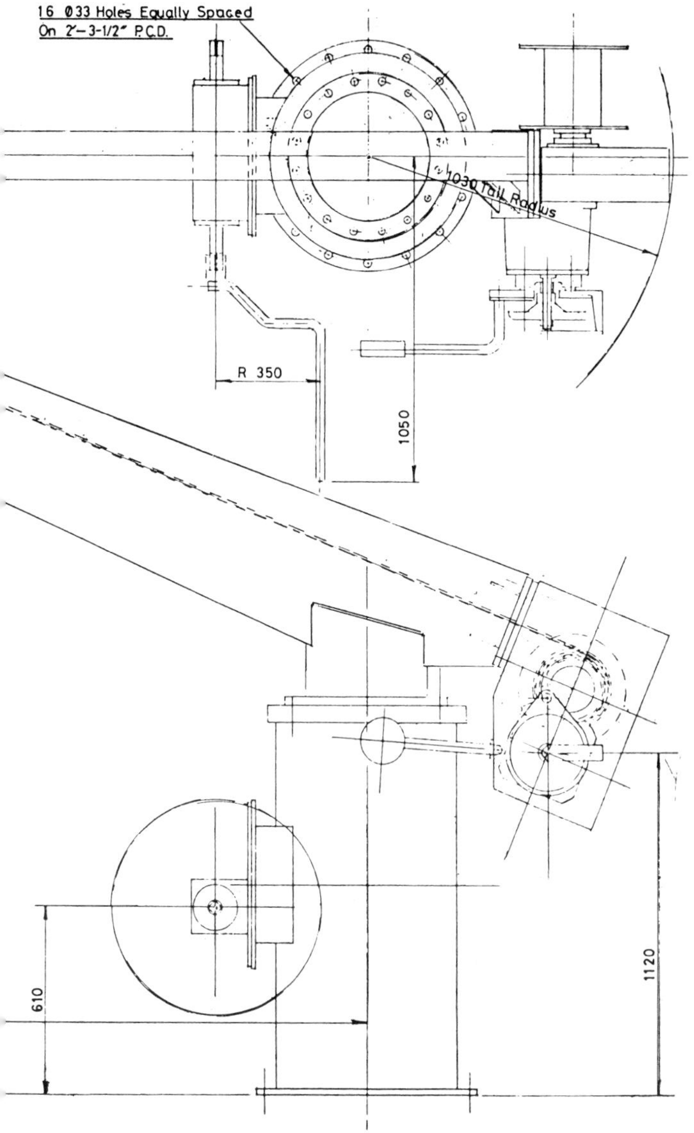

16 Ø33 Holes Equally Spaced
On 2'-3-1/2" P.C.D.
1030 Tail Radius
R 350
1050
1120
610

Free-fall

Stowage tray and tilt launcher

Approved for use with D.o.T. inflatable boats, the tilt launcher provide an alternative method for launching a D.o.T. inflatable boat on either sid of the ship and is capable of launching the boat against an adverse lis exceeding 25 degrees.

The tilt launcher is designed for use on small ships and consists of tw parallel trackways laid thwartships over the engine-room skylight or othe suitable position. A flat tray on runners is placed over the trackways. Th D.o.T. inflatable boat is secured to the tray by canvas webbing and sen house slips, the tray itself is kept in position by means of a release pi which can be easily and quickly unshipped.

To launch the boat, make fast the painter, let go the senhouse slips o the boat lashings and remove the boat cover, ensure it is all clear overside remove the release pin holding the tray. Push the tray (which is quite light along the trackway to whichever side of the ship it is desired to launch th boat. When the tray has travelled the required distance, it will tilt an launch the boat automatically. The angle of tilt required to clear th bulwarks or rails, is predetermined and stops are placed to ensure that th correct angle of tilt is not exceeded.

For recovery of the boat from the water, a small derrick or single arm o radial davit must be used in conjunction with a fore and aft sling, to lift th boat out of the water and onto the ship. Once on board the boat can b taken to and placed on the tray by hand. Bail out any water in the boa before lifting.

The stowage tray and tilt launchers may be modified if required to als allow for float free launching.

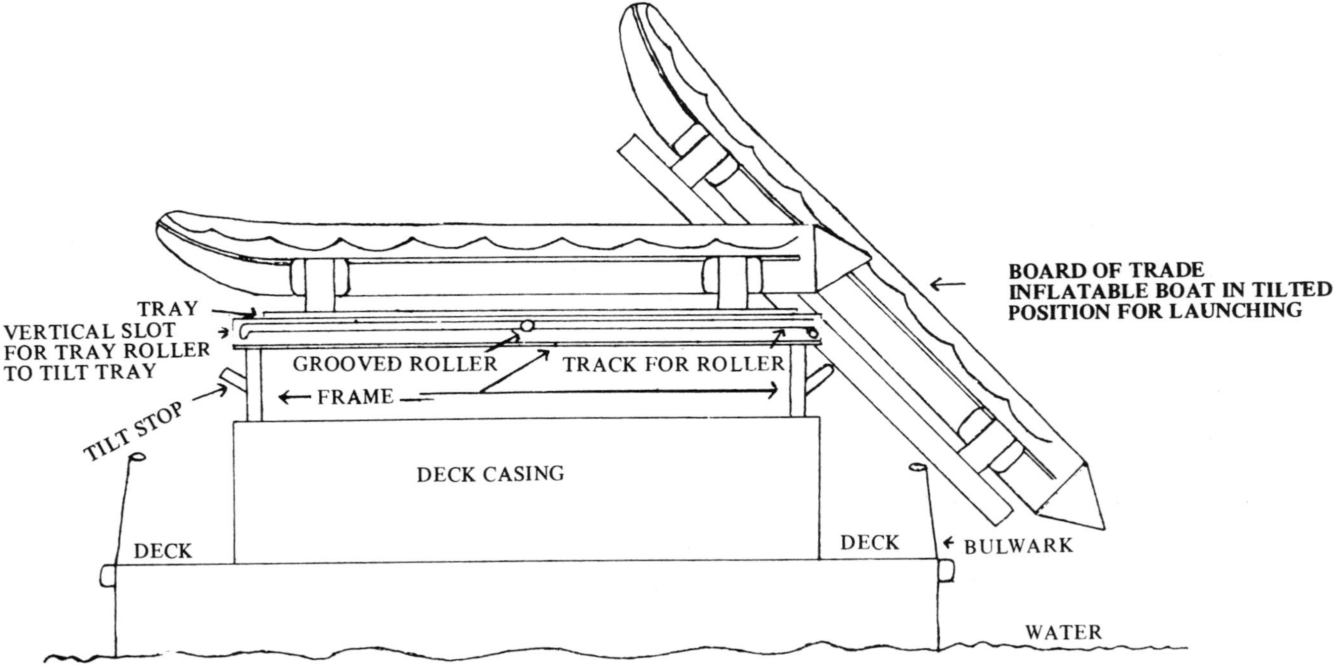

EXAMPLE

STOWAGE AND PORT AND STARBOARD LAUNCHING OF INFLATED BOAT

THE WM. CUBBINS TILT TRAY LAUNCHER FOR INFLATABLE BOATS

Freefall survival system

ONE of the biggest developments in rigid lifeboats has taken place in Norway recently where the Norwegian Maritime Directorate has initiated a project to develop a superior lifesaving system. The carriage of enclosed lifeboats had already been made compulsory by the Norwegian government but the system of launching remained inadequate.

The result of the study was the Harding freefall survival system. This, in fact, incorporates three different methods of lifeboat launching: freefall, mechanical and float-free.

In the first, and most dramatic, the rigid enclosed lifeboat is launched clear of the ship from a ramp and plunges nose-first into the sea. In the second, a gantry is employed to winch the lifeboat into the sea automatically; it can also be used to retrieve the lifeboat after exercises. In the third method, the lifeboat remains in the stowed position and is automatically released by hydrostate as the ship sinks.

In the freefall method padded seats and six-point safety belts ensure that the passengers survive the shock of impact without injury. The craft is also equipped with a sprinkler system capable of seeing it safely through an oil blaze of more than one nautical mile. Three air flasks can supply passengers and engines with 15 minutes' supply of oxygen and there are also motor-driven and manual pumps and a steering nozzle which gives increased manoeuvrability and encloses the propeller.

The complete system was the result of collaboration between Harding A/S who built the boat and A/S Nor Davit, who were responsible for the launching system.

Research has also been carried out to adapt the system for oil rigs and small boats.

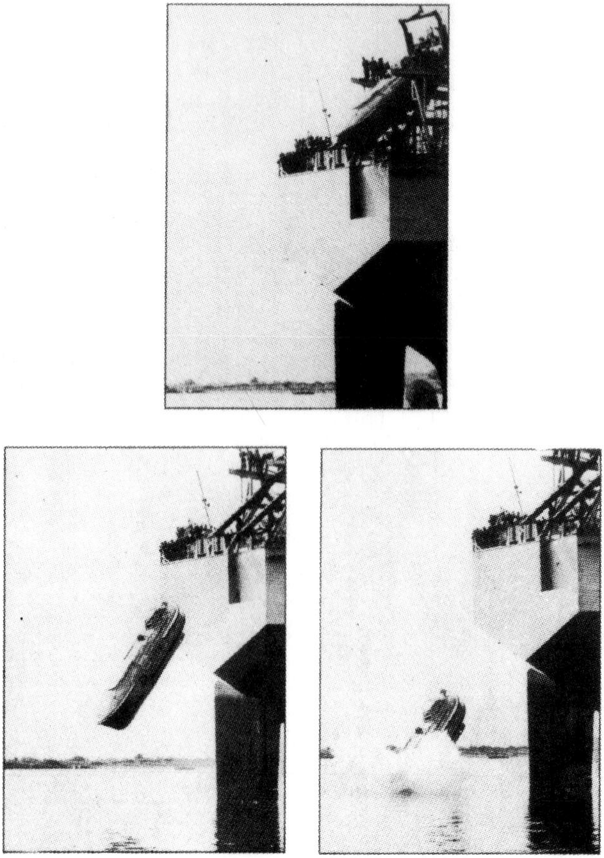

Reproduced from the "Nautical Review", February, 1980.

CHAPTER 3. LIFEBOATS UNDER 1974 & 1983 SOLAS REGULATIONS

Lifeboats are to be constructed with rigid sides and have ample stability in bad weather also sufficient freeboard when fully loaded and flooded and be sufficiently strong to allow them to be safely lowered into the water when fully loaded with persons and equipment. They are required to be fitted with internal buoyancy and may be manufactured from wood, aluminium alloy, galvanized steel or glass reinforced plastic.

Lifeboats built of wood may be either clincher *(overlapping planks)* or carvel built *(smooth sided)*. A wood lifeboat certified to carry 85 persons or more is required to be carvel built with the planks laid diagonally and in two thicknesses.

As wood lifeboats have a tendency to dry out in dry weather with the result that the seams are liable to open up, a little water should be kept in the bottom of the boat in hot weather. However, the plug should always be pulled out and the boat allowed to drain whenever there is any possibility of frost or ice.

Clincher built boats have the advantage of the fact that the edges of the overlapping planks increase the freeboard.

Aluminium and galvanised steel lifeboats are constructed in much the same way as a wood lifeboat. The fittings are of wood, for under extreme conditions, metal would burn the survivors.

They will stand up to tremendous punishment. However, great care must be taken with aluminium alloy boats in order to avoid corrosion. The buoyancy tanks must be made of aluminium alloy or expanded plastic foam substituted. Other metals should not come in contact with the boat. All paint used should have either a zinc chromate or zinc oxide base, copper and lead paints have a detrimental effect on aluminium. In the equipment, the buckets should be made of heavy duty plastic and the gripe wires must be sheathed.

Glass reinforced plastic (GRP) lifeboats are sometimes moulded in two halves and joined together by riveting onto an alloy keel frame, which has attachments for the lifting hooks. The bilge grab rail and all interior fittings are made of wood. In others, the boat is moulded in one piece and wood is only used for the bottom boards and thwarts. Buoyancy is supplied by blocks of expanded plastic foam and in some cases is actually built into the boat, though in older boats, alloy air tanks were fitted.

These boats are rot proof, fire-resistant, non-corrosive and do not sweat. Glass reinforced plastic, however, has very poor resistance to abrasion. Every effort has to be made to prevent gripes and lashings from scrubbing and the heavier gear inside the boat should be choked off as well as being lashed down so that it cannot rub with the ship's movement. Apart from this, a fibreglass lifeboat should be free of the necessity for any special care always remembering that there are some metal and wooden parts which may require painting from time to time.

Viking fibreglass lifeboats are equipped with a plastic foam buoyancy which is positioned under the side benches during construction and no attempt should be made to inspect or remove it. Authority for not removing this buoyancy is to be found above the Administration stamp situated on the bow name plate.

This form of buoyancy makes the boat unsinkable even when badly damaged. Moreover, when the boat is inspected, nothing has to be opened up or removed.

Repairs to a fibreglass boat can be achieved by using raw fibreglass and resin materials so that the surface shows little or no sign of damage. To make such repairs requires experience in handling fibreglass and polyester resin and temperatures and humidities by no means always found on the boatdeck or quayside. Any repair kits supplied by boat builders must be kept "in date."

In almost every case therefore, a far stronger and neater job can be made using ordinary skills and well known materials. The most useful fastening for light repairs, such as a tingle, is an assortment of stainless steel self tapping screws. For stronger fixing use ordinary through fastenings, rivets nuts and bolts etc. Ordinary putty or sealing compounds are satisfactory for smoothing and jointing and the job can be finished off by painting in the ordinary way.

Air tanks are normally made of muntz metal or copper and filled with either kapok contained in plastic bags or expanded plastic foam. Air tanks made of yellow metal should be wire brushed and coated with either clear varnish or linseed oil whenever the lifeboat is being overhauled.

Stretchers are required to be portable in order that they can be easily removed for the purpose of laying injured survivors on the bottom boards.

All lifeboats are required to be at least 12 inches (30cm) clear of the ship's side, when being lowered with the ship upright.

All lifeboats are to have a whaler stern so that they will rise to a following sea and can be either hove-to or rowed stern first. Except that all mechanically propelled and motor boats may have a transom stern to assist in protecting the propeller.

Names of the parts of a lifeboat.

Apron	A doubler inside the stem and sternpost to which the ends of the planking are fastened.
Bilge	The curve of the hull of the boat.
Bow sheets	The area inside the bow of the boat.
Bottom boards	Light boards in the bottom of the boat, running fore and aft. Placed there to protect the skin of the boat.
Breasthook	"V" shaped tie piece in the bow or stern.
Buoyancy tanks	Air tight tanks fitted under the side benches.
Chain plate	Eye plate, bolted on the gunwale on either side of the boat, to take the mast stay lanyards when the mast is stepped.
Cleat	Metal fitting with two horns, attached inside the gunwale for the purpose of assisting in the control of the sheets when under sail.
Deadwood	Scarphed between the apron and keelson, it strengthens the scarphs between the keel and stem and between the keel and sternpost.
Forefoot	Curved part of the stem below the waterline.
Gangboard	A plank or piece of metal in both the bow and stern through which the shank of the lifting hook passes. It prevents sideways movement of the shank.

Garboard strake	The plank next to the keel on each side of it.
Gunwale	The top edge of the hull of the boat.
Hogpiece	A broad plank placed on top of the keel and to which the garboard strake is fastened.
Keel	Strong fore and aft member on the bottom of the boat.
Keel grab rail or Bilge grab rail	A rail half the length of the boat, fitted amidships and running fore and aft, just below the waterline. Fitted for survivors to cling to on a capsized boat.
Keel plates	Plates under the keel bolted through to the sling plates.
Keelson	Strong fore and aft member in the bottom of the boat, placed on top of the timbers, in line with the keel.
Mast clamp	A steel clamp on the after side of the mast thwart, used to clamp the mast in position when it is stepped.
Mast shoe	A socket to take the heel of the mast when it is stepped.
Plug hole	A drain hole, its position is marked on the side bench.
Rising	Fore and aft hull plank inside the timbers.
Roove	A copper washer placed over the end of a copper nail from which the point has been nipped. The nail is hammered flat on the roove to give a riveted effect. Wood lifeboats are fastened with copper nails and rooves.
Rudder	A loose extension to the stern post. Used for steering.
Rudder post	A vertical rail on the after side of the sternpost, used to support the rudder.
Sheer strake	The topmost plank on each side of the hull.
Side benches	Seats running fore and aft along each side of the boat.
Stem	A vertical post forming the bow of the boat.
Sling plates	Attach the shanks of the hooks to the keel plates.
Sternpost	A vertical post forming the stern of the boat.
Sternsheets	Area in the stern of the boat, abaft the after thwart.
Stretchers	Low cross seats, sometimes known as lower thwarts.
Tabernacle	Alternative to a mast shoe. Three vertical plates, bounded on the fourth (after) side by a bolt.
Tank cleading	A vertical bulkhead or grating between the inboard side of the side benches and the bottom of the boat. Encloses the space occupied by the buoyancy tanks.
Thwarts	Cross seats at the same level as the side benches.
Thwart knees	Brackets attaching the thwarts to the hull.
Tiller	A loose handle which fits into the top of the rudder. Used to turn the rudder when steering.
Timbers	Transverse wood frames to which the planks of the hull are fastened.
Tingle	A metal patch on the outside of a boat where a repair has been effected.
Transom	A vertical board fitted athwartships in the stern of a boat in lieu of a stern post to give a flat stern.
Whaler stern	A pointed stern.
Yoke	A cross piece with two lines attached, fitted to the head of the rudder post in lieu of a tiller.

Mechanically propelled lifeboats

Mechanically propelled lifeboats like all other open lifeboats will shortly become obsolete, in as much as they will not be acceptable as life-saving equipment on ships built after 1st. July, 1986.

A mechanically propelled lifeboat has no engine and it requires manual effort to rotate the propeller. There are two generally accepted types of propelling gear, as follows:–

(a) The operators sit on thwarts facing both forward and aft, the propeller is turned by means of a fore and aft motion of vertical levers, placed between the thwarts. (Fleming Gear).

(b) The operators sit on the side benches facing inboard and propel the boat by rotating horizontal shafts which are connected to vertical fly-wheels.

The propelling gear is to be so arranged that it can be rapidly and easily made ready for use, and will not interfere with the rapid embarkation of persons into the boat. It must be capable of being operated by untrained persons and must not require adjustment to enable it to be worked by persons of different stature and must be capable of propelling the boat when it is either partially or fully loaded and when the boat is flooded. The metal part of any handle is required to be sheathed in a material other than wood, to ensure that the hands of the operators are protected in conditions of extreme cold.

The propelling gear shall be of sufficient power to enable the lifeboat to be propelled at a speed of at least 3.5 knots in smooth water over a distance of ¼ of a mile when the boat is fully loaded. It shall also be capable of propelling the boat both ahead and astern. A device is required to be fitted by means of which the helmsman can cause the boat to go ahead or astern when the propelling gear is in operation.

I.M.O. General requirements for life-saving appliances. On ships built after 1st July, 1986.

1. Paragraph 2(g) applies to all ships. With respect to ships constructed before 1st. July, 1986. Paragraph 2 (g) shall apply no later than 1st. July, 1991.

2. Unless expressly provided otherwise, all life-saving appliances prescribed in these Rules shall:

(a) be constructed with proper workmanship and materials;

(b) not be damaged in stowage throughout an air temperature range of minus 30 degrees C (–22 deg. F) to plus 65 degrees C (149 deg. F).

(c) If they are likely to be immersed in sea water during their use operate throughout the sea water range of minus 1 degree (30 deg. F) to plus 30 degrees C (86 deg. F).

(d) Where applicable be rot-proof, corrosion resistant and not be unduly affected by sea water, oil or fungal attack;

(e) where exposed to sunlight be resistant to deterioration;

(f) be of highly visible colour on all parts where this will assist detection;

(g) be fitted with retro-reflective material where it will assist in detection and in accordance with the recommendations of the Organization;

(h) If they are to be used in a sea-way be capable of satisfactory operation in that environment.

3. The Administration shall determine the period of acceptability of life-saving appliances which are subject to deterioration with age. Such life-saving appliances shall be marked with a means for determining their age or the date by which they must be replaced.

IMO. General requirements for lifeboats on ships built after 1st. July, 1986.

1. **Construction of lifeboats**

(a) All lifeboats shall be properly constructed and shall be of such form and proportions that they have ample stability in a sea-way and sufficient freeboard when loaded with their full complement of persons and equipment. All lifeboats shall have rigid hulls and shall be capable of maintaining positive stability when in an upright position in calm water and loaded with their full complement of persons and equipment and holed in any one location below the waterline, assuming no loss of buoyancy material and no other damage.

(b) All lifeboats shall be of sufficient strength to:

(i) enable them to be safely lowered into the water when loaded with their full complement of persons and equipment; and

(ii) be capable of being launched and towed when the ship is making headway at a speed of 5 knots in calm water.

(c) Hulls and rigid covers shall be fire retarding or non-combustible.

(d) Seating shall be provided on thwarts, benches or fixed chairs fitted as low as practicable in the lifeboat and constructed so as to be capable of supporting the number of persons each weighing 100kg. (220lbs) for which spaces are provided in compliance with the requirements of paragraph 2 (b) (ii) of this Regulation.

(e) Each lifeboat shall be of sufficient strength to withstand a load, without residual deflection on removal of that load:

(i) in the case of boats with metal hulls, 1.25 times the total mass of the lifeboat when loaded with its full complement of persons and equipment; or

(ii) in the case of other boats, twice the total mass of the lifeboat when loaded with its full complement of persons and equipment.

(f) Each lifeboat shall be of sufficient strength to withstand, when loaded with its full complement of persons and equipment and with, where applicable, skates or fenders in position, a lateral impact against the ship's side at an impact velocity of at least 3.5 m/s and also a drop into the water from a height of at least 3m. (9ft. 9in.)

(g) The vertical distance between the floor surface and the interior of the enclosure or canopy over 50% of the floor area shall be:

(i) not less than 1.3m (4.25ft.) for a lifeboat permitted to accommodate nine persons or less;

(ii) not less than 1.7m (5ft. 7in.) for a lifeboat permitted to accommodate 24 persons or more;

(iii) not less than the distance as determined by linear interpolation between 1.3m. and 1.7m. for a lifeboat permitted to accommodate between nine and 24 persons.

2. **Carrying capacity of lifeboats**

(a) No lifeboat shall be approved to accommodate more than one hundred and fifty persons.

(b) The number of persons which a lifeboat shall be permitted to accommodate shall be equal to the lesser of:

(i) the number of persons having an average mass of 75 kg. (165lbs), all wearing lifejackets, that can be seated in a normal position without interfering with the means of propulsion or the operation of any of the lifeboat's equipment; or

(ii) The number of spaces that can be provided on the seating arrangements in accordance with figure 1. The shapes may be overlapped as shown, provided foot-rests are fitted and there is sufficient room for legs and the vertical separation between the upper and lower seat is not less than 350 mm (14 ins.)

(iii) Each seating position shall be clearly indicated in the lifeboat.

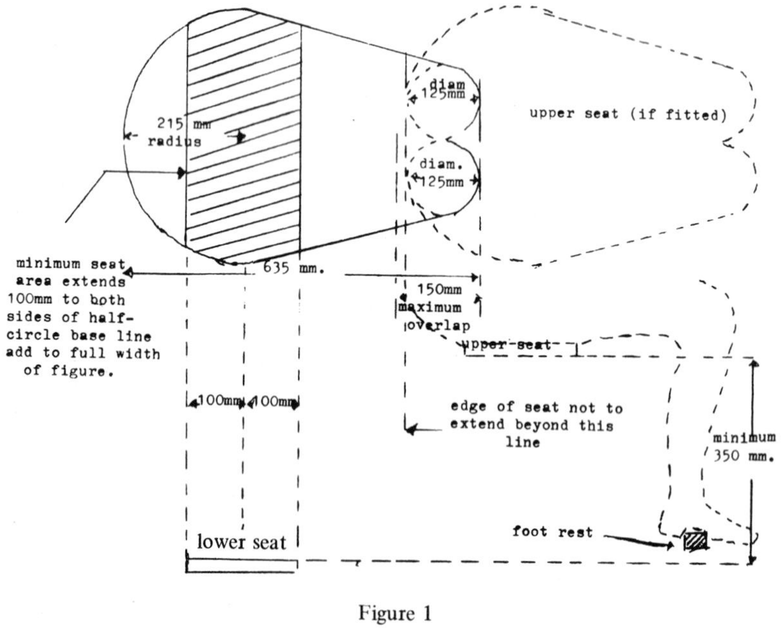

Figure 1

3. **Access into lifeboats**

(a) Every passenger ship lifeboat shall be so arranged that it can be rapidly boarded by its full complement of persons. Rapid disembarkation shall also be possible.

(b) Every cargo ship lifeboat shall be so arranged that it can be boarded by its full complement of persons in not more than 3 minutes from the time the instruction to board is given. Rapid disembarkation shall also be possible.

(c) Lifeboats shall have a boarding ladder that can be used on either side of the lifeboat to enable persons in the water to board the lifeboat. The lowest step of the ladder shall be not less than 0.4m. (16 ins.) below the lifeboat's light waterline. *(This bottom step should be weighted to prevent it floating.)*

(d) The lifeboat shall be so arranged that helpless people can be brought on board either from the sea or on stretchers.

(e) All surfaces on which persons might walk shall have a non-skid finish.

4. **Lifeboat buoyancy**

All lifeboats shall have inherent buoyancy or shall be fitted with inherently buoyant material which shall not be adversely affected by sea water, oil or oil products, sufficient to float the lifeboat, with all its equipment on board when flooded and open to the sea. Additional inherently buoyant material, equal to 280 N of buoyant force per person shall be provided for the number of persons the lifeboat is permitted to accommodate. Buoyant material, unless in addition to that required above, shall not be installed external to the hull of the lifeboat.

5. **Lifeboat freeboard and stability**

All lifeboats, when loaded with 50% of the number of persons the lifeboat is permitted to accommodate seated in their normal positions to one side of the centreline, shall have a freeboard, measured from the waterline to the lowest opening through which the lifeboat may become flooded, of at least 1.5% of the lifeboat's length or 100mm (4ins.), whichever is the greater.

6. **Lifeboat propulsion**

(a) Every lifeboat shall be powered by a compression ignition engine. No engine shall be used for any lifeboat if its fuel has a flashpoint of 43 degrees C (109 deg. F) or less (closed cup test).

(b) The engine shall be provided with either a manual starting system or a power starting system with two independent rechargeable energy sources. Any necessary starting aids shall also be provided. The engine starting systems and starting aids shall start the engine at an ambient temperature of -15 degrees C (+5 deg. F) within 2 minutes of commencing the start procedure. The starting system shall not be impeded by the engine casing, thwarts or other obstructions.

(c) The engine shall be capable of operating for not less than 5 minutes after starting from cold with the lifeboat out of the water.

(d) The engine shall be capable of operating when the lifeboat is flooded up to the centreline of the crankshaft.

(e) The propeller shafting shall be so arranged that the propeller can be disengaged from the engine. Provision shall be made for ahead and astern propulsion of the lifeboat.

(f) The exhaust pipe shall be so arranged as to prevent water from entering the engine in normal operation.

(g) All lifeboats shall be designed with due regard to the safety of persons in the water and to the possibility of damage to the propulsion system by floating debris.

(h) The speed of a lifeboat when proceeding ahead in calm water, when loaded with its full complement of persons and equipment and with all engine powered auxiliary equipment in operation, shall be at least 6 knots and at least 2 knots when towing a 25 person liferaft loaded with its full complement of persons and equipment or its equivalent. Sufficient fuel, suitable for use throughout the temperature range expected in the area in which the ship operates, shall be provided to run the fully loaded lifeboat at 6 knots for a period of not less than 24 hours.

(i) The lifeboat engine, transmission and engine accessories shall be enclosed in a fire-retarding casing or other suitable arrangements providing similar protection. Such arrangements shall also protect the engine from exposure to weather and sea. Adequate means shall be provided to reduce the engine noise. Starter batteries shall be provided with casings which form a watertight enclosure around the bottom and sides of the batteries. The battery casings shall have a tight fitting top which provides for necessary gas venting.

(j) The lifeboat engine and accessories shall be designed to limit electro-magnetic emissions so that engine operation does not interfere with the operation of radio life-saving appliances used in the lifeboat.

(k) Means shall also be provided for recharging all engine starting, radio and searchlight batteries. Radio batteries shall not be used to provide power for engine starting. Means shall be provided for recharging lifeboat batteries from the ship's power supply at a supply voltage not exceeding 55 V which can be disconnected at the lifeboat embarkation station.

(l) Water resistant instructions for starting and operating the engine shall be provided and mounted in a conspicuous place near the engine starting controls.

7. **Lifeboat fittings**

(a) All lifeboats shall be provided with not less than one drain valve fitted near the lowest point in the hull, which shall automatically open to drain water from the hull when the lifeboat is not waterborne and automatically close to prevent entry of water when the lifeboat is waterborne. Each drain valve shall be provided with a cap or plug to close the valve, which shall be attached to the lifeboat by a lanyard, a chain, or other suitable means. Drain valves shall be readily accessible from inside the lifeboat and their position shall be clearly indicated.

(b) All lifeboats shall be provided with a rudder and tiller. When a wheel or other remote steering mechanism is also provided the tiller shall be capable of controlling the rudder in case of failure of the steering mechanism. The rudder shall be permanently attached to the lifeboat.

The tiller shall be permanently installed on, or linked to, the rudder stock; however, if the lifeboat has a remote steering mechanism, the tiller may be removable and securely stowed near the rudder stock. The rudder and tiller shall be so arranged as not to be damaged by operation of the release mechanism or the propellor.

(c) Except in the vicinity of the rudder and propeller, a buoyant life-line shall be becketed around the outside of the lifeboat.

(d) Lifeboats which are not self-righting when capsized shall have suitable handholds on the underside of the hull to enable persons to cling to the lifeboat. The handholds shall be fastened to the lifeboat in such a way that, when subjected to an impact sufficient to cause them to break away from the lifeboat, they break away without damaging the lifeboat.

(e) All lifeboats shall be fitted with sufficient watertight lockers or compartments to provide for the storage of the small items of equipment, water and provisions required by paragraph 8 of this Regulation. Means shall be provided for the storage of collected rain water.

(f) Every lifeboat to be launched by a fall or falls shall be fitted with a release mechanism complying with the following:

(i) The mechanism shall be so arranged that all hooks are released simultaneously;

(ii) The mechanism shall have two release capabilities as follows:

1. a normal release capability which will release the life-boat when it is waterborne or when there is no load on the hooks;

2. an on-load release capability which will release the lifeboat with a load on the hooks. This release shall be so arranged as to release the lifeboat under any conditions of loading from no-load with the lifeboat waterborne to a load of 1.1 times the total mass of the lifeboat when loaded with its full complement of persons and equipment. This release capability shall be adequately protected against accidental or premature use;

3. the release control shall be clearly marked in a colour that contrasts with its surroundings;

4. The mechanism shall be designed with a factor of safety of 6 based on the ultimate strength of the materials used, assuming the mass of the lifeboat is equally distributed between the falls.

(g) Every lifeboat shall be fitted with a release device to enable the forward painter to be released when under tension.

(h) Every lifeboat shall be provided with a permanently installed earth connection and arrangements for adequately siting and securing in the operating position the antenna provided with the portable radio apparatus.

(i) Lifeboats intended for launching down the side of a ship shall have skates and fenders as necessary to facilitate launching and prevent damage to the lifeboat.

(j) A manually controlled lamp visible on a dark night with a clear atmosphere at a distance of at least two miles for a period of not less than 12 hours shall be fitted to the top of the cover or enclosure. If the light is a flashing light, it shall initially flash at a rate of not less than 50 flashes per minute over the first two hours of operation of the 12 hour operating period.

(k) A lamp or source of light shall be fitted inside the lifeboat to provide illumination for not less than 12 hours to enable reading of survival and equipment instructions; however, oil lamps shall not be permitted for this purpose.

(l) Unless expressly provided otherwise, every lifeboat shall be provided with effective means of bailing or be automatically self-bailing.

(m) Every lifeboat shall be so arranged that an adequate view forward, aft and to both sides is provided from the control and steering position for safe launching and manoeuvring.

8. **Lifeboat equipment**

All items of lifeboat equipment, whether required by this paragraph or elsewhere in this Regulation, with the exception of boat hooks which shall be kept free for fending off purposes, shall be secured within the lifeboat by lashings, storage in lockers or compartments, storage in brackets or similar mounting arrangements or other suitable means. The equipment shall be secured in such a manner so as not to interfere with any abandonment procedures. All items of lifeboat equipment shall be as small and of as little mass as possible and shall be packed in a suitable and compact form. Except where otherwise stated, the normal equipment of every lifeboat shall consist of:

(i) sufficient buoyant oars to make headway in calm seas. Thole pins, crutches or equivalent arrangements shall be provided for each oar provided. Thole pins or crutches shall be attached to the boat by lanyards or chains;

(ii) two boat hooks;

**(iii) a buoyant bailer and two buckets;

(iv) a survival manual;

(v) a binnacle containing an efficient compass which is luminous or provided with suitable means of illumination. In a totally enclosed lifeboat, the binnacle shall be permanently fitted at the steering position; in any other lifeboat, it shall be provided with suitable mounting arrangements;

*(vi) a sea-anchor of adequate size fitted with a shock resistant hawser and a tripping line which provides a firm hand grip when wet. The strength of the sea-anchor, hawser and tripping line shall be adequate for all sea conditions;

(vii) two efficient painters of a length equal to not less than twice the distance from the stowage position of the lifeboat to the waterline in the lightest sea-going conditions or 15m (49ft.) whichever is the greater. One painter attached to the release device required by paragraph 7 (g) of this Regulation, shall be placed at the forward end of the lifeboat and the

***Buckets should have lanyards spliced onto the handles.*

**Normally the sea-anchor hawser shall be at least 3 times the length of the lifeboat and be provided with a swivel. It is to be protected from chafing at the fairlead and the tripping line should be at least 11.5 feet (3.5m.) longer than the hawser.*

other shall be firmly secured at or near the bow of the lifeboat ready for use;

(viii) two hatchets, one at each end of the lifeboat;

(ix) watertight recepticals containing a total of 3 litres (5.3 pints) of fresh water for each person the lifeboat is permitted to accommodate, of which one litre (1.8 pints) per person may be replaced by de-salting apparatus capable of producing an equal amount of fresh water in 2 days; fresh water stored in tanks is to be frequently changed.

(x) a rust proof dipper with lanyard;

*(xi) a rust-proof graduated drinking vessel;

(xii) a food ration totalling not less than 10,000 kj for each person the lifeboat is permitted to accommodate; these rations shall be kept in air-tight packaging and be stowed in a watertight container;

(xiii) four rocket parachute flares;

(xiv) six hand flares;

(xv) two buoyant smoke signals;

(xvi) one waterproof electric torch suitable for morse signalling together with one spare set of batteries and one spare bulb in a waterproof container;

(xvii) one daylight signalling mirror with instructions for its use for signalling to ships and aircraft;

N.B. The mirror is used to reflect the sun's rays; the angle between the sun and the ship or aircraft must be acute enough for the mirror to be able to reflect the rays in the desired direction. To use the mirror, hold the top and bottom of the mirror between the index finger and thumb of the right hand, across the palm of the hand, polished side outward and with the hole at the bottom left hand corner, look through the hole at the vessel or air-craft you are trying to attract. Holding the sight at the full length of its cord away from the mirror with the left hand, look through the hole in the sight also. Now keep angling the mirror, trying to get a reflection of the cross on the mirror onto the back of the sight, keeping the vessel or air-craft in view through the holes all the time. Each time you succeed in reflecting the cross onto the back of the sight, you will have flashed a light that will be visible to your target and which, if seen, may well result in your rescue. The reflection produced should be visible well over 5 miles (8km). on a sunny day.

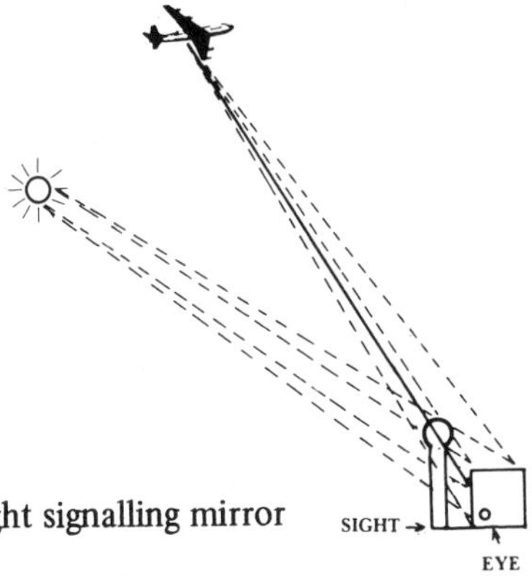

Using the daylight signalling mirror

**U.K. Regulations require 3 drinking vessels to be provided.*

(xviii) one copy of the life-saving signals prescribed by the Regulations, on a waterproof card or in a waterproof container;

N.B. A copy of these signals is reproduced on pages 313 to 316.

(xix) one whistle or equivalent sound signal;

(xx) a first-aid outfit in a waterproof case capable of being closed tightly after use;

(xxi) six doses of anti-seasickness medicine and one seasickness bag for each person;

(xxii) a jack-knife to be kept attached to the boat with a lanyard;

(xxiii) three tin openers;

(xxiv) two buoyant rescue quoits, attached to not less than 30m (100ft.) of buoyant line;

(xxv) a manual pump

(xxvi) one set of fishing tackle;

(xxvii) sufficient tools for minor adjustments to the engine and its accessories;

(xxviii) portable fire extinguishing equipment suitable for extinguishing oil fires;

(xxix) a searchlight capable of effectively illuminating a light coloured object at night having a width of 18m (60ft.) at a distance of 180m. (585ft.) for a total period of 6 hours and of working for not less than 3 hours continuously;

(xxx) an efficient radar reflector;

(xxxi) thermal protective aids sufficient for 10% of the number of persons the lifeboat is permitted to accommodate or two, whichever is the greater;

(xxxii) in the case of ships engaged on voyages of such nature and duration that in the opinion of the Administration, the items specified in sub-paragraphs (xii) and (xxvi) are unnecessary, the Administration may allow these items to be dispensed with.

9. **Lifeboat markings**

(a) The dimensions of the lifeboat and the number of persons it is permitted to accommodate shall be marked on it in clear permanent characters.

(b) The name and port of registry of the ship to which the lifeboat belongs shall be marked on each side of the lifeboat's bow in block capitals of the Roman alphabet.

(c) Means of identifying the ship to which the lifeboat belongs and the number of the lifeboat shall be marked in such a way that they are visible from above.

N.B. In all lifeboats the buckets are normally of a two gallon (0.9 litre) size and are galvanized. However, heavy duty plastic buckets will be supplied in aluminium boats. For lifeboats up to 9m. (30ft.) in length a circular or square mouth sea-anchor is optional. Boats over 9m (30ft.) in length are to be supplied with square mouth or folding sea-anchors, these anchors have galvanized iron spreaders across the mouth and an ash wood spreader on the upper edge.

In lifeboats provided with oil lamps and matches, should the box become wet, wipe it on your sleeve, this should dry it sufficiently to enable a match to strike.

Where lifeboats are intended for launching down the ship's side, all openings in the ship's side over which the lifeboat will pass are required to have permanent fenders provided at the opening, to ensure that the lifeboat cannot be caught in the opening when being lowered or raised.

Pea whistles are not an acceptable whistle.

The manual pump should be bolted in place and the cover is to be easily removable for cleaning, the handle sheathed to prevent it burning the hands of the operator in arctic conditions. The pump is to be self-priming at a height of 1.2m (4ft.) with a strum or strainer fitted to the suction. Causes of failure to function:– a blocked strainer, clean it. A leak in the suction pipe, change it with the discharge pipe. Ensure the interior of the pump is clean. Prime if necessary by pouring water down the discharge pipe.

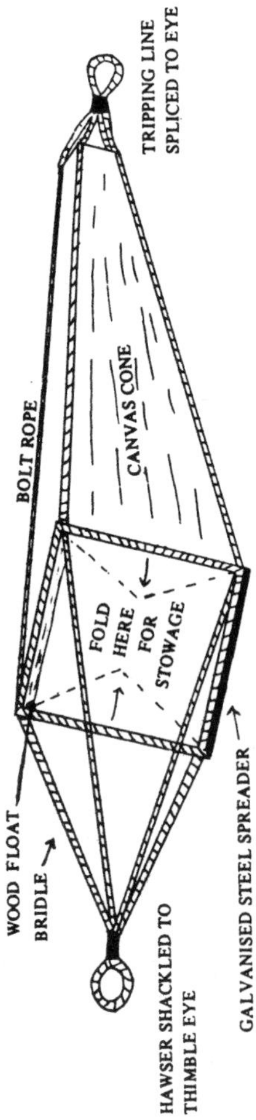

COLLAPSIBLE SQUARE MOUTH SEA ANCHOR

MEDIUM SIZE FOR BOATS 22 TO 30 FEET LONG (6.7 to 9.1m)

LENGTH OF SIDE OF MOUTH 24 INCHES (61cm.)

LENGTH OF CANVAS CONE 4 FEET (1.2m)

IMO. Requirements for Partially enclosed lifeboats on ships built after 1st. July, 1986.

1. Partially enclosed lifeboats shall comply with the requirements for all lifeboats and in addition shall comply with the following:–

2. Every partially enclosed lifeboat shall be provided with effective means of bailing or be automatically self-bailing.

3. Partially enclosed lifeboats shall be provided with permanently attached rigid covers over not less than 20% of the length of the lifeboat from the stem and not less than 20% of the lifeboat from the aftermost part of the lifeboat. The lifeboat shall be fitted with a permanently attached foldable canopy which together with the rigid covers completely encloses the occupants of the lifeboat in a weatherproof shelter and protects them from exposure. The canopy shall be so arranged that;

(a) it is provided with adequate rigid sections to permit erection of the canopy;

(b) it can easily be erected by not more than two persons;

(c) it is insulated to protect the occupants against heat and cold by means of not less than two layers of material separated by an air gap or other equally efficient means; means shall be provided to prevent accumulation of water in the air gap;

(d) its exterior is of a highly visible colour and its interior is of a colour which does not cause discomfort to the occupants;

(e) it has entrances at both ends and on each side, provided with efficient adjustable closing arrangements which can be easily and quickly opened and closed from inside or outside so as to permit ventilation but exclude sea water, wind and cold; means shall be provided for holding the entrances securely in the open and closed position;

(f) with the entrances closed, it admits sufficient air for the occupants at all times;

(g) it has means for collecting rain water;

(h) the occupants can escape in the event of the lifeboat capsizing.

4. The interior of the lifeboat shall be of a highly visible colour.

5. The radiotelegraph installation required by these Regulations shall be installed in a cabin large enough to accommodate both the equipment and the person using it. No separate cabin is required if the construction of the lifeboat provides a sheltered space to the satisfaction of the Administration.

Self righting partially enclosed lifeboats

1. Self-righting partially enclosed lifeboats shall comply with the Regulations for partially enclosed lifeboats and in addition with the following requirements.

2. **Enclosure**

(a) Permanently attached rigid covers shall be provided extending over not less than 20% of the length of the lifeboat from the stem and not less than 20% of the length of the lifeboat from the aftermost part of the lifeboat.

(b) The rigid covers shall form two shelters. If the shelters have bulkheads they shall have openings of sufficient size to permit easy access by persons wearing an immersion suit or warm clothes and a lifejacket. The interior height of the shelters shall be sufficient to permit persons easy access to their seats in the bow and stern of the lifeboat.

(c) The rigid covers shall be so arranged that they include windows or translucent panels to admit sufficient daylight to the inside of the lifeboat with the openings or canopies closed so as to make artificial light unnecessary.

(d) The rigid covers shall have railings to provide a secure handhold for persons moving about the exterior of the lifeboat.

(e) Open parts of the lifeboat shall be fitted with a permanently attached foldable canopy so arranged that:

(i) it can be easily erected by not more than two persons in not more than 2 minutes;

(ii) it is insulated to protect the occupants against cold by means of not less than two layers of material separated by an air gap or other equally efficient means.

(f) The enclosure formed by the rigid covers and canopy shall be so arranged:

(i) as to allow launching and recovery operations to be performed without any occupant having to leave the enclosure;

(ii) that it has entrances at both ends and on each side, provided with efficient adjustable closing arrangements which can be easily and quickly opened and closed from inside or outside so as to permit ventilation but exclude sea water, wind and cold; means shall be provided for holding the entrances securely in the open and in the closed position;

(iii) that with the canopy erected and all entrances closed, sufficient air is admitted for the occupants at all times;

(iv) that it has means for collecting rain water;

(v) that the exterior of the rigid covers and canopy and the interior of that part of the lifeboat covered by the canopy is of a highly visible colour. The interior of the shelters shall be of a colour which does not cause discomfort to the occupants;

(vi) that it is possible to row the lifeboat.

3. **Capsizing and re-righting**

(a) A safety belt shall be fitted at each indicated seating position. The safety belt shall be so designed as to hold a person of a mass of 100kg (220lbs) securely in place when the lifeboat is in a capsized position.

(b) The stability of the lifeboat shall be such that it is inherently or automatically self-righting when loaded with its full or a partial complement of persons and equipment and the persons are secured with safety belts.

4. **Propulsion**

(a) The engine and transmission shall be controlled from the helmsman's position.

(b) The engine and engine installation shall be capable of running in any position during capsize and continue to run after the lifeboat returns to the upright or shall automatically stop on capsizing and be easily restarted after the lifeboat returns to the upright and the water has been drained from the lifeboat. The design of the fuel and lubricating systems shall prevent the loss of fuel and the loss of more

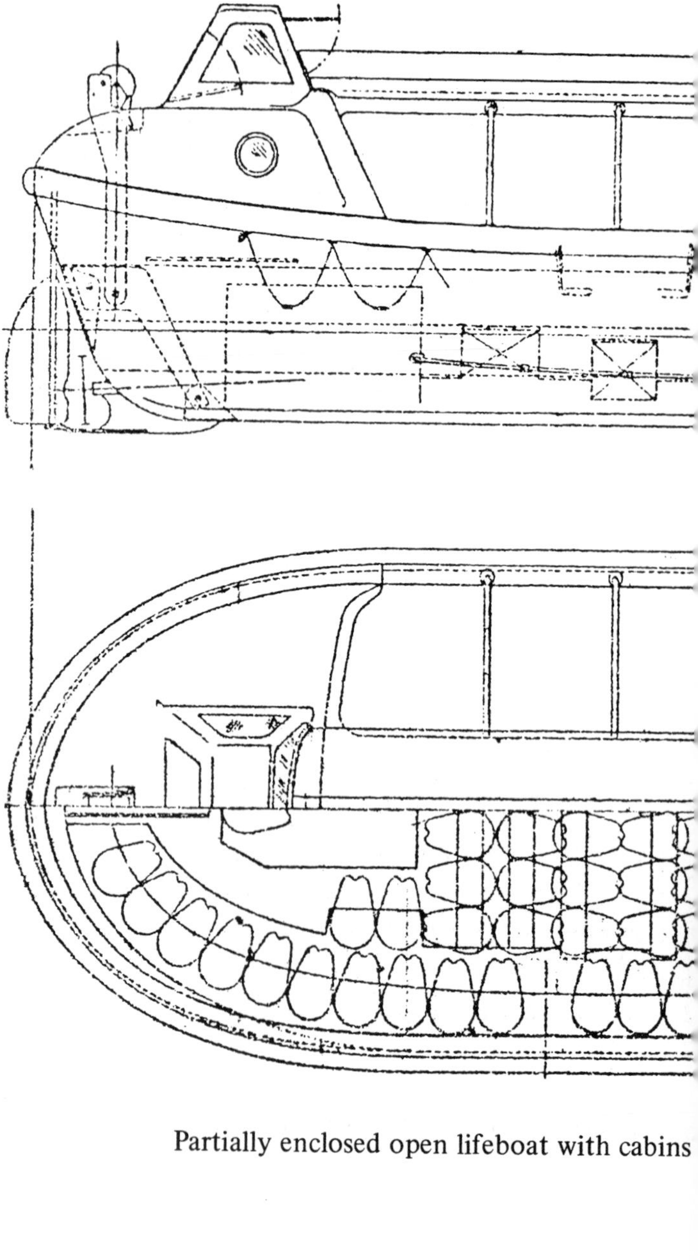

Partially enclosed open lifeboat with cabins

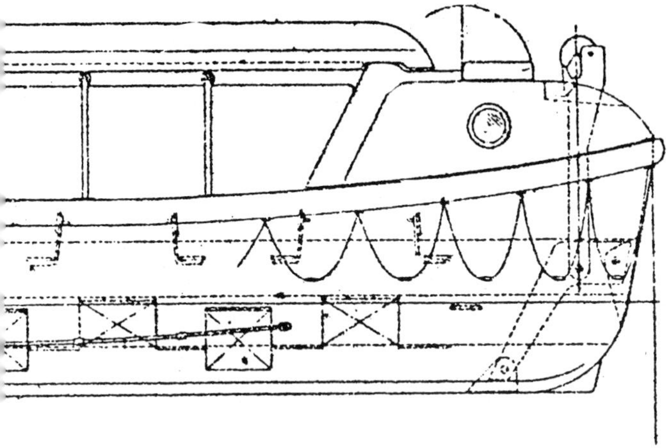

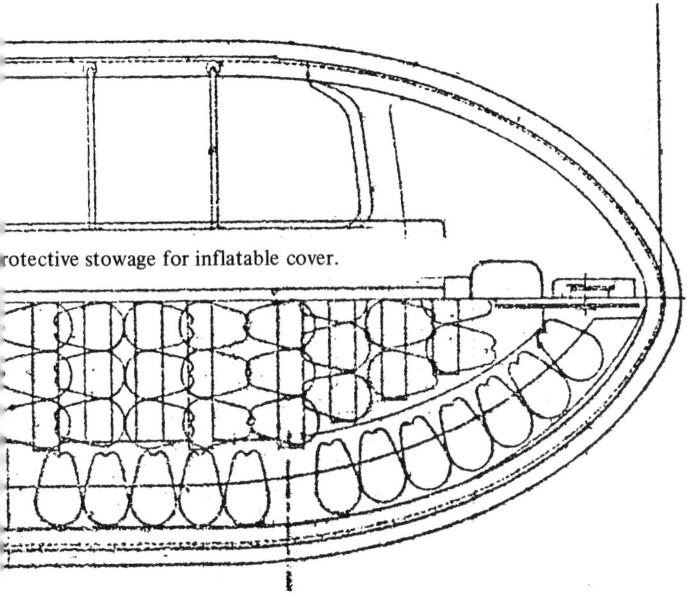

rotective stowage for inflatable cover.

ɔwn inflatable cover seating 144 persons.

than 250ml (½ pint) of lubricating oil from the engine during capsize.

(c) Air-cooled engines shall have a duct system to take in cooling air from, and exhaust it to, the outside of the lifeboat. Manually operated dampers shall be provided to enable cooling air to be taken in from, and exhausted to, the interior of the lifeboat.

5. **Construction and fendering**

(a) Notwithstanding sub-paragraph (f) in paragraph 1 of the IMO General requirements for lifeboats on page 179, a self-righting partially enclosed lifeboat shall be so constructed and fendered as to ensure that the lifeboat renders protection against harmful accelerations resulting from an impact of the lifeboat, when loaded with its full complement of persons and equipment, against the ship's side at an impact velocity of not less than 3.5 m/s.

(b) The lifeboat shall be automatically self-bailing.

N.B. The author, at the time of going to press, has been unable to obtain a line drawing of a proto-type Partially Enclosed Self-Righting Lifeboat.

IMO. Requirements for Totally Enclosed Lifeboats on ships built after 1st. July, 1986.

1. Totally enclosed lifeboats shall comply with the IMO. Requirements for all lifeboats and in addition shall comply with the following:—

2. **Enclosure**

The enclosure shall be so arranged that:

(a) it protects the occupants against heat and cold;

(b) access to the lifeboat is provided by hatches that can be closed to make the lifeboat watertight;

(c) hatches are positioned so as to allow launching and recovery operations to be performed without any occupant having to leave the enclosure;

(d) access hatches are capable of being opened and closed from both inside and outside and are equipped with means to hold them securely in open positions;

(e) it is possible to row the lifeboat;

(f) it is capable, when the lifeboat is in the capsized position with the hatches closed and without significant leakage, of supporting the entire mass of the lifeboat, including all equipment, machinery and its full complement of persons;

(g) it includes windows or translucent panels on both sides and to admit sufficient daylight to the inside of the lifeboat with the hatches closed so as to make artificial light unnecessary;

(h) its exterior is of a highly visible colour and its interior of a colour, which does not cause discomfort to the occupants;

(i) handrails provide a secure handhold for persons moving about the exterior of the lifeboat, and aid embarkation and disembarkation;

(j) persons have access to their seats from an entrance without having to climb over thwarts or other obstructions;

(k) The occupants are protected from the effects of dangerous sub-atmospheric pressures which might be created by the lifeboat engine;

3. **Capsizing and re-righting**

(a) A safety belt shall be fitted at each indicated seating position. The safety belt shall be designed to hold a person of a mass of 100kg. (220lbs) securely in place when the lifeboat is in a capsized position.

(b) The stability of the lifeboat shall be such that it is inherently or automatically self-righting when loaded with its full or a partial complement of persons and equipment and all entrances and openings are closed watertight and the persons are secured with safety belts.

(c) The lifeboat shall be capable of supporting its full complement of persons and equipment when the lifeboat is in the damaged condition of being holed in any one location below the waterline, assuming no loss of buoyancy material and no other damage and its stability shall be such that in the event of capsizing, it will automatically attain a position that will provide an above water escape for its occupants.

(d) The design of all engine exhaust pipes, air ducts and other openings shall be such that water is excluded from the engine when the lifeboat capsizes and re-rights.

4. **Propulsion**

(a) The engine and transmission shall be controlled from the helmsman's position.

(b) The engine and engine installation shall be capable of running in any position during capsize and continue to run after the lifeboat returns to the upright or shall automatically stop on capsizing and be easily restarted when the lifeboat returns to the upright. The design of the fuel and lubricating systems shall prevent the loss of fuel and the loss of more than 250ml. (½ pint) of lubricating oil from the engine during capsize.

(c) Air-cooled engines shall have a duct system to take in cooling air from and exhaust it to, the outside of the lifeboat. Manually operated dampers shall be provided to enable cooling air to be taken in from and exhausted to the interior of the lifeboat.

5. **Construction and fendering**

Notwithstanding sub-paragraph (f) in paragraph 1 of the IMO. General Requirements for Lifeboats on page 175, a totally enclosed lifeboat shall be so constructed and fendered as to ensure that the lifeboat renders protection against harmful accelerations resulting from an impact of the lifeboat, when loaded with its full complement of persons and equipment against the ship's side at an impact velocity of not less than 3.5 m/s.

6. **Free-fall lifeboats**

A lifeboat arranged for free-fall launching shall be so constructed that it is capable of rendering protection against harmful accelerations resulting from being launched, when loaded with its full complement of persons and equipment from at least the maximum height at which it is designed to be stowed above the waterline with the ship in its lightest sea-going condition, under unfavourable conditions of up to 10 degrees of trim and with the ship listed not less than 20 degrees either way.

Lifeboats with a self contained air support system

In addition to the foregoing requirements a lifeboat with a self-contained air support system shall be so arranged that when proceeding with all entrances and openings closed, the air in the lifeboat remains safe and breathable and the engine runs normally for a period of not less than 10 minutes. During this period the atmosphere pressure inside the lifeboat shall never fall below the outside atmospheric pressure nor shall it exceed it by more than 20 mb. The system shall have visual indicators to indicate the pressure of the air supply at all times.

Fire-protected lifeboats

1. In addition to complying with the above requirements, a fire-protected lifeboat when waterborne shall be capable of protecting the number of persons it is permitted to accommodate when subjected to a continuous oil fire that envelops the lifeboat for a period of not less than 8 minutes.

2. **Water spray system**

A lifeboat which has a water spray fire-protection system shall comply with the following:–

(a) water for the system shall be drawn from the sea by a self-priming motor pump. It shall be possible to turn "on" and turn "off" the flow of water over the exterior of the lifeboat;

(b) the sea water intake shall be so arranged as to prevent the intake of flammable liquids from the sea surface;

(c) the system shall be arranged for flushing with fresh water and allowing complete drainage.

A "Watercraft" totally enclosed G.R.P. lifeboat has been tested under the most rigorous conditions in the presence of IMO., and many others. The results of the tests showed that the temperature inside the craft did not exceed a maximum of 54 degrees C (130 deg. F.) reducing to 42.8 degrees C (109 deg. F.) amidships at head level.

The "Watercraft-Schat" Survival System enables the crew to evacuate and to be afloat totally enclosed and independent of the outside atmosphere. The craft is fully protected by means of a waterspray system which enables it to go through an oil fire up to a distance of 1 mile. The craft is extremely manoeuvrable with a speed of over 6 knots and is fuelled for at least 24 hours operation. Embarkation of the craft is through two large watertight openings which permit all persons to be seated, with doors closed ready for lowering in less than 60 seconds (during which time the engine can be hydraulically started for immediate getaway). Immediate lowering is effected by pulling and maintaining tension on the control wire which passes through the canopy top adjacent to the control position from which the craft is driven and the engine controlled. The craft descends at a controlled speed and being suspended from two points, will not spin. The descent can be halted at any point by releasing the control wire. The boat release system is operated by a single lever which releases both hooks simultaneously when the craft is waterborne.

The craft is constructed of glassfibre reinforced plastic with fire retardant additives and pigmented international orange. It is totally enclosed with two large inward opening access doors which are hinged to permit the lower half to be clipped into the top half for ventilation. The doors will also hinge and stow under the roof. A large armour plated glass escape hatch is fitted at each end of the cover and this will allow access to the lifting hooks, if necessary. A similar hatch is fitted on top of the cover immediately above the helmsman's position, which permits him to obtain maximum vision when standing. An automatic opening and closing ventilator is also provided, on top of the cover. Four rowing ports with removable watertight covers are fitted. A compressed-air system is incorporated which operates a continuous watershed for 10 minutes, with an air supply sufficient for the personnel as well as the engine running at full throttle. A positive pressure is built up inside the craft preventing the entry of fumes and gases. Air is not re-circulated. Seat belts are provided and the ventilators close automatically in the event of the craft capsizing. The diesel engine is completely enclosed in a glassfibre casing with a hinged lid and can be restarted immediately after righting, if the craft should capsize. Cooling is by means of a completely enclosed fresh water system The exhaust system is fitted with a non-return clack valve to ensure that water does not enter the engine.

The rowing of totally enclosed lifeboats is a somewhat difficult operation, due to the fact that the oarsmen are unable to see the blades of their oars in most cases. In the smaller enclosed lifeboats, paddles are sometimes substituted for oars due to the difficulty of handling oars in the confined space.

“WATERCRAFT Totally Enclosed Glassfibre Lifeboat under way

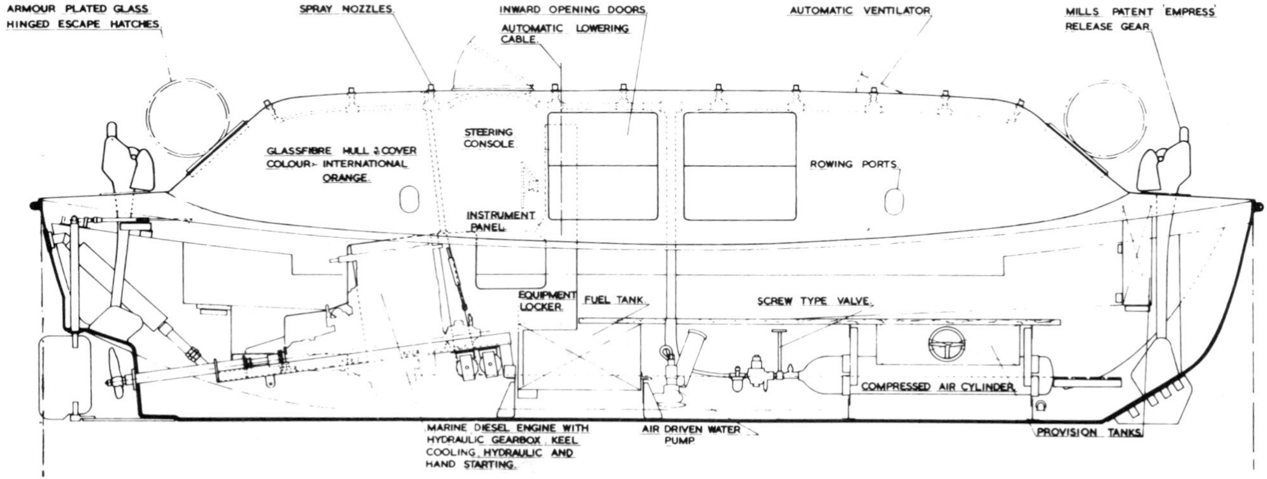

TYPICAL GENERAL ARRANGEMENT OF A WATERCRAFT G.R.P. TOTALLY ENCLOSED SELF-RIGHTING LIFEBOAT

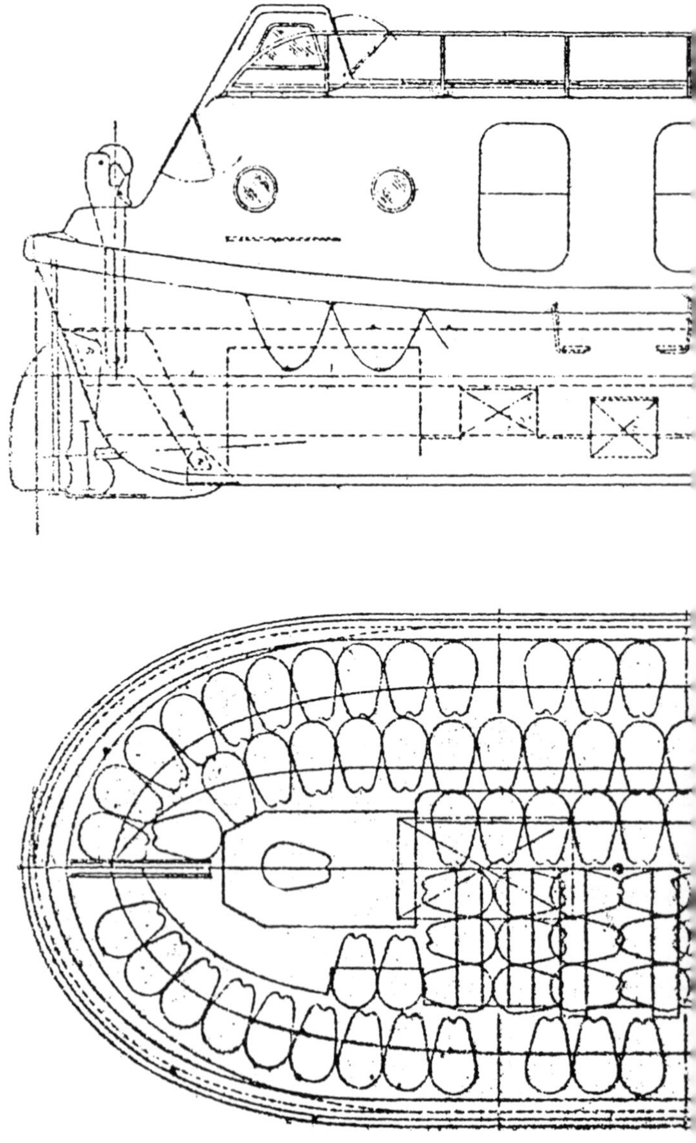

"WATERCRAFT" Totally enclosed lifeboat seating 144 per

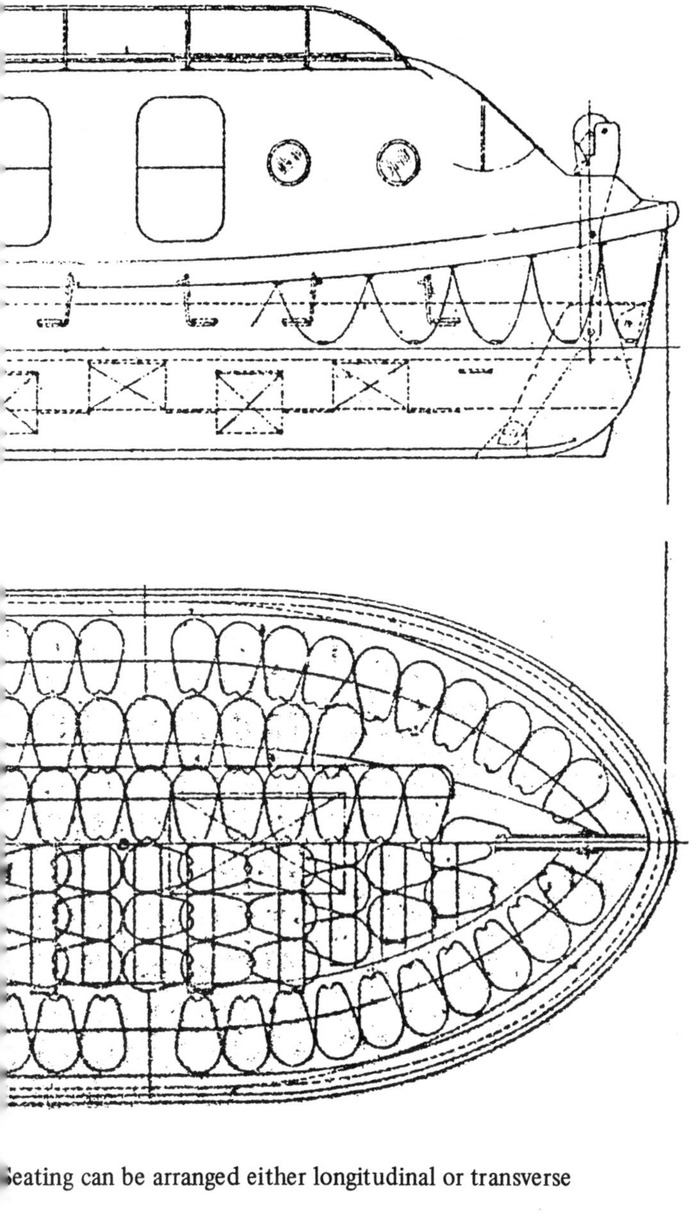

Seating can be arranged either longitudinal or transverse

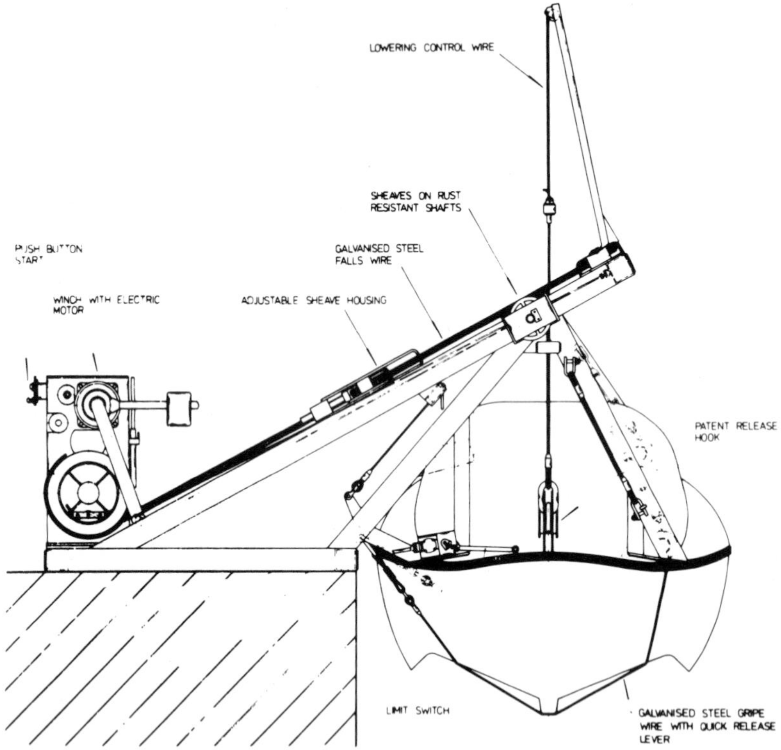

A 21.3 ft. (6.50m) totally enclosed Viking Watercraft G.R.P. Self-righting Lifeboat attached to Schat Davits on a typical off-shore platform installation.

RUCKER SURVIVAL CAPSULES

The Whittaker Survival System consists of:

1. A 14, 21, 36/38 or 50/54 man survival capsule.
2. holding and launching platform.
3. Platform electrical system.
4. Winch.

Engineered to match the 50/54 man Capsule, the PL50/54 Platform s specially designed to support the Capsule firmly, allow unobstructed ccess, and upon retrieval, provide a method of guiding it smoothly nd safely into the stowed position. Limited by a centrifugal brake, the aunching system lowers the Capsule at approximately 135 feet per ninute on a 113,000 lb. capacity wire rope. The single point design nakes launch and recovery during training a simple procedure.

)FFICIAL BURN TESTS DEMONSTRATE FITNESS OF VHITTAKER SURVIVAL SYSTEMS TO SAVE OFFSHORE CREWMEN.

Totally engulfing flames 200 feet high. A pall of black smoke rising ,000 feet from the surface of a specially built test pond. This was the earing environment a 28-man Whittaker survival capsule withstood s a safe haven for human life during its officially required fire tests.

During the test the temperature outside the capsule reached 1986° F. s flames from 975 gallons of kerosene roared around it for 7 minutes nd 35 seconds. During the test the capsule was held totally immobile nd was therefore unable to benefit from such heat reducing effects as vind, wave and capsule motion.

Inside environment: During the test, inside air temperature average vas 95° F.

Whittaker survival systems are normally only used on oil-rigs in lieu of totally enclosed lifeboats. They are approved by the U.S. Coast Guard for installation on non-self-propelled drilling rigs, self-propelled drilling riggs, fixed structures and artificial islands.

WHITTAKER SURVIVAL SYSTEM
CAPSULE ARRANGEMENT

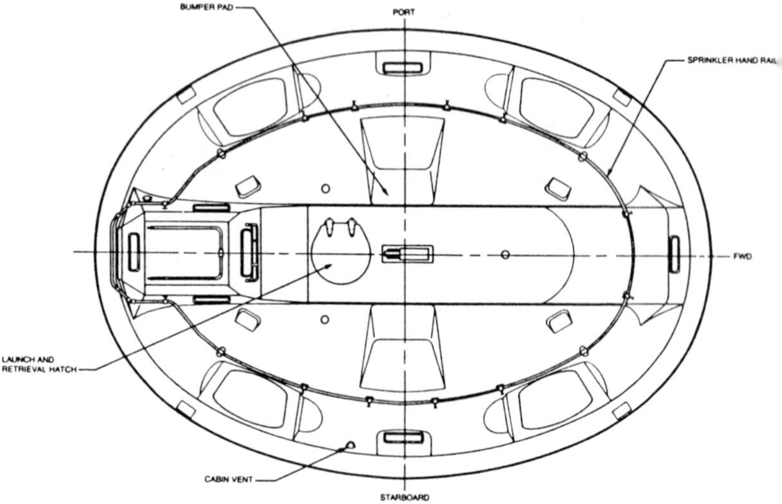

TOP VIEW

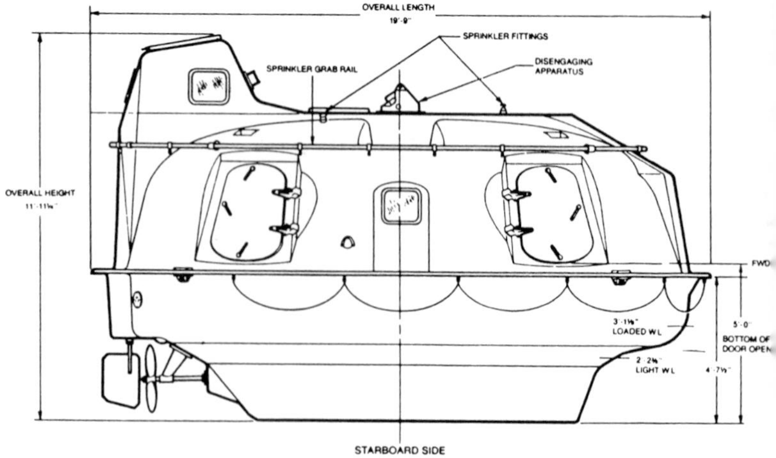

STARBOARD SIDE

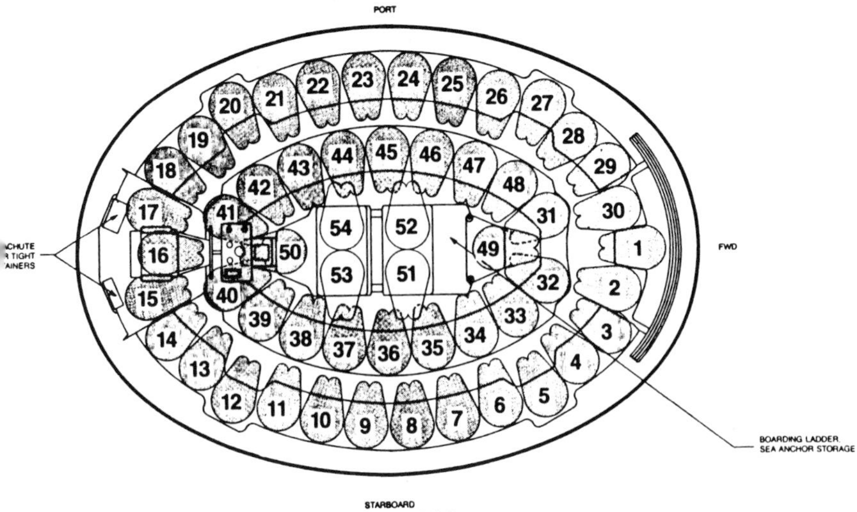

SEATING ARRANGEMENT

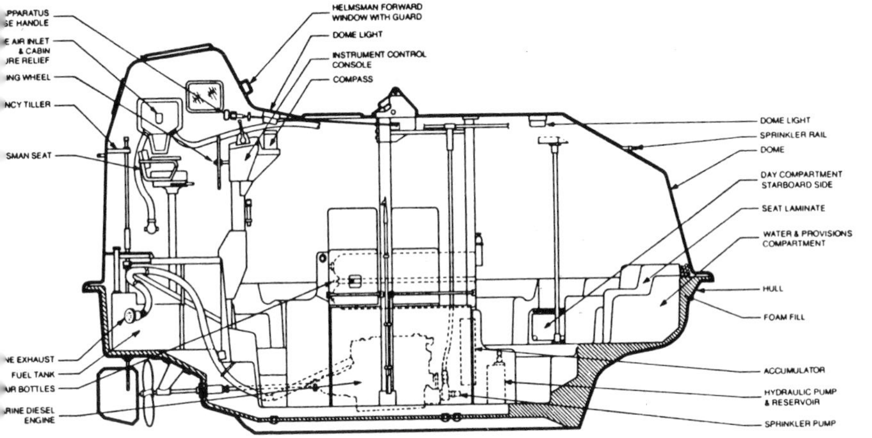

INTERIOR ARRANGEMENT

50/54-MAN LAUNCH AND RECOVERY SYSTEM

U.S.C.G. Approved

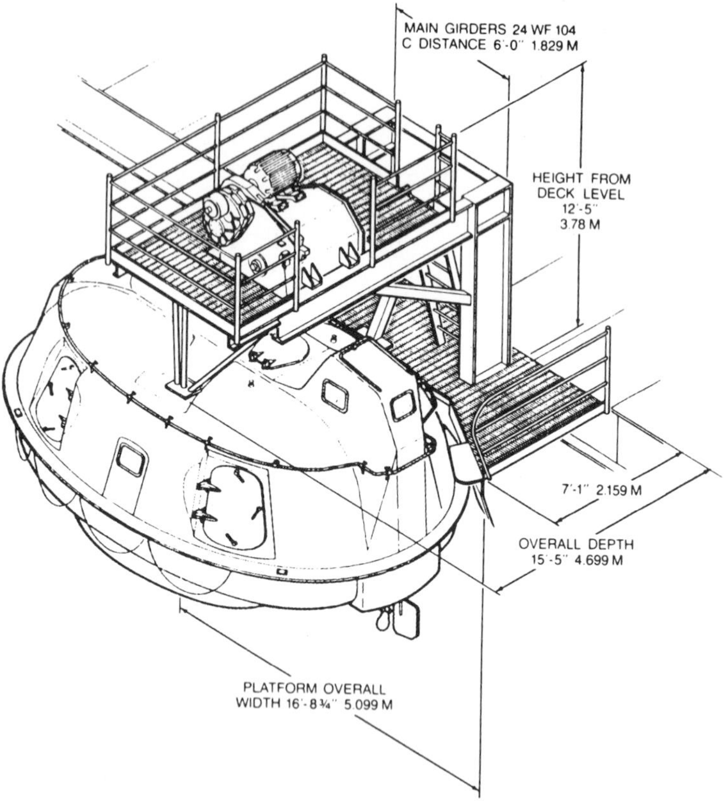

RESCUE BOATS ON SHIPS BUILT AFTER 1st JULY 1986

1. **General requirements**

(a) Except as provided by this Regulation, all rescue boats shall comply with the requirements for all lifeboats on pages 179 to 187, except that they shall not be required to be provided with watertight lockers or compartments for the storage of small items of equipment or means for the storage of rainwater, a permanently attached earth connection and antenna securing requirements, a manually controlled lamp attached to the top of the cover or enclosure, a source of light inside the boat, the equipment as listed, or be marked with the required lifeboat markings. Unless the rescue boat is also a lifeboat.

(b) Rescue boats may be of either rigid or inflated construction or a combination of both and shall:

- (i) be not less than 3.8m (13ft) and not more than 8.5m (27.9ft) in length;
- (ii) be capable of carrying at least five seated persons and a person lying down.

(c) Rescue boats which are a combination of rigid and inflated construction shall comply with the appropriate requirements of this Regulation to the satisfaction of the Administration.

(d) Unless the rescue boat has adequate sheer, it shall be provided with a bow cover extending for not less than 15% of its length.

(e) Rescue boats shall be capable of manoeuvring at speeds up to 6 knots and maintaining that speed for a period of at least 4 hours.

(f) Rescue boats shall have sufficient mobility and manoeuvrability in a sea-way to enable persons to be retrieved from the water, marshal liferafts and tow the largest liferaft carried on the ship when loaded with its full complement of persons and equipment or its equivalent at a speed of at least 2 knots.

(g) A rescue boat shall be fitted with an inboard engine or outboard motor. If it is fitted with an outboard motor, the rudder and tiller may form part of the engine. Notwithstanding the General requirements for all lifeboats, petrol driven outboard engines with an approved fuel system may be fitted in rescue boats provided the fuel tanks are specially protected against fire and explosion.

(h) Arrangements for towing shall be permanently fitted in rescue boats and shall be sufficiently strong to marshal or tow liferafts as required by sub-paragraph (f) above.

(i) Rescue boats shall be fitted with weathertight stowage for small items of equipment.

2. **Rescue boat equipment**

(a) All items of rescue boat equipment, with the exception of boat hooks which shall be kept free for fending off purposes, shall be secured within the rescue boat by lashings, storage in lockers or compartments, storage in brackets or similar mounting arrangements, or other suitable means. The equipment shall be secured in such a manner so as not to interfere with any launching or recovery procedures. All items of rescue boat equipment shall be as small and of as little mass as possible and shall be packed in suitable and compact form.

(b) The normal equipment of every rescue boat shall consist of:

(i) sufficient buoyant oars or paddles to make headway in calm seas. Thole pins, crutches or equivalent arrangements shall be provided for each oar. Thole pins or crutches shall be attached to the boat by lanyards or chains;

(ii) a buoyant bailer;

(iii) a binnacal containing an efficient compass which is luminous or provided with a suitable means of illumination;

(iv) a sea-anchor and tripping line with a hawser of adequate strength not less than 10m (32.5ft) in length;

(v) a painter of sufficient length and strength, attached to the release device and placed in the forward end of the rescue boat;

(vi) one buoyant line, not less then 50m (164ft) in length, of sufficient strength to tow a liferaft as required by sub-paragraph 1. (f) above.

(vii) one waterproof electric torch suitable for morse signalling together with one set of spare batteries and one spare bulb in a waterproof container;

(viii) one whistle or equivalent sound signal;

(ix) a first-aid outfit in a waterproof case capable of being closed tightly after use;

(x) two buoyant rescue quoits, attached to not less than 30m (97.5ft) of buoyant line;

(xi) a searchlight capable of effectively illuminating a light coloured object at night having a width of 18m (58.5ft) at a distance of 180m (585ft) for a total period of 6 hours and of working for at least 3 hours continuously;

(xii) an efficient radar reflector;

(xiii) thermal protective aids sufficient for 10% of the number of persons the rescue boat is permitted to accommodate or two, whichever is the greater.

(c) In addition to the equipment required by sub-paragraph 2 (b) above, the normal equipment of every rigid rescue boat shall include:

(i) a boat hook;

(ii) a bucket;

(iii) a knife or hatchet.

(d) In addition to the equipment required by sub-paragraph 2 (b) above, the normal equipment of every inflated rescue boat shall consist of:

(i) a buoyant safety knife;

(ii) two sponges;

(iii) an efficient manually operated bellows or pump;

(iv) a repair kit in a suitable container for repairing punctures;

(v) a safety boat hook.

3. **Additional requirements for inflated rescue boats**

(a) Notwithstanding the requirements for all lifeboats, inflated rescue boats shall not be required to be fire retarding or non-combustible or to be of sufficient strength to withstand a load without residual deflection on removal of the load.

(b) An inflated rescue boat shall be constructed in such a way that, when suspended by its bridle or lifting hook:

(i) it is of sufficient strength and rigidity to enable it to be lowered and recovered with its full complement of persons and equipment;

(ii) it is of sufficient strength to withstand a load of 4 times the mass of its full complement of persons and equipment at an ambient temperature of 20 degrees C (68 deg.F) plus or minus 3 degrees C (6 deg.F) with all relief valves inoperative:

(iii) it is of sufficient strength to withstand a load of 1.1 times the mass of its full complement of persons and equipment at an ambient temperature of minus 30 degrees C (–22 deg.F), with all relief valves operative.

(c) Inflated rescue boats shall be so constructed as to be capable of withstanding exposure:

(i) when stowed on an open deck on a ship at sea;

(ii) for 30 days afloat in all sea conditions.

(d) In addition to carrying the marking required to be marked on all lifeboats, inflated rescue boats shall be marked with a serial number, the maker's name or trade mark and the date of manufacture.

(e) The buoyancy of an inflated rescue boat shall be provided by either a single tube sub-divided into at least five separate compartments of approximately equal volume or two separate tubes neither exceeding 60% of the total volume. The buoyancy tubes shall be so arranged that in the event of any one of the compartments being damaged, the intact compartments shall be able to support the number of persons which the rescue boat is permitted to accommodate, each having a mass of 75kg. (165 ibs), when seated in their normal positions with positive freeboard over the rescue boat's entire periphery.

(f) The buoyancy tubes forming the boundary of the inflated rescue boat shall on inflation provide a volume of not less than $0.17m^3$ for each person the rescue boat is permitted to accommodate.

(g) Each buoyancy compartment shall be fitted with a non-return valve for manual inflation and means for deflation. A safety relief valve shall also be fitted unless the Administration is satisfied that such an appliance is unnecessary.

(h) Underneath the bottom and on vulnerable places on the outside of the inflated rescue boat, rubbing strips shall be provided to the satisfaction of the Administration.

(i) Where a transom is fitted it shall not be inset by more than 20% of the overall length of the rescue boat.

(j) Suitable patches shall be provided for securing the painters fore and aft and the becketed lifelines inside and outside the boat.

(k) The inflated rescue boat shall be maintained at all times in a fully inflated condition.

DUNLOP INFLATABLE RESCUE BOAT.

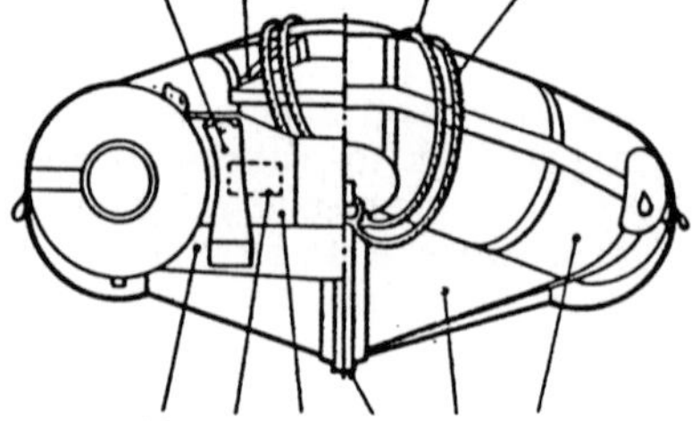

LOOSE EQUIPMENT IS STOWED IN ITEM 21

CALLED UP ON ASSEMBLY DRAWINGS BUT ITEMISED FOR CLARITY

Item No.	No. off per Unit	Description
1	1 Set	Hull Tubes & Assembly
2	1	Floor Assembly
3	1	Transom Assembly
4	1	Bow Board Assembly
5	1	Keel Assembly
6	1	Deck Assembly
7	1	Thwart Assembly
8	2	Inner Lifeline Assembly
9	2	Outer Lifeline Assembly
10	1	Dodger Assembly
11	1	Side Rubbing Strake
12	1	Bottom Rubbing Strake
13	2	Aft Rubbing Strake
14	3	Stowage Pocket Assembly
15	1 Set	Steering/Sculling Patch
16	1	Painter Patch
17	1	Tank Lashing Assembly
18	1	Painter Line
19	1 Set	Transom Patches
20	1	Drogue
21	1	Stowage Box Assembly
22	2	Support Pocket
23	2	Paddles
24	2	Oars
25	3	Stowage Patch
26	1	Rowlock/Oarflap Assembly

Item No.	No. off per Unit	Description
27	1	Rowlock/Righting Line Assembly
28	1	Identification Label
40	2	Rescue Quoit Assembly
41	1	Drain Tube Assembly
42	1	Safety Knife Assembly
43	5	Deflate Plug Assembly
44	5	S.R.T.U. Valve Assembly
45	4	Fuel Line Securing Cord
46	1	Stretcher Bar
47	1	Dodger Support Tube
48	1	Belly Band Assembly (Mid.)
49	1	Belly Band Assembly (For'd)
50	2	Ring Bolt
59	1	Fuel Tank Fastening Line
60*	1	Searover Manual Assembly
61	5	S.R.T.U. Valve Plug (Inc. in item 44)
62	1	Set of Leak Stoppers
63	1	Six Leg Sling
64	1	Bailer
65	1 Set	Spare Batteries and Bulb
66	1	Electric Torch
67	2	Sponges
68	1	Repair Kit
69	1	Bellows
70	1	Fire Extinguisher

DUNLOP INFLATABLE RESCUE BOATS

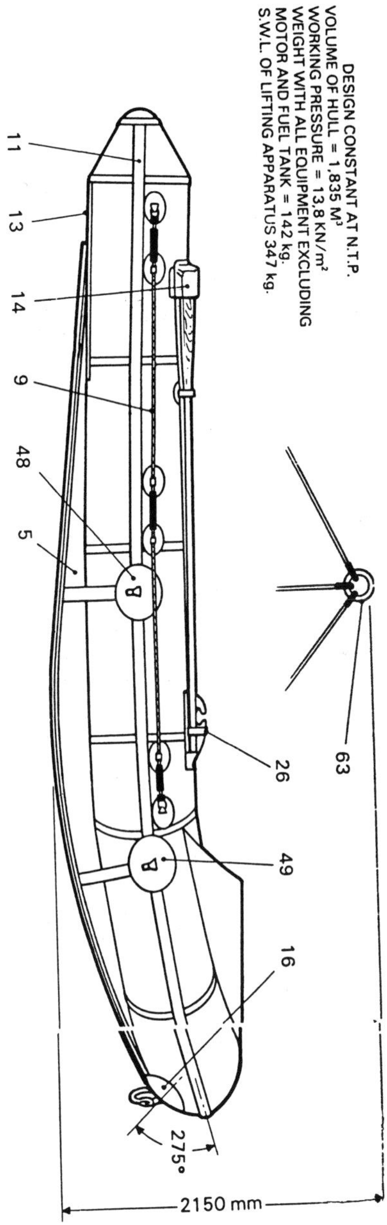

DUNLOP INFLATABLE RESCUE BOATS

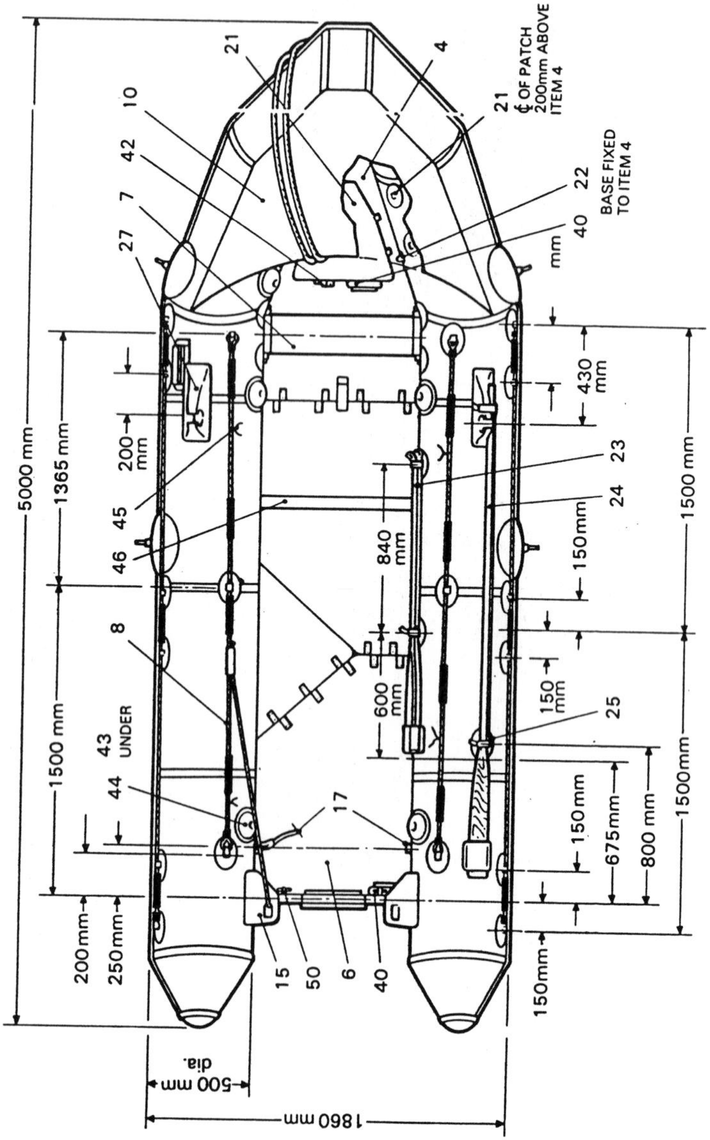

A "DUNLOP" 10 man D.o.T. Inflatable boat under way with its full complement 10 men and powered by a 10 hp outboard motor. This boat also has approval when fitted with central console steering and a 25 hp engine for North Sea stand-by use.

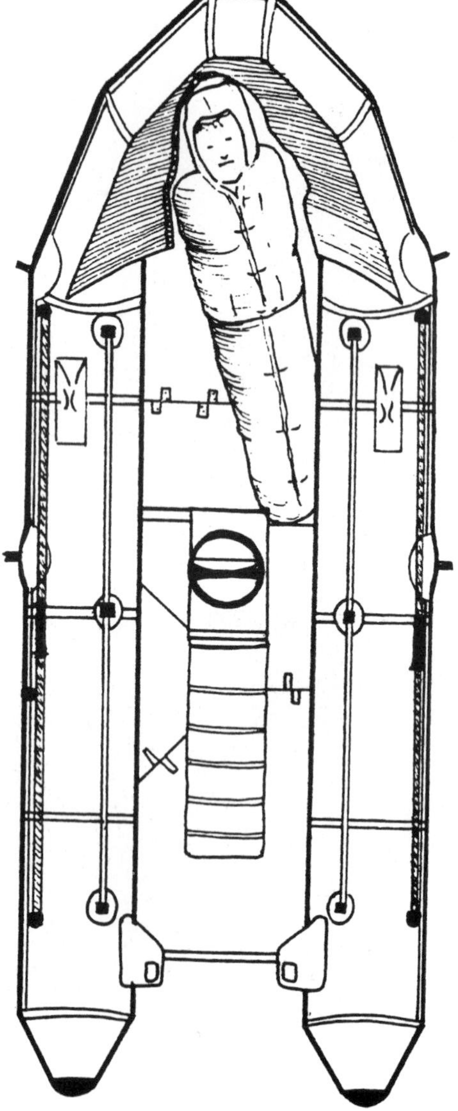

STRETCHER CASE IN DUNLOP RESCUE BOAT

"WATERCRAFT" Rescue boat under way with "jet" propulsion. When recovering, simply hook recovery hooks onto the uprights.

Whether on a ship or an offshore platform, this high speed rescue boat is a great asset for picking up survivors or for man overboard situations.

"OSBORNE" Fast Semi-rigid Rescue Boat

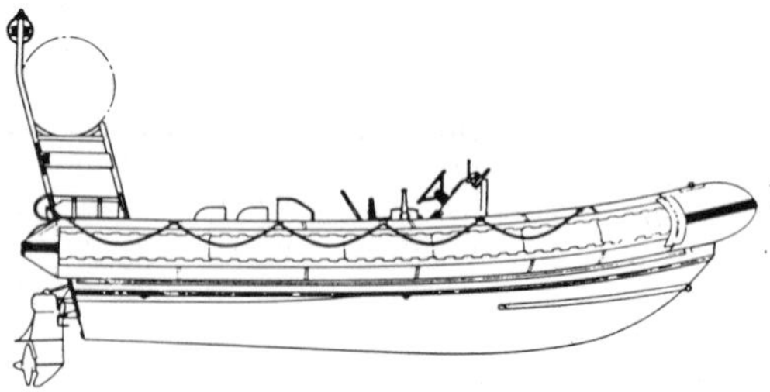

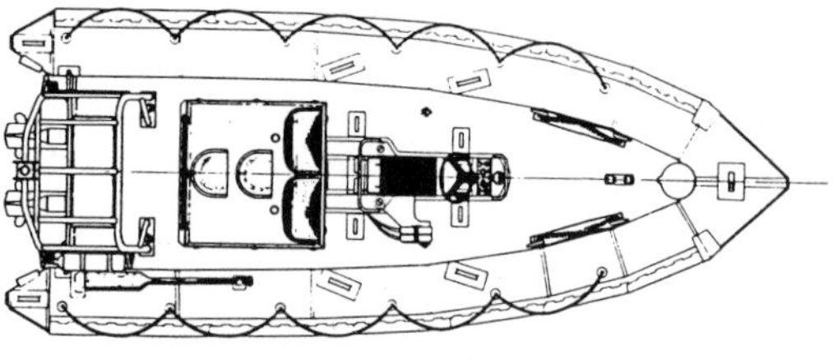

STANDARD EQUIPMENT

- 2 × 70 hp Volkswagen Piranha Normally Aspirated Inboard Diesel Engines
- 2 × Souter/Enfield Mk II Z Drives
- 1 × Hinging/Lift off GRP Moulded Engine Case
  Waterproof Electrics and Instrumentation
- 1 × Kelvin Hughes 'Husun' VHF R/T in special waterproof casing
- 1 × Righting 'A' Frame Roll Bar and Navigation Lights
- 1 × Righting Air Bag and set of Inflation equipment
- 1 × Set Heavy Duty Abrasion Cladding P & S
- 4 × S/S Lighting Eyes
- 1 × Set Lighting Strops (to suit single point suspension)
- 1 × Sea Anchor and Bucket Storage
  Signwriting on Collar P & S (customer requirements)
- 2 × Pannier Stowage Bags P & S
  Hydraulic Steering
- 1 × Compass
- 1 × Search Lights (Waterproof & Shockproof)
- 3 × Paddles C/W Stowage
- 3 × Mooring Cleats
- 2 × Dry Powder Fire Extinguishers
- 1 × Drogue/Bow Fairlead
- 2 × M.O.B. Emergency Throwing Quoits and Lines
- 1 × Inflation Bellows
- 1 × Knife and Stowage
- 1 × Calibrated Fuel Tank Dipstick
- 2 × Mooring Warps
- 1 × Bow Hauling Out Eye
- 2 × Stern Hauling Out Eyes
- 2 × Engine Hours Metres
- 1 × Bilge Pump

OPTIONAL EXTRAS

- ☐ Steel Shipping/Holding Cradle
- ☐ Boat Launching Trailer
- ☐ Terylene 'All-over' Boat Cover
- ☐ Terylene 'All-over' Console Cover
- ☐ Additional Moulded handholds/footholds
- ☐ Helicopter Stretcher assembly C/W Collar Storage Framework

Osborne Rescue Boats Ltd reserve the right to modify the standard specification without notice.

LIFEBOAT AND RESCUE BOAT ENGINES.

Diesel engines only are approved for duties in lifeboats and rescue boats, except that outboard motors when attached to inflated boats and Class C boats may be petrol engines. Diesel engines are renowned for their reliability and ease of starting in cold weather to which may be added that the risk of fire is considerably reduced. Air-cooled engines are normally fitted in open and partially enclosed lifeboats and rescue boats, for which they are particularly suited owing to their comparative light weight and simplicity, and because air-cooling permits the running of the engine indefinately for maintenance whilst the boat is in the davits.

However, air-cooled engines are not really suited to totally enclosed boats. Totally enclosed boats are normally fitted with water-cooled engines, which, in order to enable them to run in the davits for short periods, are cooled by means of a sealed fresh water system, which in turn is cooled when the boat is afloat by means of a heat exchanger placed outside the boat and which runs parallel to the keel along the bottom of the boat. On ships liable to trade in areas where frost may be expected, the sealed fresh water system should be protected from frost by anti-freeze.

Each engine is supplied with a set of spares, tools and an instruction card. Engines will normally be started by hand, however in the larger lifeboats, the engine may be too powerful for hand starting, in which case at least two alternative starting devices are required to be fitted. A hand-wound spring motor together with either an hydraulic starting system or an electric starter motor will be fitted to the engine. Smaller engines, will in addition to manual starting be fitted with a spring-wound starter motor or other suitable starting device.

In order to run or even start, an engine must have a supply of both fuel and air in the correct mixture and the oil should be up to the mark on the dip-stick. Water or an air-lock in the fuel system or a clogged air intake or exhaust will prevent the engine starting. Lack of sufficient oil in the sump will cause the engine to over-heat and seize up. Dirt or water in the fuel will cause the engine to stop.

The engine, except outboard motors, will have a gear box attached and the boat may be propelled either ahead or astern simply by pushing a lever either forward or aft, according to the direction desired, the central position of the lever being neutral.

Although not very successful when fitted to lifeboats, some rescue boats are fitted with jet-propulsion. The jet-propulsion unit is driven by the boat's diesel engine. Twin jet units take in water through a grill in the bottom of the boat; this water is impelled by vanes to give it velocity and is expelled in twin jets. The boat may be steered by either the wheel and rudder, or by changing the direction of the jets of expelled water or both. The boats can be impelled ahead, astern, sideways or as required. There are no propellers to foul and the boat can be navigated anywhere there is sufficient water to float it. The boat can be held against the ship's side simply by impelling the boat sideways and taken away by reversing the direction of the impelled water.

To manually start a motor lifeboat engine

1. Check that the gear lever is in neutral.
2. Prime the fuel system if necessary.
3. Check that the engine is free to turn without obstruction.
4. Turn throttle control lever to almost vertical or "Fast".
5. Move the de-compression lever towards the fly-wheel. Fit starting handle.
6. Turn engine slowly from 3 to 20 turns to prime combustion chamber and lubricating system.
7. Crank the engine really fast. When speed is obtained, return the de-compression lever to the firing position. Continue to crank until the engine fires.
8. Remove starting handle and reduce engine speed as required.
9. To stop the engine:- Turn throttle control anti-clockwise and hold it until the engine stops. Or if fitted, Pull the remote stopping control.

   After starting:- Check the oil pressure gauge and with a water-cooled engine the overboard discharge of cooling water.

PRIMING THE FUEL SYSTEM ON A LISTER ENGINE

(a) Fill fuel tank or connect fuel supply.

(b) Slacken each bleed screw A on top of the filter body and in the outlet banjo union. Tighten each bleed screw when a full air free flow of fuel is obtained working from the fuel tank.

(c) Slacken bleed screw B on fuel pump(s) nearest the tank first; tighten when all air has been displaced from fuel at each pump.

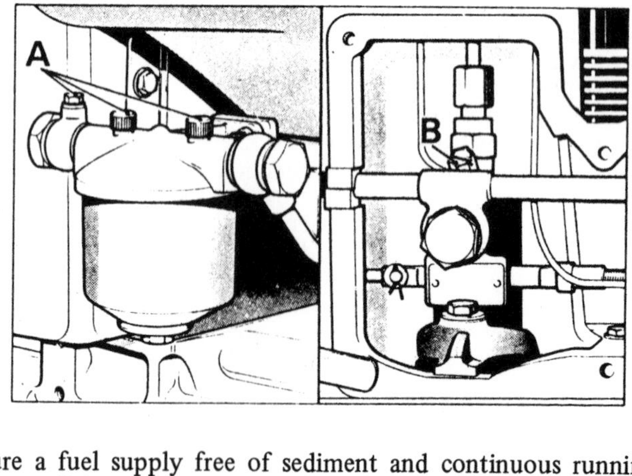

To ensure a fuel supply free of sediment and continuous running in an emergency, all lifeboat and rescue boat fuel tanks should be thoroughly cleaned out annually.

LISTER ST2 AIR COOLED ENGINE WITH HAND STARTING

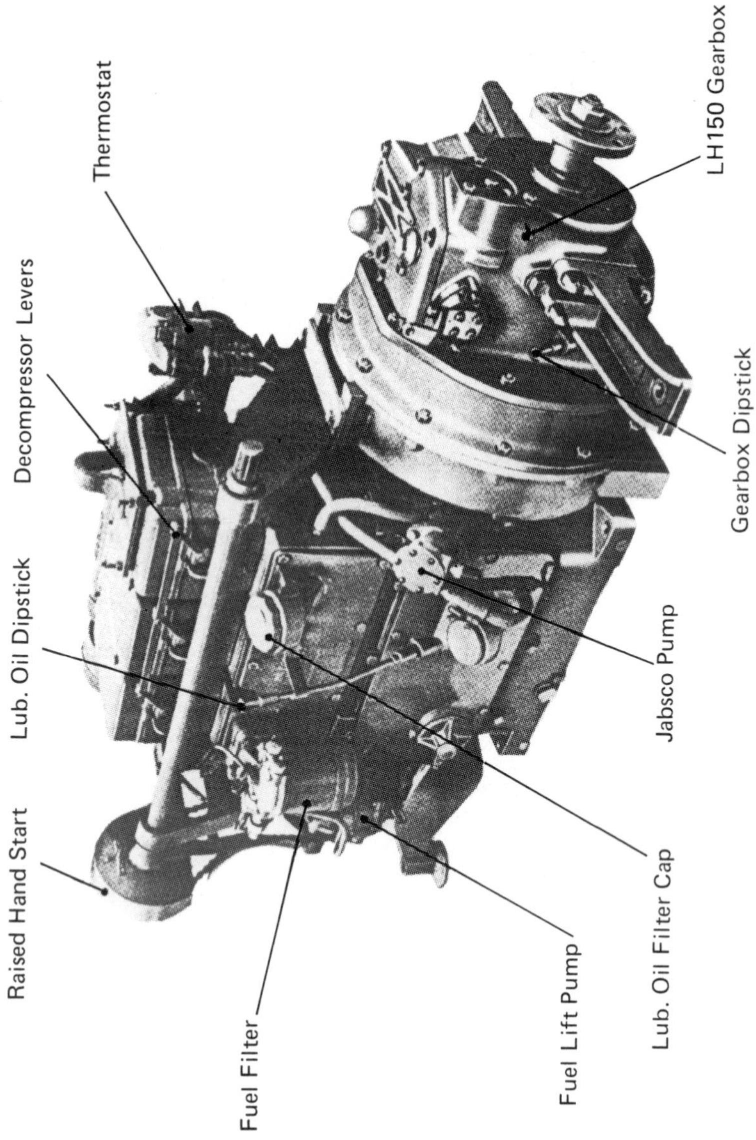

LISTER STW3 WATER COOLED ENGINE

DEPARTMENT OF TRANSPORT MERCHANT SHIPPING NOTICE
No. **M.1165**

LIFEBOAT ENGINES AND OTHER COMPRESSION IGNITION ENGINES USED IN AN EMERGENCY

Notice to Shipowners, Superintendants and Chief Engineers

This notice supersedes Notice M. 843

(1) Ships' motor lifeboats, emergency generators and emergency fire pumps are required to be put into service quickly in the event of an emergency and to be operated under a wide variety of climatic conditions. It is, therefore, essential that the correct type of fuel and grade of lubricating oil are used to enable the engine to be started and run whatever the ambient temperature.

(2) With regard to lubricating oil, the problem is not generally acute. The amount of running is relatively small and consequent renewal of the oil is infrequent, whilst the procurement of suitable multigrade lubricating oils rarely presents a problem.

(3) The selection of fuel oil, however, requires careful attention when low temperature operation is considered. Not only can the fuel become more viscous at lower temperatures but there is also the problem associated with the formation of wax crystals which can stop the flow of fuel to the engine. Special fuel oils for low temperature operation are readily available in areas of the world where such temperatures are regularly experienced. However, when ships ply between different temperature zones, it is necessary to make sure that any fuel taken on board for emergency purposes would be suitable for use at the lowest ambient temperature the ship is likely to encounter in service. Suitable fuels are not always available from stock in the United Kingdom and, therefore, orders should be placed with suppliers well in advance of the date by which the fuel will be required.

(4) The Merchant Shipping (Passanger Ship Construction and Survey) Regulations 1984, The Merchant Shipping (Cargo Ship Construction and Survey) Regulations 1984 and the Merchant Shipping (Cargo Ship Construction and Survey) Regulations 1981 (Amendment) Regulations 1984 permit fuel oil of a lower flash point—ie. not lower than 43C. (closed test)—to be used in emergency generators. Machinery such as lifeboat engines and emergency fire pumps would be considered to fall within the same category and consequently be permitted to use the lower flash point fuel. However, it is pointed out that the flash point of the fuel is not necessary a guide to its suitability for use at low temperatures. It is, therefore, essential in all cases to specify the lowest temperature for which the fuel has to be suitable.

(5) Requirements for particular applications are as follows:

(a) *Emergency Generators and Emergency Fire Pumps*
The engine should be capable of being readily started in its cold condition at a temperature of 0°C. If this is impractical

or if temperatures below 0° C. are likely to be encountered provision must be made for heating the engine so that it will start readily. The fuel oil provided should be suitable for use at 0° C. or the lowest anticipated ambient temperature in the space containing the engine whichever is the lower.

(b) *Lifeboat Engines*

The engine shall be capable of being started, using starting aids if necessary, at an ambient temperature of —15° C. within two minutes of commencing the start procedure. The fuel oil supplied should be suitable for use at —15° C., unless the ship carrying the lifeboat is constantly engaged in voyages in the Tropics when a more appropriate fuel oil may be supplied.

Department of Transport
Marine Directorate
London WC1V 6LP
March 1985

DOWTY HYDROJET 300

he hydrojet 300 is one of a range of Dowty high
erformance water propulsion units intended for
narine use and is the result of development and
xperience in this field over many years.
he hydrojet unit incorporates a well proven and
obust axial flow type pump and a special rotating
ozzle steering system. Single and two stage
ersions permit direct coupling to diesel engines up
o 300 horsepower.
)owty hydrojet units, modified to suit particular
nstallation requirements, are also produced for
mphibious military vehicles.

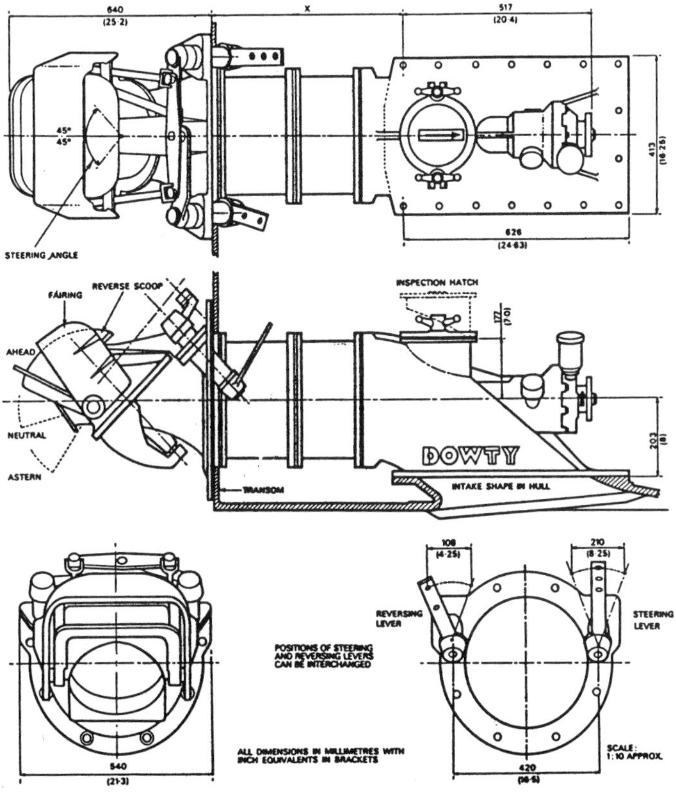

Photograph showing water intakes and jet expulsion orifices on a "Watercraft" G.R.P. Lifeboat

꞉HAPTER IV LIFERAFTS.

꞉MO. General requirements for all liferafts on ships ꞉uilt after 1st. July 1986

. **Construction of liferafts**

(a) Every liferaft shall be so constructed as to be capable of withstanding exposure for 30 days afloat in all sea conditions.

(b) The liferaft shall be so constructed that when it is dropped into the water from a height of 18m. (58·5ft.), the liferaft and its equipment will operate satisfactorily, if the liferaft is to be stowed at a height of more than 18m. (58·5ft.) above the waterline in the lightest sea-going condition, it shall be of a type which has been satisfactorily drop-tested from at least that height.

(c) The floating liferaft shall be capable of withstanding repeated jumps on to it from a height of at least 4·5m. (14·5ft.) above its floor both with and without the canopy erected.

(d) The liferaft and its fittings shall be so constructed as to enable it to be towed at a speed of 3 knots in calm water when loaded with its full complement of persons and equipment and with one of its sea-anchors streamed.

(e) The liferaft shall have a canopy to protect the occupants from exposure which is automatically set in place when the liferaft is launched and waterborne. The canopy shall comply with the following:

(i) it shall provide insulation against heat and cold by means of either two layers of material separated by an air gap or other equally efficient means. Means shall be provided to prevent accumulation of water in the air gap.;

(ii) its interior shall be of a colour that does not cause discomfort to the occupants;

(iii) each entrance shall be clearly indicated and be provided with efficient adjustable closing arrangements which can be easily and quickly opened from inside and outside the liferaft so as to permit ventilation but exclude sea water, wind and cold. Liferafts accommodating more than eight persons shall have at least two diametrically opposite entrances.

(iv) it shall admit sufficient air for the occupants at all times, even with the entrances closed;

(v) it shall be provided with at least one viewing port;

(vi) it shall be provided with means for collecting rain water;

(vii) it shall have sufficient headroom for sitting occupants under all parts of the canopy.

꞉. **Minimum carrying capacity and mass of liferafts**

(a) No liferaft shall be approved which has a carrying capacity of less than six persons.

(b) Unless the liferaft is to be launched by an approved

appliance and is not required to be portable, the total mass o the liferaft, its container and its equipment shall not be more than 185kg. (416 lbs.).

3. **Liferaft fittings**

(a) Lifelines shall be securely becketed around the inside and outside of the liferaft.
(b) The liferaft shall be provided with arrangements for adequately siting and securing in the operating position the antenna provided with the portable radio apparatus.
(c) The liferaft shall be fitted with an efficient painter of length equal to not less than twice the distance from the stowed position to the waterline in the lightest sea-going condition or 15m (48.75ft) whichever is the greater.

4. **Davit launched liferafts**

(a) In addition to the above requirements, a liferaft for use with an approved launching appliance shall:
(i) when the liferaft is loaded with its full complement of persons and equipment, be capable of withstanding a lateral impact against the ship's side at an impact velocity of not less than 3.5 m / s and also a drop into the water from a height of not less than 3m. (9.75ft.) without damage that will effect its function;
(ii) be provided with means for bringing the liferaft alongside the embarkation deck and holding it securely during embarkation.
(b) Every passenger ship davit launched liferaft shall be so arranged that it can be rapidly boarded by its full complement of persons.
(c) Every cargo ship davit launched liferaft shall be so arranged that it can be boarded with its full complement of persons in not more than 3 minutes from the time the instruction to board is given.

5. **Equipment**

(a) The normal equipment of every liferaft shall consist of:
(i) one boyant rescue quoit, attached to not less than 30m. (97.5ft.) of boyant line;
(ii) one knife of the non-folding type having a buoyant handle and a lanyard attached and stowed in a pocket on the exterior of the canopy near the point at which the painter is attached to the liferaft. In addition, a liferaft which is permitted to accommodate 13 persons or more shall be provided with a second knife which need not be of a non-folding type;
(iii) for a liferaft which is permitted to accommodate not more than 12 persons, one buoyant bailer. For a liferaft which is permitted to accommodate 13 persons or more, two buoyant bailers;

(iv) two sponges;

(v) two sea-anchors each with a shock resistant hawser and tripping line, one being spare and the other permanently attached to the liferaft in such a way that when the liferaft inflates or is waterborne it will cause the liferaft to lie oriented to the wind in the most stable manner. The strength of each sea-anchor and its hawser and tripping line shall be adequate for all sea conditions. The sea-anchors shall be fitted with a swivel at each end of the line and shall be of a type which is unlikely to turn inside out between its shroud lines;

(vi) two buoyant paddles;

(vii) three tin openers. (Safety knives containing special tin-opener blades are satisfactory for this requirement);

(viii) one first-aid outfit in a waterproof case capable of being closed tightly after use;

(ix) one whistle or equivalent sound signal;

(x) four rocket parachute flares;

(xi) six hand flares;

(xii) two buoyant smoke signals;

(xiii) one waterproof electric torch suitable for morse signalling together with one spare set of batteries and one spare bulb in a waterproof container;

(xiv) an efficient radar reflector;

(xv) one daylight signalling mirror with instructions on its use for signalling to ships and aircraft; *(For instructions as to its use, see footnote to sub-paragraph (xvii) on page 185).*

(xvi) one copy of the life-saving signals on a waterproof card or in a waterproof container; *(See pages 313 to 316).*

(xvii) one set of fishing tackle;

(xviii) a food ration totalling not less than 10,000 kj for each person the liferaft is permitted to accommodate; these rations shall be kept in airtight packaging and be stowed in a watertight container;

(xix) watertight receptacles containing a total of 1.5 litres (2.7 pints) of fresh water for each person the liferaft is permitted to accommodate, of which 0.5 litres (0.9 pints) per person may be replaced by a de-salting apparatus capable of producing an equal amount of fresh water in 2 days.

Some liferafts supply sealing lids for opened water tins.

(xx) one rustproof drinking vessel;

(xxi) six doses of anti-seasickness medicine and one seasickness bag for each person the liferaft is permitted to accommodate;

(xxii) instructions on how to survive;

(xxiii) instructions for immediate action;

(xxiv) thermal protective aids for 10% of the number of persons the liferaft is permitted to accommodate or two, whichever is the greater.

H

(b) The marking required on liferafts equipped in accordance with sub-paragraph (a) above shall be "SOLAS A PACK" in block capitals of the Roman alphabet.

(c) In the case of passenger ships engaged on short international voyages of such a nature and duration that, in the opinion of the Administration, not all the items specified in sub-paragraph (a) above are necessary, the Administration may allow the liferafts carried on any such ships to be provided with the equipment specified in sub-paragraphs (a) (i) to (vi) inclusive, (a) (viii), (a) (ix), (a) (xiii) to (xvi) inclusive and (a) (xxi) to (xiv) inclusive and one half of the equipment specified in sub-paragraphs (a) (x) to (xii) inclusive. The marking required on such liferafts shall be "SOLAS B PACK" in block capitals of the Roman alphabet. A liferaft container and liferaft approved for use on pleasure craft shall be marked "DOT(UK) approved".

(d) Where appropriate the equipment shall be stowed in a container which, if it is not an integral part of, or permanently attached to, the liferaft, shall be stowed and secured inside the liferaft and be capable of floating in water for at least 30 minutes without damage to its contents.

HALF LIFERAFT CONTAINER USED AS SEA ANCHOR.

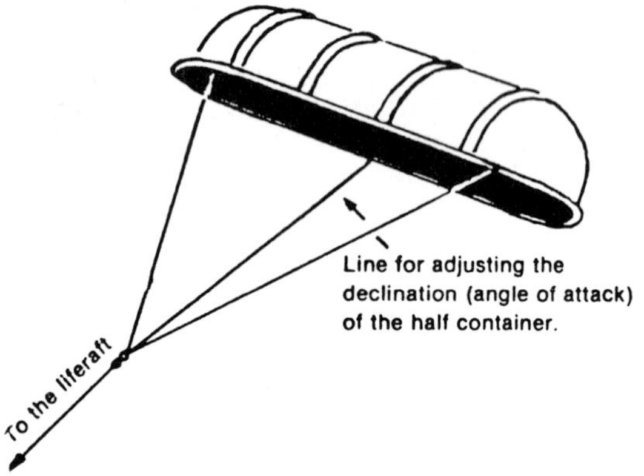

Reproduced from "Siglingamal"

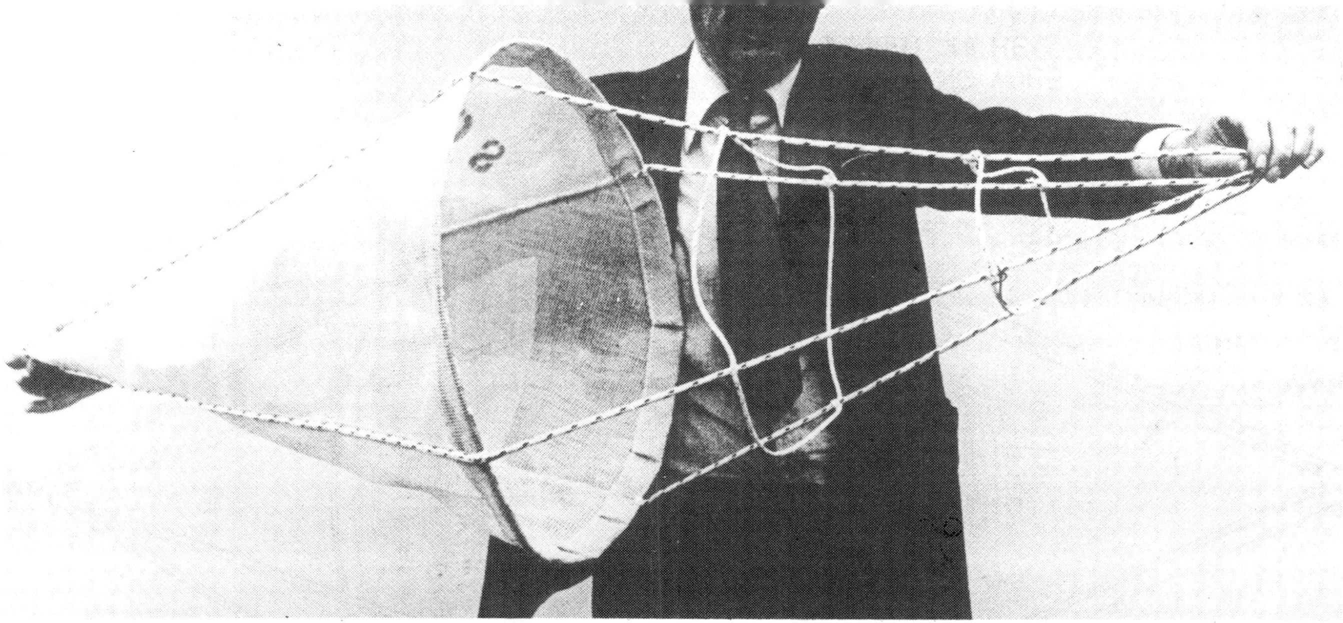

Approved by the D.o.T. The B.M.T. perforated drogue differs from the Icelandic type in that the net on the shroud lines is reduced to two circles of line and a stiff wire grommet is incorporated into the mouth, to keep it open. The opening at the rear is closed, being replaced by perforated material. In addition to its function as a sea anchor, it would appear to be ideal for the purpose of netting plankton.
(Photograph reproduced from N.M.I. R127)

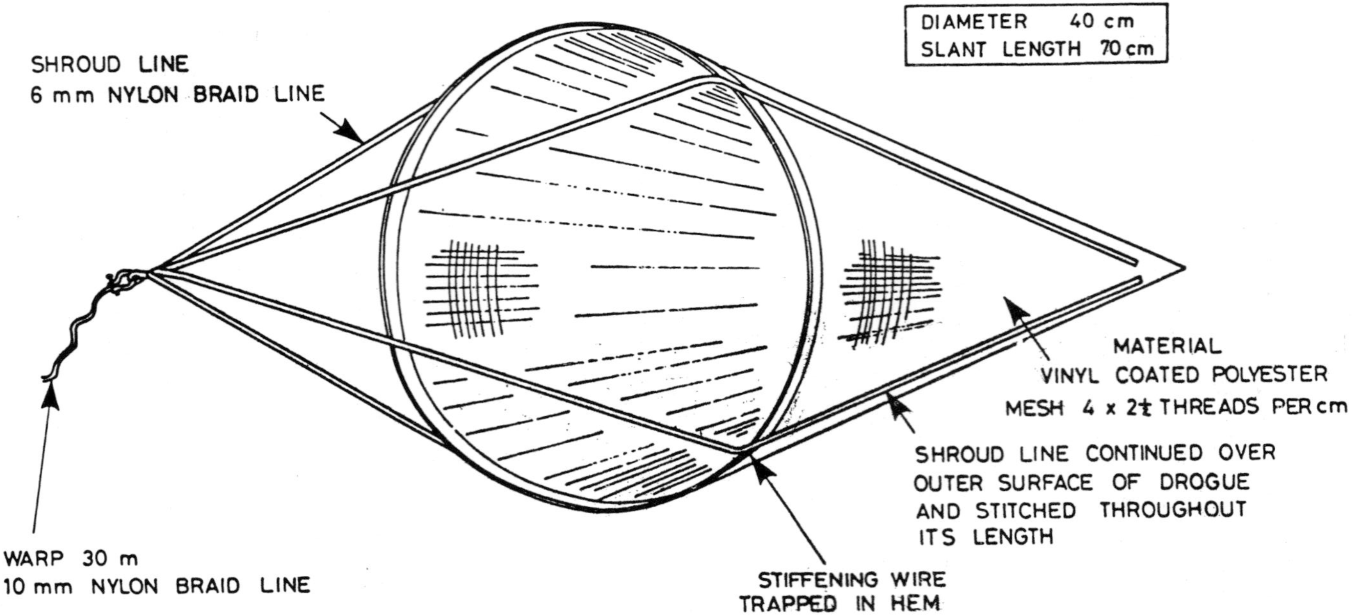

PROPORTIONS OF N.M.I. PERFORATED DROGUE

BEAUFORT AIR-SEA EQUIPMENT LIMITED

INTEGRAL BAILER

An integral "through-the-floor" baling device is fitted in the floor to allow the liferaft to be baled, rapidly, without the need to open the canopy entrance flaps. When not in use, the baler is housed in a fabric cover, complete with draw cord, which is attached to the floor around the periphery of the baler.

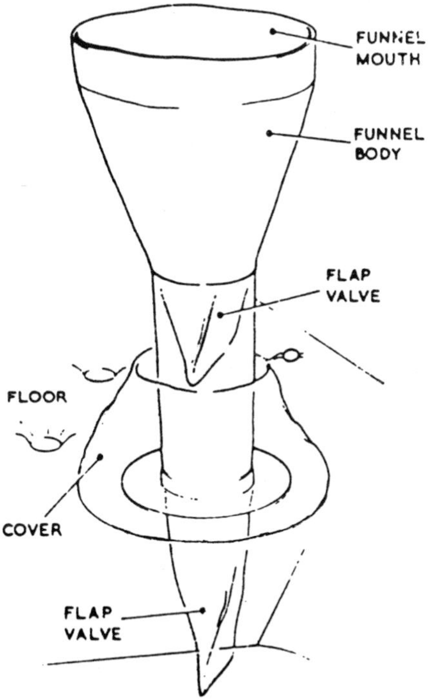

Normally, the hand topping-up bellows or pump supplied in the equipment of every inflatable liferaft, can be adapted in reverse for use as a pump. Put the pump or bellows in the water and lead the hose through the door or into the funnel of the integral bailer, then pump.

RFD Emergency Packs for Marine Liferafts

Complying with the requirements of D.O.T. and to SOLAS 1974 requirements

Typical Pack

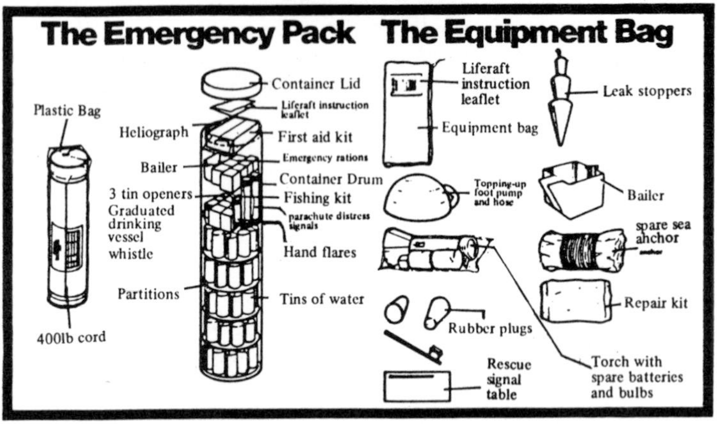

SOLAS defines 3 scales of equipment. D.O.T. defines these as packs A, B & C. The packs listed, together with the standing equipment in the raft comply with these requirements.

		D.O.T. PACK A									B	C
		4	6	8	10	12	15	16	20	25	All sizes	
a	RL&Q *	1	1	1	1	1	1	1	1	1	1	1
b	Knife & Bailer	1	1	1	1	1	2	2	2	2	4–12=1 15–25=2	4–12=1 15–25=2
c	Sponges *	2	2	2	2	2	2	2	2	2	2	2
d	Sea Anchors	2	2	2	2	2	2	2	2	2	2	2
e	Paddles *	2	2	2	2	2	2	2	2	2	2	2
f	Repair Kit	1	1	1	1	1	1	1	1	1	1	1
g	Pump	1	1	1	1	1	1	1	1	1	1	1
h	Safety Tin Openers	3	3	3	3	3	3	3	3	3	–	–
	First Aid Kit	1	1	1	1	1	1	1	1	1	–	–
j	Drinking Vessel	1	1	1	1	1	1	1	1	1	–	–
k	Torch c/w B & B	1	1	1	1	1	1	1	1	1	1	–
l	Heliograph	1	1	1	1	1	1	1	1	1	–	–
m	Parachute Flare	2	2	2	2	2	2	2	2	2	1	–
n	Hand Flare	6	6	6	6	6	6	6	6	6	3	–
o	Fishing Kit	1	1	1	1	1	1	1	1	1	–	–
p	Rations	4	6	8	10	12	15	16	20	25	–	–
q	Water (Litres)	6	9	12	15	18	22.5	24	30	37.5	–	–
r	Anti-Seasickness Tabs.*	24	36	48	60	72	90	96	120	150	–	–
s	Immediate Action Leaflet	1	1	1	1	1	1	1	1	1	1	1
t	Rescue Signal Table	1	1	1	1	1	1	1	1	1	1	1

Notes:

(1) Ration normally comprises 2/6 Kg biscuit and 1/6 Kg glucose tablets. Total energy 10.46 Kj or 2,500 Kcal.

(2) The sponges also serve a dual purpose in protecting the paddles.

(3) Where necessary local variations are made to meet the requirements of overseas governments.

(4) * Denotes item is within liferaft but not in the pack.

Refer to latest edition of The Merchant Shipping (Life-saving Appliances) Rules for rules governing type of pack required.

RFD Inflatables Limited

6. **Float free arrangements for liferafts**

(a) **Painter system**

The liferaft painter system shall provide a connection between the ship and the liferaft and shall be so arranged as to ensure that the liferaft when released and, in the case of the inflatable liferaft, inflated is not dragged under by the sinking ship.

(b) If a weak link is used in the float free arrangement, it shall:

(i) not be broken by the force required to pull the painter from the lifecraft container;

(ii) if applicable, be of sufficient strength to permit the inflation of the lifecraft;

(iii) break at a force of 2.2kN plus or minus 0.4kN.

(c) **Hydrostatic release units**

If a hydrostatic release unit is used in the float free arrangements, it shall:

(i) be constructed of compatible materials so as to prevent malfunction of the unit. Galvanising or other forms of metallic coating on parts of the hydrostatic release unit shall not be accepted;

(ii) automatically release the liferaft at a depth of not more than 4m. (13ft.);

(iii) have drains to prevent the accumulation of water in the hydrostatic chamber when the unit is in its normal position;

(iv) be so constructed as to prevent release when seas wash over the unit;

(v) be permanently marked on its exterior with its type and serial number;

(vi) be provided with a document or identification plate stating the date of manufacture, type and serial number;

(vii) be such that each part connected to the painter system has a strength of not less than that required for the painter.

THE DUNLOP HYDROSTATIC RELEASE UNIT

INTRODUCTION

Dunlop Hydrostatic Release Units are designed to automatically free inflatable liferafts from their stowages after they have become submerged. The normal method of stowing inflatable liferafts is by a lashing and a quick release slip, but there are occasions when manual release is impossible because the vessel sinks very rapidly. Therefore, the use of the Dunlop H.R.U. is a very desirable safety measure.

NOTE
On current models, the ring on item 1 is no longer fitted and item 2 fits directly into the base plate. Therefore, when installing, item 1 should be positioned at 90° to that shown in the sketch.

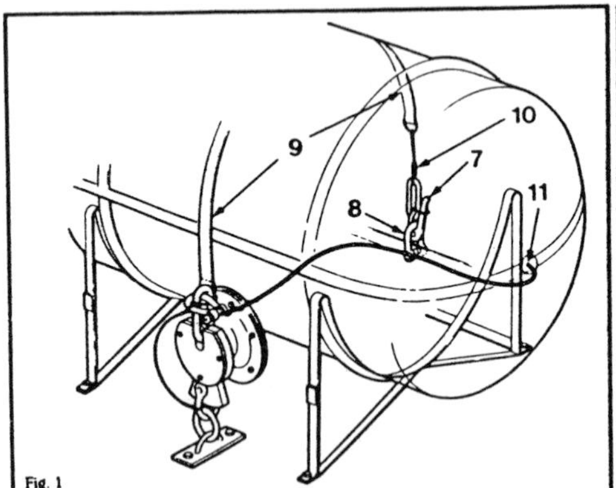

Fig. 1

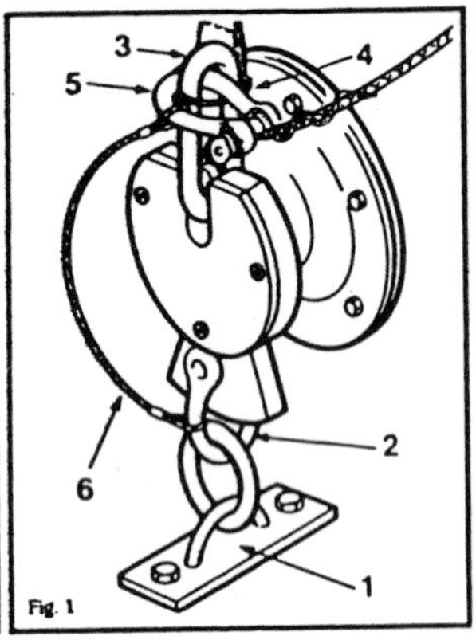

Fig. 1

INSTALLATION see Fig. 1

The system should be installed as follows:

(1) Securely fix the metal cradle in the required position*.

(2) Weld or bolt the base plate (1) to the ship's deck so that it lies in the direction shown and, when subsequently fitted, the H.R.U. will be at normal to the deck. On existing cradles the crossbar on the side without the footbar must be removed.

(3) Fit the H.R.U. to the plate using shackle (2).

(4) Place the liferaft container in the cradle making sure that the top half is uppermost and that the black weathertight seal is horizontal and not at an angle.

(5) Fit the Hoop ring (3) through the buckle (4) of the lashing strap and hold in place with shackle (5).

(6) Make sure that the weak link (6) is in position as shown in the sketch and is not under tension and not obstructed in any way.

(7) Secure the senhouse slip (7) to the crossbar of the cradle with shackle (8).

(8) Pass the lashing strap (9) over the container and lash its buckle to the senhouse slip with the cordage provided (10) making sure that sufficient tension is achieved to hold the container securely and tightly in the cradle.

(9) Fasten the free end of the painter line to the upper shackle (5) using a five tuck splice. Make sure that the painter is undisturbed at the point where it emerges from the container (11) so that the antiwicking arrangement is not impaired.

*Important. The location of the cradle must be such that the senhouse slip, which is used for manual release, is readily accessible and is not obstructed in any way.

OPERATION see Fig. 2

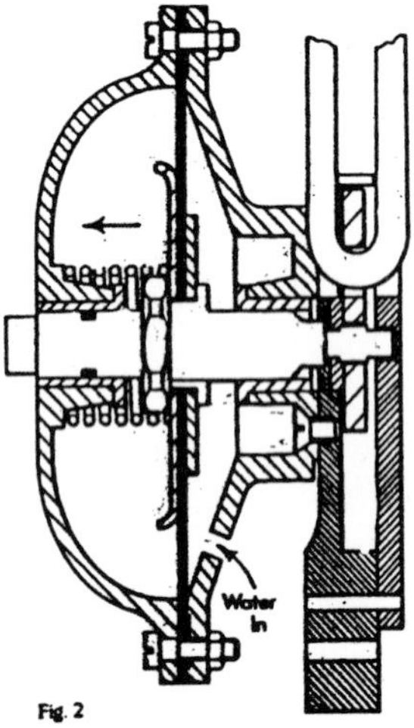

Fig. 2

When the unit is submerged, water fills one chamber and pressure in excess of the spring loading causes the diaphragm to move sideways, which withdraws the plunger from engagement with the hook. Tension of the associated lashing causes the hook to rotate thus releasing the hoop ring and allowing the liferaft to float free.

Experience shows that H.R.U. and the senhouse slip should be contained on the same side of the liferaft container, as shown in the diagram on page 231 and not on opposite sides as shown in the above diagram (fig. 7).

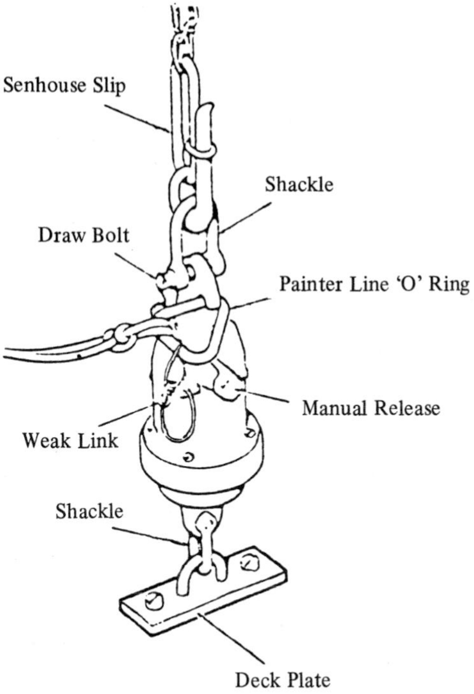

The correct method of attaching a senhouse slip and a hydrostatic release unit to the lashing on an inflatable liferaft container.
(Reproduced by permission of Berwin Engineering Ltd.)

HOW IT WORKS:

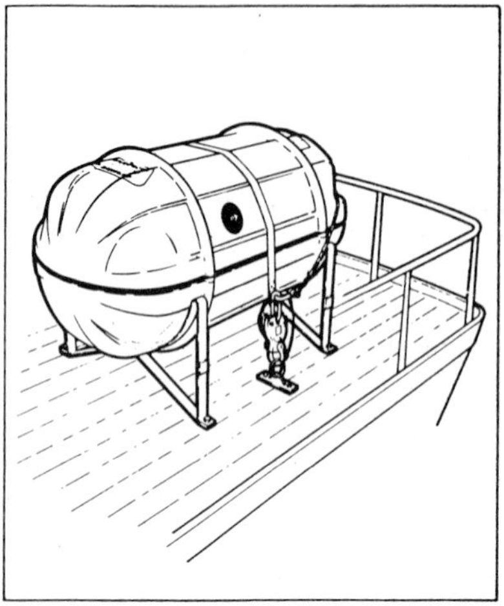

A liferaft on a typical stowage showing the painter line made fast to the weak link system of the H.R.U.
Note that the location is such that the senhouse slip is readily accessible and that the H.R.U. is on the seaward side of the vessel.

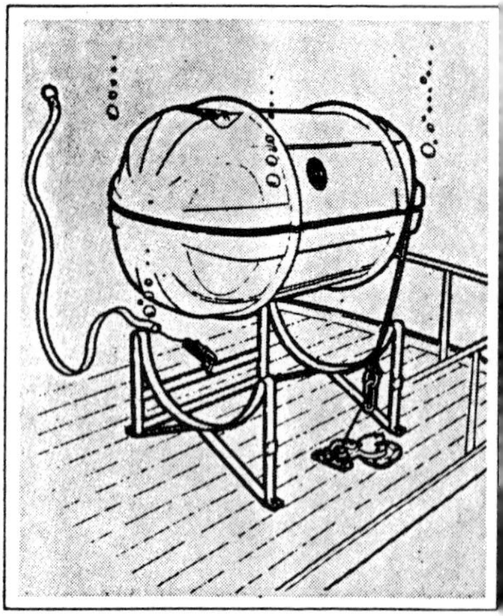

When submerged to a depth of 1.5 to 3.7 metres (5 to 12 feet), the H.R.U. operates automatically. Freed of its lashing strap, the container rises to the surface paying out the inflation painter line as it goes but retaining it to the H.R.U. by means of the weak link.

It is this combination of the rising container and the sinking vessel which provides the painter tension which activates the gas cylinder and starts inflation once all the free painter line is paid out.

As the vessel continues to sink, the weak link will break – thus releasing the liferaft from the vessel.
The weak link of 227 ± 45 Kgf (500 ± 100 lbf) breaking strength ensures that a raft is not dragged under by the sinking vessel before it has time to inflate to its full dimensions.

THE DUNLOP HYDROSTATIC HRU RELEASE SYSTEM

Maintenance on board

1. Ensure that the water inlet hole is free from obstructions. Do not probe with sharp tools or damage to the diaphragm may result.

Do not paint

2. Periodically examine the weak link for fraying or deterioration.
3. Ensure that the painter remains properly connected at all times.
4. Periodically check the tension on the lashing strap and make sure that the tensioning cordage is intact and securely tied off.

IMO. Requirements for inflatable liferafts on ships built after 1st July 1986

1. Inflateable liferafts shall comply with the requirements for all liferafts and, in addition, shall comply with the following:-

2. **Construction of inflatable liferafts**

(a) The main buoyancy chamber shall be divided into not less than two separate compartments, each inflated through a non-return inflation valve on each compartment. The buoyancy chambers shall be so arranged that in the event of any one of the compartments being damaged or failing to inflate, the intact compartments shall be able to support, with positive freeboard over the lifecraft's entire periphery, the number of persons which the lifecraft is permitted to accommodate, each having a mass of 75kg. (165 lbs.) and seated in their normal positions.
(b) The floor of the liferaft shall be waterproof and shall be capable of being sufficiently insulated against cold either:

(i) by means of one or more compartments that the occupants can inflate, or which inflate automatically and can be deflated and re-inflated by the occupants; or
(ii) by other equally efficient means not dependent on inflation.

(c) The liferaft shall be inflated with a non-toxic gas. Inflation shall be completed within a period of one minute at an ambient temperature of between 18 and 20 degrees C (64 to 68 deg.F). After inflation the liferaft shall maintain its form when loaded with its full complement of persons and equipment.
(d) Each inflatable compartment shall be capable of withstanding a pressure equal to at least three times the working pressure and shall be prevented from reaching a pressure exceeding twice the working pressure either by means of relief valves or by a limited gas supply. Means shall be provided for fitting the topping-up air pump or bellows so that the working pressure can be maintained.

3. **Carrying capacity of inflatable liferafts**

The number of persons which a liferaft shall be permitted to accommodate shall be equal to the lesser of:
(a) the greatest whole number obtained by dividing by 0.096 the volume, measured in m^3 of the main buoyancy tubes (which for

this purpose shall include neither the arches nor the thwarts if fitted) when inflated; or
(b) the greatest whole number obtained by dividing by 0.372 the inner horizontal cross-sectional-area of the liferaft measured in m^2(which for this purpose may include the thwart or thwarts, if fitted) measured to the innermost edge of the buoyancy tubes; or
(c) the number of persons having an average mass of 75kg. (165 lbs.), all wearing lifejackets, that can be seated with sufficient comfort and headroom without interfering with the operation of any of the liferaft's equipment.

Access into inflatable liferafts

(a) At least one entrance shall be fitted with a semi-rigid boarding ramp to enable persons to board the liferaft from the sea so arranged as to prevent significant deflation of the liferaft if the ramp is damaged. In the case of a davit launched liferaft having more than one entrance, the boarding ramp shall be fitted at the entrance opposite the bowsing lines and embarkation facilities.
(b) Entrances not provided with a boarding ramp shall have a boarding ladder, the lowest rung of which shall be situated not less than 0.4m (16in.) below the liferaft's light waterline.
(c) There shall be means inside the liferaft to assist persons to pull themselves into the liferaft from the ladder.

Stability of inflatable liferafts

(a) Every inflatable liferaft shall be so constructed that when fully inflated and floating with the canopy uppermost, it is stable in a sea-way.
(b) The stability of the liferaft when in the inverted position shall be such that it can be righted in a sea-way and in calm water by one person.
(c) The stability of the liferaft when loaded with its full complement of persons and equipment shall be such that it can be towed at speeds of up to 3 knots in calm water.

Inflatable liferaft fittings

(a) The strength of the painter system including its means of attachment to the liferaft, except the weak link, shall be not less than 10.0kN for a liferaft permitted to accommodate nine persons or more, and not less than 7.5kN for any other liferaft. The liferaft shall be capable of being inflated by one person.
(b) A manually controlled lamp visible on a dark night with a clear atmosphere at a distance of at least two miles for a period of not less than 12 hours shall be fitted to the top of the liferaft canopy. If the light is a flashing light it shall flash at a rate of not less than 50 flashes per minute for the first two hours of operation of the 12 hour operating period. The lamp shall be powered by a sea-activated cell or a dry chemical cell and shall light automatically when the liferaft inflates. The cell shall be of a type

that does not deteriorate due to damp or humidity in the stowed liferaft.

(c) A manually controlled lamp shall be fitted inside the liferaft capable of continuous operation for a period of at least 12 hours. It shall light automatically when the liferaft inflates and be of sufficient intensity to enable reading of survival and equipment instructions.

7. **Containers for inflatable liferafts**

(a) The liferaft shall be packed in a container that is:

(i)so constructed as to withstand hard wear under conditions encountered at sea;

(ii) inherently buoyant when packed with the liferaft and its equipment;

(iii) as far as practicable watertight, except for drain holes in the container bottom.

(b) The liferaft shall be packed in its container in such a way as to ensure, as far as possible, that the waterborne liferaft inflates in an upright position on breaking free from its container.

(c) The container shall be marked with:

(i) The maker's name or trade mark;

(ii) its serial number;

(iii) name of the approving authority and the number of persons it is permitted to carry;

(iv) SOLAS;

(v) type of emergency pack enclosed;

(vi) date when last serviced;

(vii) length of painter;

(viii) maximum permitted height of stowage above waterline (depending on drop test height and length of painter;

(ix) launching instructions.

8. **Markings on inflatable liferafts**

The liferaft shall be marked with:

(a) maker's name or trade mark;

(b) its serial number;

(c) date of manufacture (month and year);

(d) name of approving authority;

(e) name and place of servicing station when it was last serviced;

(f) number of persons it is permitted to accommodate over each entrance in characters of a colour contrasting with that of the liferaft not less than 100m (4in) in height.

9. **Davit launched inflatable liferafts**

(a) In addition to complying with the above requirements a liferaft for use with an approved launching appliance shall, when suspended from its lifting hook or bridle, withstand a load of:

(i) 4 times the mass of its full complement of persons and equipment, at an ambient temperature and a stabalized liferaft temperature of 20 degrees C (68 deg.F), plus or minus

3 degrees C (5deg.F.) with all relief valves inoperative; and
(ii) 1.1 times the mass of its full complement of persons and equipment at an ambient temperature of minus 30 degrees C (-22 deg F) with all relief valves operative.

(b) Rigid containers for liferafts to be launched by a launching appliance shall be so secured that the container or parts of it are prevented from falling into the sea during and after inflation and launching of the contained liferaft.

Additional equipment for inflatable liferafts

(a) In addition to the equipment required for all liferafts, every inflatable liferaft shall be provided with:

(i) one repair outfit for repairing punctures in buoyancy compartments;

(ii) one topping-up pump or bellows.

(b) The knives required to be carried shall be safety knives.

The equipment will be in a sealed watertight container lashed to the interior of the liferaft, should it fall out, it is easily reclaimable. Equipment for immediate use may be supplied in a separate draw-string bag.

Never in any circumstances interfere with the deflation plugs on the outside of the liferaft. They are placed there for the use of the manufacturer or his agent and are used to draw off the last of the gas by suction, after the liferaft has been tested and before packing.

Many inflatable liferafts are now being constructed with an integral "through the floor" bailing device.

Department of Transport Merchant Shipping Notice
No. **M.1135**

LIFE-SAVING APPLIANCES

MISUSE OF INFLATABLE LIFERAFTS/BOATS

Notice to Shipowners, Masters and Skippers

This notice supersedes Notices M. 726 and M. 1008

(1) Cases have occurred where items of inflatable equipment which formed part of the statutory life-saving equipment of a ship have been used for purposes not connected with an emergency or with the saving of life. For instance the Department has discovered that inflatable boats have been used as floating platforms to assist in the painting of ships.

(2) The coated fabric from which the inflatable equipment is made is resistant to most forms of chemical attack but can be damaged by certain paints which could render the equipment unfit for its statutory purpose.

(3) The Department takes a serious view of the use of inflatable equipment for miscellaneous duties, and attention is drawn to the provision of Section 430 of the Merchant Shipping Act, 1894, in which penalties are prescribed for wilfully rendering statutory life-saving appliances unfit for service during a voyage, and for failure to keep such appliances at all times fit and ready for use.

(4) If at any time a liferaft can be recovered after emergency use it should, if possible, be maintained in a condition ready for further use until such time as it can be properly serviced and re-packed.

Department of Transport
Marine Directorate
London WC1V 6LP
June 1984

LIFEGUARD. FORTIES LIFERAFTS 4/6/8, Persons.

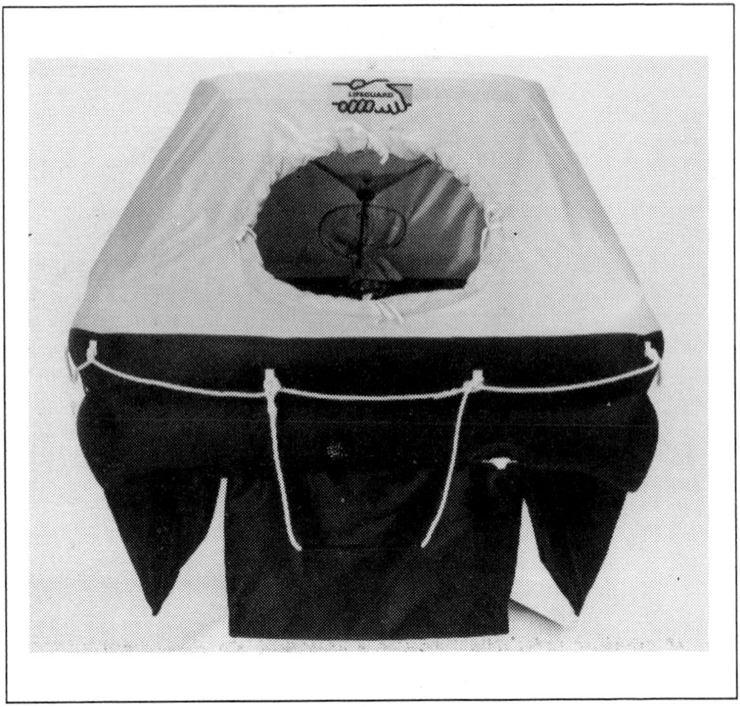

Lifeguard's new liferaft unitary construction design is the latest additional safety feature. The unitary construction design removes the risk of separation between the two buoyancy chambers but at the same time continues to provide independant air compartments. This type of construction is also combined with seamless angle joints so that the two features together reduce the air holding seam run by as much as 64% so further increasing the safety margin against leakage. Another design feature is the placing of the pressure relief valves on the exterior of the raft thus avoiding any risk of asphyxia when the pressure relief valves are blowing off.

These liferafts also have the advantage of circular entrances similar to those required by the Icelandic Government.

When fitted with RORC/ORC pack type 'R', Forties liferafts comply with RORC/ORC Regulations.

The liferafts are provided with either a soft valise or a rigid container. A valise should be installed in a dry, rat-proof location, whereas a container may be mounted on deck.

AVON INFLATABLES LIMITED
Construction Diagram

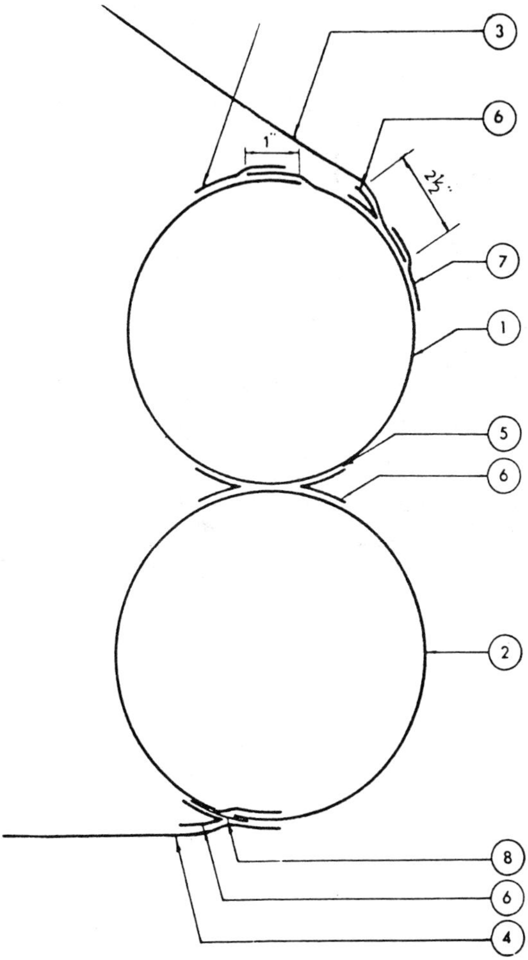

1.	TOP BUOYANCY TUBE	5.	ADHESIVE K.B.63
2.	BOTTOM BUOYANCY TUBE	6.	HINGE
3.	CANOPY (INCLUDING INNER ENTRANCE)	7.	1" WIDE TAPE CUT FROM CANOPY FABRIC
4	FLOOR	8.	TAPE I.P. No. 498

MODIFIED RAFT CONSTRUCTION 4, 6 and 8 MAN FOR ALL RAFTS MANUFACTURED AFTER DECEMBER, 1976.

BEAUFORT LIFERAFT TYPE RBM

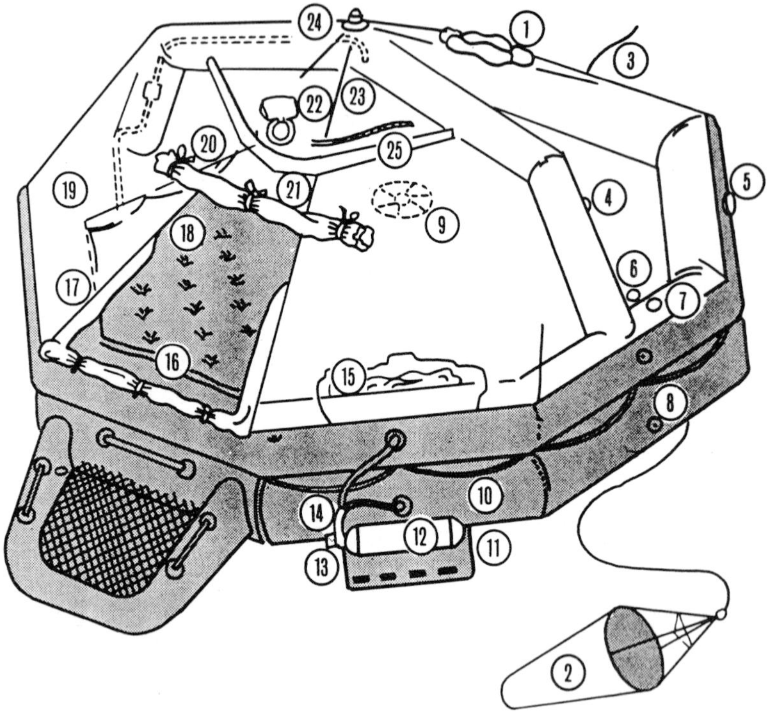

1. Observation Port
2. Sea Anchor
3. Painter Line
4. Topping-up/Deflation Valve (Arch)
5. Pressure Relief Valve
6. Topping-up/Deflation Valve (Floor)
7. Topping-up/Deflation Valve (Chambers)
8. Deflation Point
9. Integral Baler
10. Lifeline
11. Stabilising Pocket
12. CO_2 Cylinder
13. Operating Head
14. Inflation Hose Assembly
15. Drawstring Bag (Emergency Pack)
16. Boarding Handle
17. Knife Pocket
18. Battery Pocket
19. Canopy
20. Outer Entrance Cover
21. Rainwater Collector Tube
22. Rescue Line and Quoit
23. Interior Light
24. Exterior Light
25. Handline

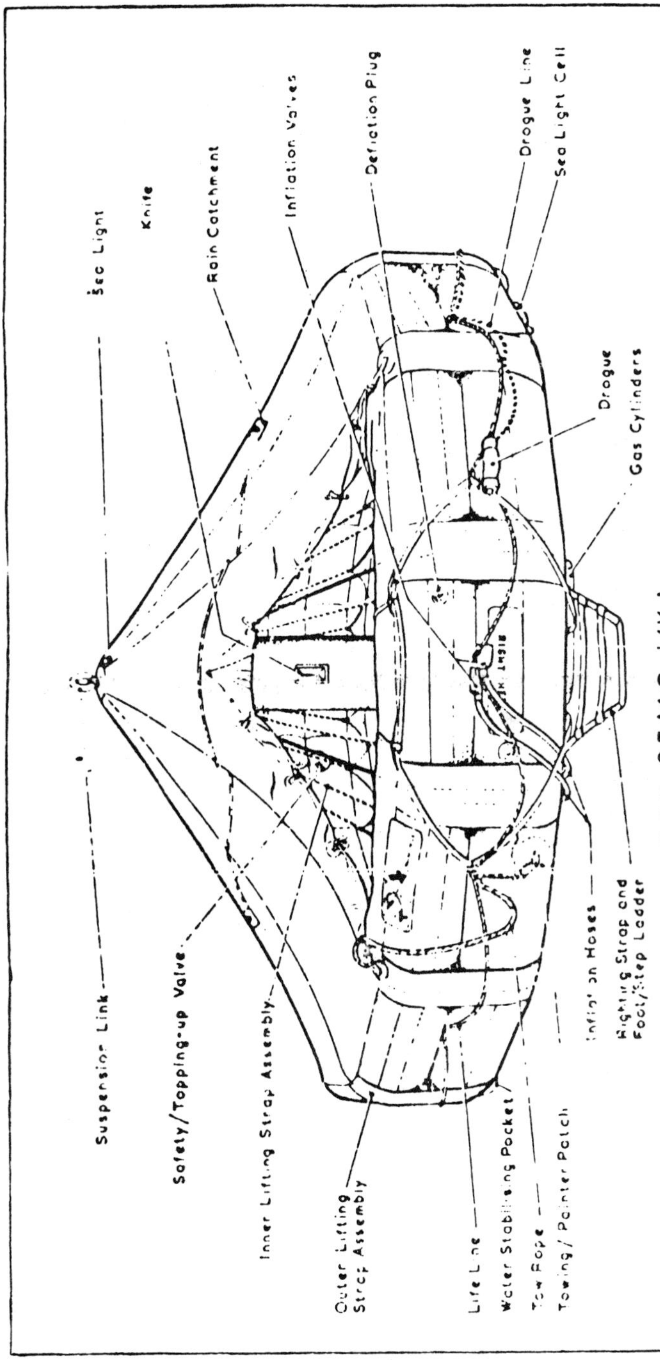
Sea Light
Knife
Rain Catchment
Inflation Valves
Deflation Plug
Drogue Line
Sea Light Cell
Drogue
Gas Cylinders
Righting Strap and Foot/Step Ladder
Inflation Hoses
Towing/Painter Patch
Tow Rope
Water Stabilising Pocket
Life Line
Outer Lifting Strap Assembly
Inner Lifting Strap Assembly
Safety/Topping-up Valve
Suspension Link

R.F.D. 25 M.C. MK.I.

BEAUFORT
DAVIT LAUNCHED TYPE Q LIFERAFT
Available in-25 man size only

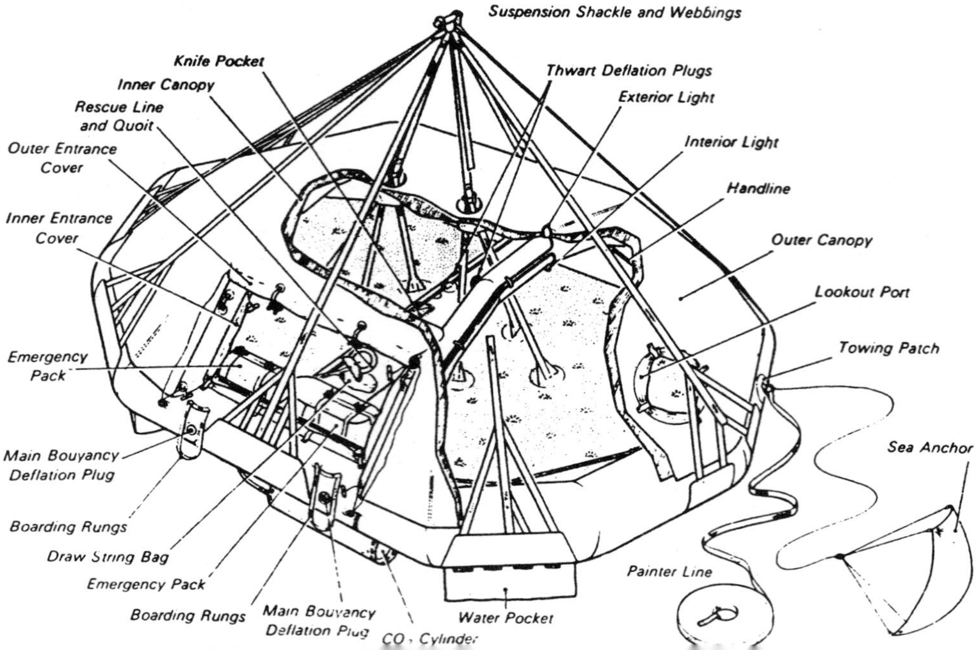

NOTES ON INFLATABLE LIFERAFT STOWAGE

When a liferaft comes aboard, perhaps after being serviced. Care must be taken not to damage the container in any way. Do not roll it as this could cause the gas bottle to shift. Do not throw it down onto the deck as this could damage the container and allow water to seep in. Do not drop it. Always handle the container by means of the handling lines if fitted. G.R.P. containers and valises must always be stowed with the lettering uppermost and never on end, to ensure that the vibration of the ship does not cause the gas bottle to shift, and because some G.R.P. containers have drain holes underneath, the joining seam is to be horizontal. Do not walk, stand or jump on stowed liferaft containers as this could crack the container.

On occasion, extra straps are fitted to G.R.P. containers to secure them in transit. These must be removed. Do not remove any strap whatsoever unless it is clearly stated on that strap that, that strap is to be removed before stowage on board, because some containers have bands around them which are held together with a weak link that breaks from pressure when the liferaft is inflated.

G.R.P. containers are usually cylindrical in shape, however, some of the smaller rafts provided to yachts and fishing vessels may be box shaped. Containers may be stowed singly in cradles or in tiers on chutes and are normally secured in place with nylon strips and sen-house slips and usually attached to hydrostatic release units. Make the end of the painter securely fast to the hydrostatic release unit, or the ship, if no unit is attached. Do this immediately the container is stowed, then it will not be forgotten. Do not pull any extra painter out of the container as this would break the watertight seal. When stowed in cradles, they normally require to be lifted and thrown overside when abandoning ship. When stowed in chutes, they can be released separately.

A fabric valise should be stowed either on a shelf or preferably in a box with collapsible sides (so that it cannot jam) that it just fits. The lid should be secured with a wood peg through a hasp and staple, to ensure that the valise is not thrown out in severe weather. They must be stowed so that they will not be damaged by the weather, trapped sea water, copper or copper alloys, heat of the boilers or engine room, sparks or fumes from the funnel, oil or to be attacked by rats. (Rats will reduce an inflatable liferaft to tatters overnight. Stow the box on chocks with chicken wire underneath to guard against rat attack.

Inflation of the liferaft must be automatic on the pulling of the painter. The painter has a dual role of acting as both painter and operating cord. It is attached to a firing mechanism on the gas bottle, so that when it is almost fully pulled out, a tug will fire the gas bottle and the raft inflates, the end of the painter being firmly made fast to the liferaft. It sometimes happens that on tugging the painter to fire the gas bottle on a launched liferaft, only half the raft inflates. This is probably due to the fact that some liferafts have two gas bottles. Give the painter another hard tug which will probably fire the second gas bottle and inflate the raft fully.

In order to ensure inflation in low temperatures, the CO_2 gas used for inflation of the liferaft contains a small percentage of nitrogen to act as an anti-freeze agent and ensure that all the CO_2 is discharged, although the liferaft will take somewhat longer to inflate in freezing conditions.

RFD INFLATABLE LIFERAFTS

Notice to Owners of and Masters of Merchant Ships, Owners and Skippers of Fishing Vessels and Yachtsmen

1. RFD Inflatables Ltd, have introduced a new method of sealing the Glass Reinforced Plastic (GRP) containers in their MK 7A range of inflatable liferafts. The system consists of a number of separate 13 mm wide nylon bands which are tensioned to secure the two sections of the container together. It replaces the previous method which used an extruded container seal with an adhesive. To maintain watertightness a 100 mm wide self-adhesive tape is secured around the outside to seal the joint between the two sections of the container.

2. To standardise container securing arrangements RFD have extended the fitting of this new system to their MK4, MK5 and MK6 range of liferaft.

3. In an attempt to prevent the nylon banding being inadvertently mistaken for transit strapping it has been partially covered with a self-adhesive tape. The tape is marked with an illustration of a cross on a pair of scissors indicating that the band should not be cut. Despite this precaution it has been reported that some ships are removing the bands as soon as the liferafts have been secured in their stowage racks.

4. RFD Inflatables Ltd. have recently issued a service bulletin, No.13/82, to all their approved service stations instructing them to check that the bands are not removed when the liferafts have been replaced on board a vessel. In addition the self-adhesive warning tape will in future be extended to completely cover each individual band on the container (see sketch overleaf).

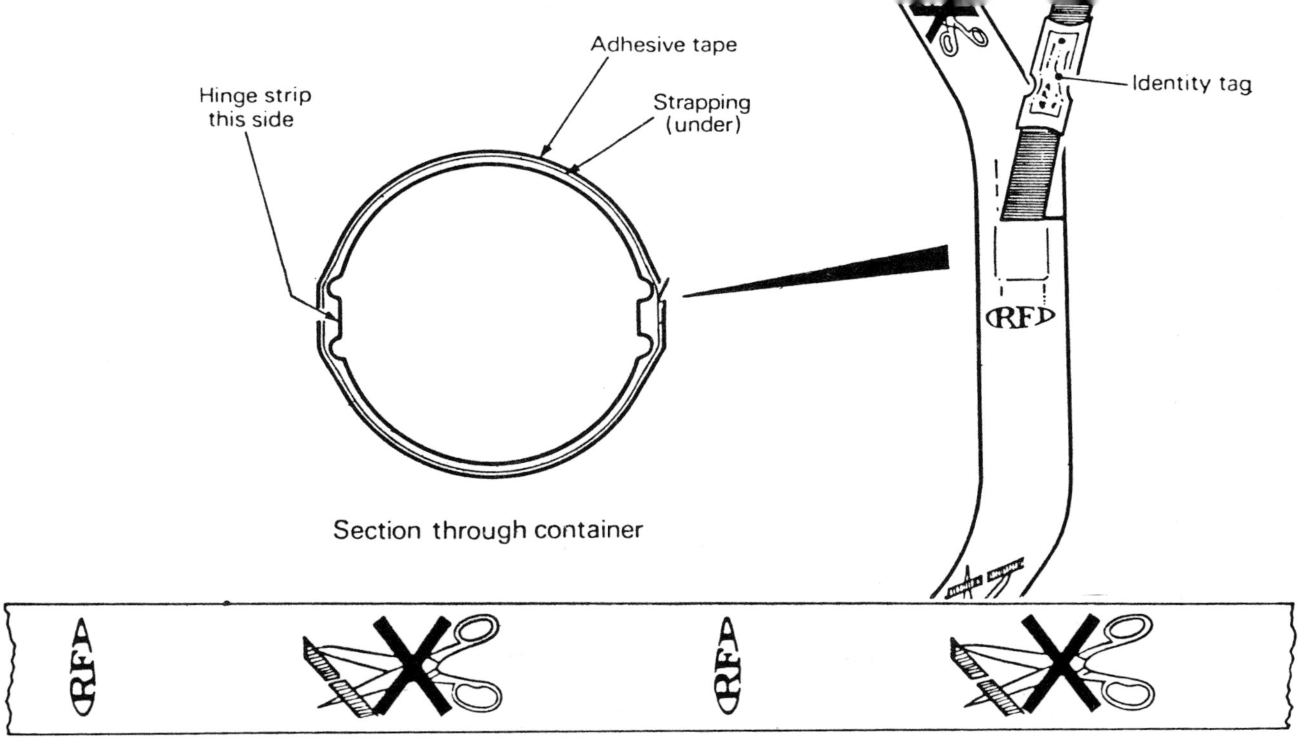
Adhesive tape
Hinge strip this side
Strapping (under)
Identity tag
RFD
Section through container
RFD
RFD
Self-adhesive tape

On some ferries and passenger ships, a rocket launched inflatable liferaft may be situated on the stern of the vessel. In the event of a person falling overboard, the rocket launcher can be electrically fired from the bridge and the liferaft deployed astern, in the hope that it may be of assistance to the person in the water and to act as a marker.

Again, on some hovercraft in particular, an electrical release system operated from the bridge, can release and launch the liferafts in an emergency. Electrical release systems are incorporated into the manual and hydro-static systems required by the Regulations.

RFD/Plumett Liferaft Launching System

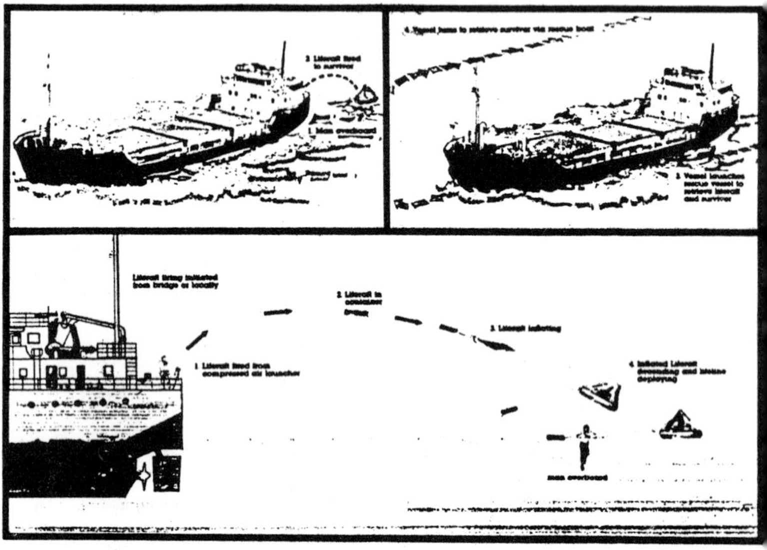

This compressed air-launching system has been especially developed for 'firing' a liferaft to a man overboard. The system can be installed on an oil rig, a ship or used in conjunction with coastal rescue from a shore station.

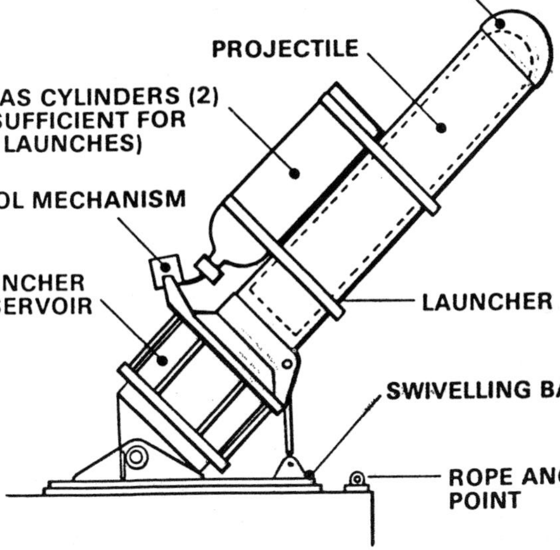

The liferaft is the RFD 'Seasava' (up to 4-person capacity) having a single buoyancy tube, single floor, self-erecting canopy, fitted with stabilising water pockets and inflation system.

METHOD OF OPERATION

1. Elevation will normally be pre-set, dependent on installation.
2. Aim projectile by turning launcher on swivel base.
3. Activate launcher by pushing button.
4. Air is released from the cylinder into the launcher reservoir. At pre-determined pressure in the reservoir the projectile will be launched automatically. The time between activating the system and automatic launch is approx. 3 to 4 seconds.
   As the projectile moves away from the launcher the retaining cord pays out, the cord end being attached to the launcher base.
6. At a pre-determined point in the trajectory of the projectile, the CO_2 cylinder attached to the liferaft is automatically triggered and whilst the projectile continues on its way, the liferaft commences to inflate. During this inflation phase the system is decelerating and descending to the water. To obviate the risk of damage to the liferaft due to possible snatch loads (caused by the rapid deceleration) a shock absorber is fitted between the retaining line and the liferaft attachment point.
7. The survivor can either swim to the liferaft, or wait for it to drift down to him, and then board the liferaft.
8. The liferaft can either be set adrift to await recovery by the stand-by vessel, or be pulled back to the vessel/rig/shore (whichever is applicable) by using the retaining cord.

EXTRACTS FROM "SIGLINGAMAL" for DECEMBER 1980 o the RESULTS OF TRIALS WITH INFLATABLE LIFERAFTS I ICELANDIC WATERS.

Many aspects of liferaft design and construction are being tested an modified, both in Iceland and abroad. Recently, trials have taken plac on improving the entrance openings and means of access from the sea Methods of improving the stability have also been examined. Thes consisted of enlarging ballast pockets and increasing their distance fron the centre of the raft to attain the maximum righting moment. The sea anchor system was also examined and improvements were made to it design and construction. The reports also specify the new requirement for the design, production and equipment of inflatable liferafts, for us on Icelandic vessels.

In February 1980 the Icelandic Directorate of Shipping carried ou full-scale trials of inflatable liferafts, under adverse weather condition in the waters off the north-west coast of Iceland, some additional trial also took place in March 1980.

The aim was to gain knowledge and experience, which would lead t an improvement in the design and manufacture of liferafts and thei associated equpment. An additional requirement was to improve th stability of the liferaft, and increase the efficiency of the sea-ancho systems. Other important functions were to test the canopy strength improve methods of closing the entrances and redesign the boardin facilities.

Seven 10 man rafts of various makes were tested between 2nd February 1980 and 17th. March 1980. Trials lasted for several days o several occasions and were mainly carried out under adverse weathe conditions. Winds over 80 knots were encountered, with temperature close to freezing and wave heights of up to 10 m. (33 feet). The rafts wer loaded with varying numbers of sandbags placed in various positions Different versions of sea-anchors were used and modified entrances, se water pockets and ladders were also tested.

One 10 man RFD MK6 liferaft, modified with seven long ballas pockets each of 22 litres capacity, was left at sea in a hurricane o 25th February and was found still afloat on 3rd. March 1980. This inflatable liferaft by then had low pressure in the upper buoyancy tube as it was damaged by abrasion in a small spot and from there gas escaped slowly, most likely caused by one of the sand ballast bags which had been secured inside the raft when it was launched.

The general condition of the liferaft when it was found was the following:-

The lower buoyancy tube was undamaged, also the bottom, the upper buoyancy tube was worn at a small spot, where it was leaking, the arch supporting the canopy was undamaged and the canopy on one side of the arch was undamaged, but on the other side the canopy was torn, and a part of it missing. The upper buoyancy tube was pumped up with air after recovery, and it kept the air reasonably well in spite of the slow leakage through the worn fabric.

Generally the inflatable liferaft was in such a condition, that most likely it would have been possible to retain its full buoyancy, if men had been topping up the pressure in the upper tube by using equipment provided in the liferaft.

This inflatable liferaft had already been fitted with the strengthening V-strips between the buoyancy tubes, as it is now intended to require. The canopy on the other hand was not of the new type specified in the following requirements.

Regarding each of the special items tested the following should be noted.

1. Sea-water ballast pockets.

The purpose of the sea-water ballast pockets below the bottom of the inflated liferafts is to increase their stability, if partly lifted out of the water. It was difficult to judge separately the effect of these sea-water ballast pockets, as several other factors have also great influence on the stability of the liferaft, e.g., the sea anchors, the number of persons in the liferaft and where they sit.

Although it is believed to be generally acceptable that an increase of the volume of the sea-water ballast pockets has considerable effect towards increased stability, as soon as the liferaft is partly lifted out of the water. Furthermore the distance of the centre of gravity of the volume of the sea water pockets from the centre of the bottom of the liferaft has to be the biggest possible.

The effect of the sea water pockets on the stability was tested by trying to capsize manually an empty but inflated 10 man liferaft in calm sea. After the size of the sea water pockets had been increased and moved as far as possible to the outer-side of the lower buoyancy tubes, it proved to be very difficult, really nearly impossible, for 3 men in the water to capsize the raft, but with the existing smaller size of sea-water pockets it was possible to turn it over.

Therefore it is reasonable to believe that increased size of the sea water pockets, and increased distance from the centre will also increase the stability in a sea-way.

Furthermore the form of the sea-water pockets appeared to be important for their effectivity. The square form of the pockets, without stiffenings, seemed more easily to fold together when lifted out of the water, than the proposed V-form. Also the size of the openings near the top of the sea water pockets is important. Furthermore it is essential that there is a space (opening) between the sea water pockets preventing air being trapped below the bottom of the liferaft, after it has been lifted partly out of the water. These items are further explained in the following requirements.

2. Sea anchors.

These trials have shown clearly that sea anchors fastened to the inflatable liferafts are one of the most important factors increasing their stability in a sea-way. The existing types of sea-anchors now supplied with the liferafts were tested. Most of them broke off liferafts or were torn by the pulling forces.

Very often they also got entangled and did thus lose their effect. After some tests with different modifications of the sea anchor equipment it was found that, by increasing the diameter of the opening at the after end of the sea anchors, and increasing their length considerably, so as to make them more cylindrical, the pulling force from the sea anchor was more smooth.

When a breaking wave pushed the liferraft forward, this new type of sea anchor pulled softly on the line to the liferaft, but however still wit sufficient pulling force to reduce the risk of capsizing the liferaft, an this pulling force even sometimes raised the liferaft again, if it ha capsized.

It was at the beginning not known why the sea anchors got entangled but the fact was that the after end of the sea anchors moved forward and got in between the shroud lines at the forward end and partly o fully closed the forward opening of the sea anchor. By dragging the se anchors slowly and either pulling or slackening the line irregularly, i was found that these variations in the pulling speed caused the se anchor to be entangled as explained above. To prevent this from happening, a net was fitted between the shroud lines at the forward en of the sea anchor. After this modification, closing the space between th shroud lines by the net, the sea anchor never got entangled. This an other items regarding the sea anchors are further explained in th following requirements.

3. The canopies of the liferafts

The strength of the canopies of the inflatable liferafts and the closin of the entrance openings were specially studied during the trials. Th existing entrance opening types were tested, as well as a new Icelandi type, based on the entrance openings commonly used on glacier-tents The fastening patches for the sea anchor lines on the liferaft are place such that the entrance openings are at a right angle to the direction o drifting,* which is more or less the same as the wind direction. Th purpose being to reduce the stress on the entrance closing device Existing entrance openings are usually of triangular shape provide with double fabric shutters tied down on the outside and tied up on th inside. The experience has shown that water and wind penetrat through this closing device and it appears that the canopy failing usually originate at the corners of the triangular entrance openings, in spite of strong reinforcements provided there.

**N.B. It is pointed out that this arrangement would preclude a throug draught in the tropics.*

Therefore trials were made with circular entrance-openings and a sleeve type closing device, which could be drawn together by a string and tied up inside the canopy.* This type has the advantage that:-

1) the opening has no corners, which have proved a weak point for tearing;

2) the canopy can be glued all around to the upper buoyancy compartment;

3) no loose fabric (flaps) remains on the outside which wind and weather could penetrate and tear off;

4) a man on watch can keep his head alone outside the canopy looking for ships and aircraft, and still the canopy of the liferaft is closed.

During the trials at sea this closing device never failed though the existing type of closing device was torn open.

**N.B. It is pointed out that this arrangement would preclude anyone outside the liferaft opening the entrance.*

4. Device for entering into an inflatable liferaft

Former experience has shown that it can be difficult for a person to imb aboard an inflatable liferaft. Therefore certain trials were made n this subject. Specially the equipment fitted for making this more asy. In this connection it was found necessary to try to find out if the rcular entrance opening and the sleeve type of closing would make ntering any more difficult than the existing closing device. Climbing on oard out of the water was tried by persons in full oil-skins, with fejackets as well as without.

Results from these tests showed that circular entrance openings and eeve type closing caused no more difficulty than the existing type of ntrances.

On the other hand, the following requirements include proposals for ne design for a net-rope ladder, its fastening to the centre of the raft, nd the necessity that its lowest step reaches well below the bottom of he raft, and that step to be fitted with a weight e.g., lead thread, to keep submerged. Without this weight, the rope ladder tends to surface and loat to the side, which makes it very difficult to get a knee or a foot into he lowest step. This net-rope ladder is further described in the ollowing requirements.

5. Ventilation openings

As with the circular entrance opening and sleeve type closing device, proved necessary to fit ventilating openings on the canopy, but it hould be possible to keep these openings fully or partly closed.

These trials have shown that some details of the inflatable liferafts nd their equipment can be developed further. The resulting odifications should however, not be considered as final solutions. urther testing and experience will no doubt bring forward further evelopments. On the other hand it is believed that the results from the ests here mentioned are sufficiently important to be put into use as soon s possible.

XTRACTS FROM THE ICELANDIC REQUIREMENTS FOR NFLATABLE LIFERAFTS . 1981.

ONLY APPLICABLE TO ICELANDIC VESSELS SUPPLIED FOR INFORMATION ONLY

1.06 An inflatable liferaft intended for 9 persons or more, shall be rovided with two entrance openings. An approved type of rope-ladder t each entrance shall be fitted and fastened to the centre floor of the raft nd lead out through the openings and for at least 50 centimetres (20 ns.) below the bottom of the raft.

The ladder shall be designed for easy climbing and a good handgrip.

The three lowest steps on the ladder shall be of stiff, rapid sinking, eavy material. At least two further steps at the outside of the buoyancy ompartment and reaching inside the liferaft, shall be of stiff material.

Removable side fastenings for the ladder shall be placed on the upper) buoyancy compartment on each side of the ladder to keep it pen (stretched) when in use.

1.07. For added stability, all inflatable liferafts intended for 6 persons r more shall be fitted with at least five separate sea-ballast pockets of

V-type with lead weight or similar at the lower edge to keep them oper The sea-ballast pockets are to be positioned on the underside of th bottom along its edges except where the inflation bottle is placed. Se Fig. 2. To prevent air being trapped underneath the raft there ar to be openings between the sea-ballast pockets to ensure that air ca escape freely.

1.08. An inflatable liferaft shall be carried in an approved containe constructed to withstand hard wear under conditions met with at sea

The liferaft in its container shall be inherently buoyant so that it wi float free if fitted with a hydrostatic release equipment.

All containers for inflatable liferafts shall be insulated to preven forming of moisture or white frost on the inside of the containers.

1.09. There must be a reasonable margin of buoyancy, if the liferaft i damaged or partially fails to inflate. To achieve this, the buoyanc chambers shall be so arranged that they are divided into an even numbe of separate compartments. Half of these compartments must be capabl of supporting out of the water the number of persons which the liferaf is permitted to accommodate. A reinforcing strip of strong material (V strip) shall be glued over all joints between the buoyancy compartment both inside and out to safeguard against failure.

2.05. All entrance openings shall be of circular form. Rafts for 9 persons or more shall have two opposite entrance openings, 80 centimetres (32 ins.) in diameter. Rafts for less than 9 persons may have one entrance opening, 70 centimetres (28 ins.) in diameter. The canopy material shall extend down to the (upper) buoyancy compartment and be joined (glued) to it all the way round the liferaft.

The entrance opening shall be provided with a sleeve for closing, 65 centimetres (26 ins.) long, and fitted with an approved door closing by a string of similar material, operàted from inside the liferaft.

The above mentioned closing device shall be on all new inflatable liferafts, except when the Director of Shipping has approved another type proposed by the manufacturer.

2.07. On polygonal rafts the sea anchor patch shall be placed on a corner of the raft or secured by a bridle, so either a corner of the raft is turned into the wind or if a straight part is into the wind, the raft is prevented from yawing.

Th entrance openings should always be at right angles to the sea anchor lines.

The sea anchor shall be made of strong approved material, and a spare sea anchor is to be provided of the same specification. A net is to be placed between the shroud lines, to prevent them from fouling (see fig. 3.) The shroud lines shall be securely fastened (sewn) to the sea anchor at the opening where it should be reinforced, e.g. with a protecting cloth sewn over the shroud lines. The sea anchor lines shall be of plaited nylon or equivalent material, and at least 35 metres (114 ft.) long. The sea anchor shall be made of synthetic cloth with a tensile strength of at least 140kp / 5cm, in both directions, and with a tear strength of the material in both directions at least 8kp. On a sea anchor, on a liferaft intended for 9 persons or more, the bigger opening shall be 60cm (24 ins.) in diameter and the smaller opening 18cm (6 ins.) in diameter and its length shall be 120cm. (48 ins.). On a sea anchor for a

A sketch of an inflatable liferaft showing the suggested circular shape of an entrance opening and water ballast pockets.
Electric light
Retro-reflective tape
Life line
Circular entrance opening provided with a sleeve-type closing device
Stretchers for the rope ladder
80
Rope ladder
22
Line to the sea anchor
Swivel
The "lazy" line to the sea anchor
50 (min)
44
Righting strap
Inflated step
Openings at top of water ballast pockets
Opening between water ballast pockets to allow air to escape
Water ballast pockets
SIGLINGAMÁLASTOFNUN RÍKISINS
(DIRECTORATE OF SHIPPING)
HAMARSHÚS TRYGGVAGATA, P. O. BOX 484, REYKJAVÍK
Metric measurements (cm)

Fig. 1

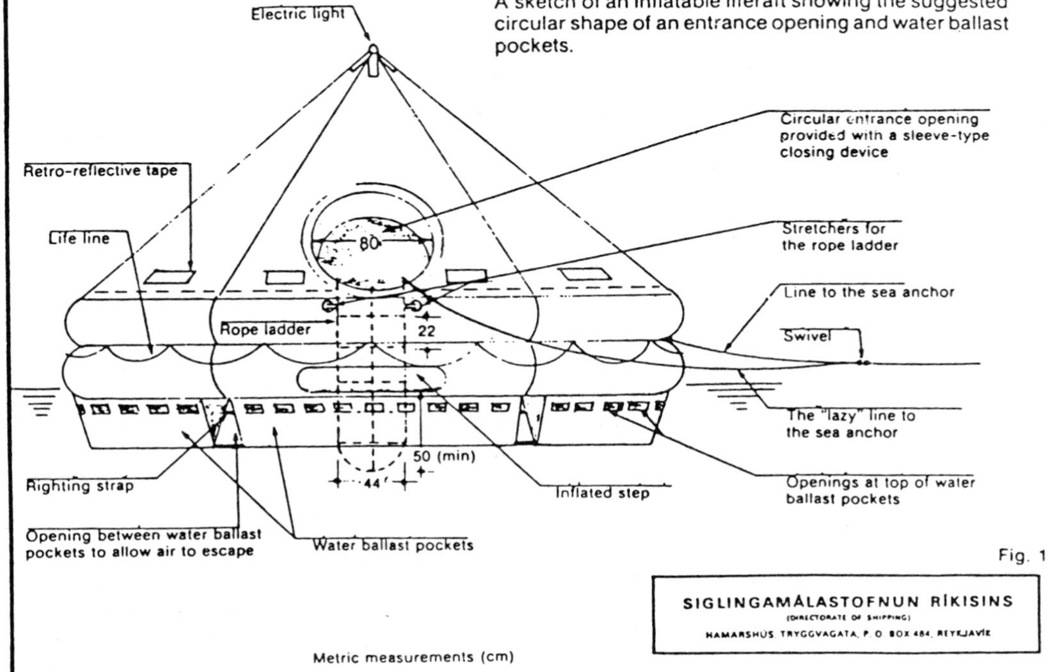

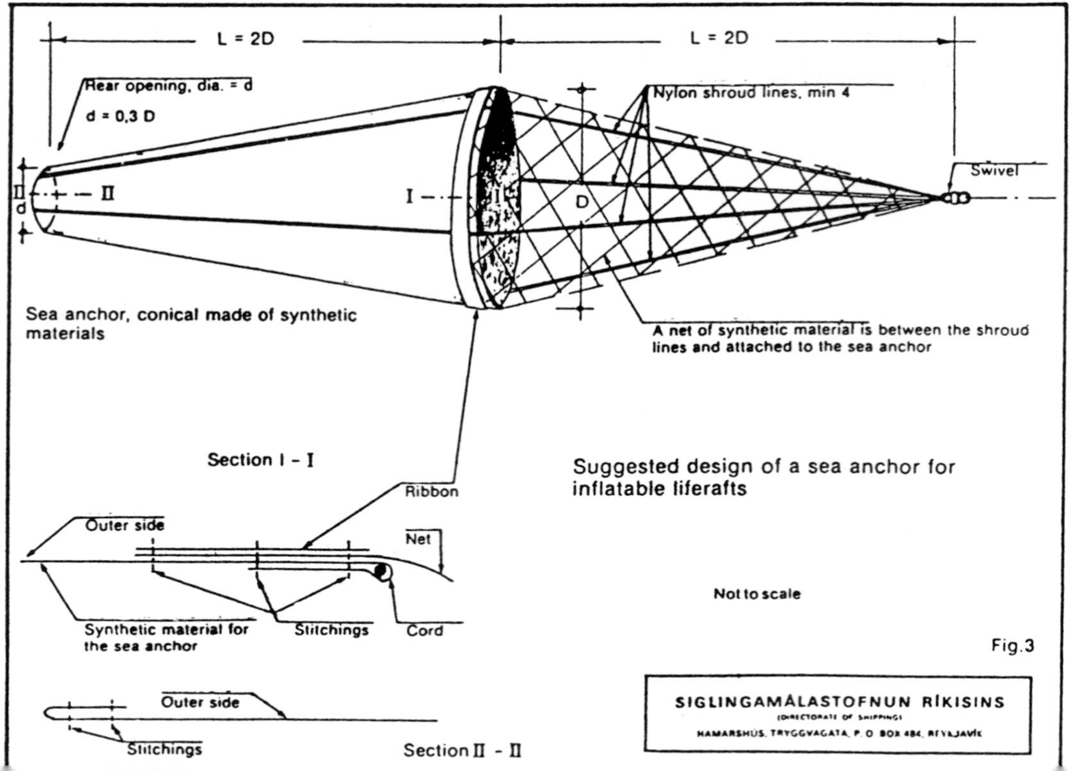

Suggested design of a sea anchor for inflatable liferafts

Fig.3

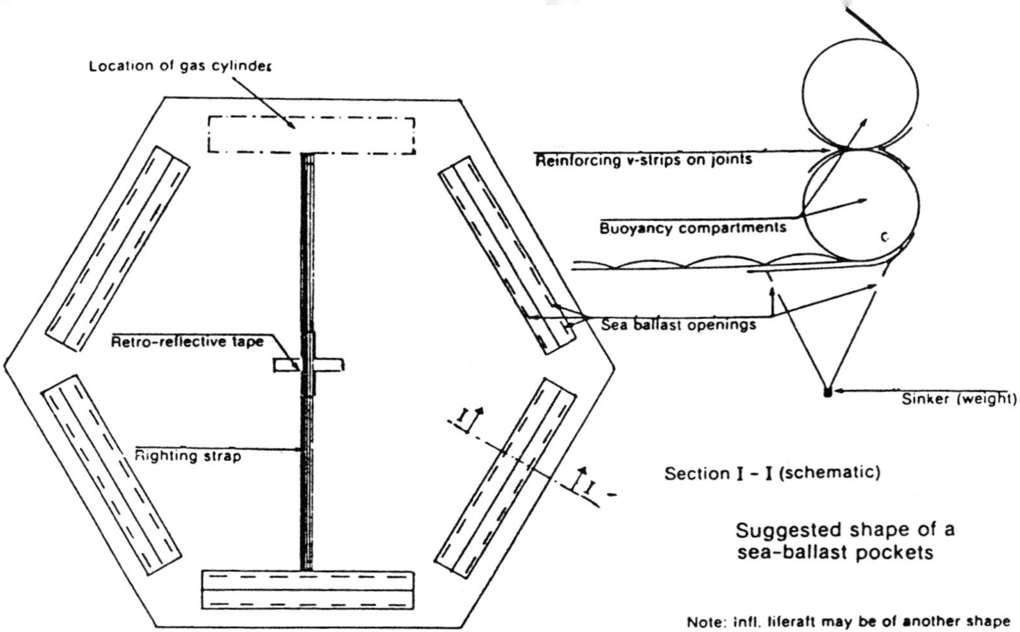

Bottom view of an inflatable liferaft

Not to scale

Section I - I (schematic)

Suggested shape of a sea-ballast pockets

Note: infl. liferaft may be of another shape than shown (hexagon).

Fig. 2

SIGLINGAMÁLASTOFNUN RÍKISINS
(DIRECTORATE OF SHIPPING)
HAMARSHÚS, TRYGGVAGATA, P. O BOX 484, REYKJAVÍK

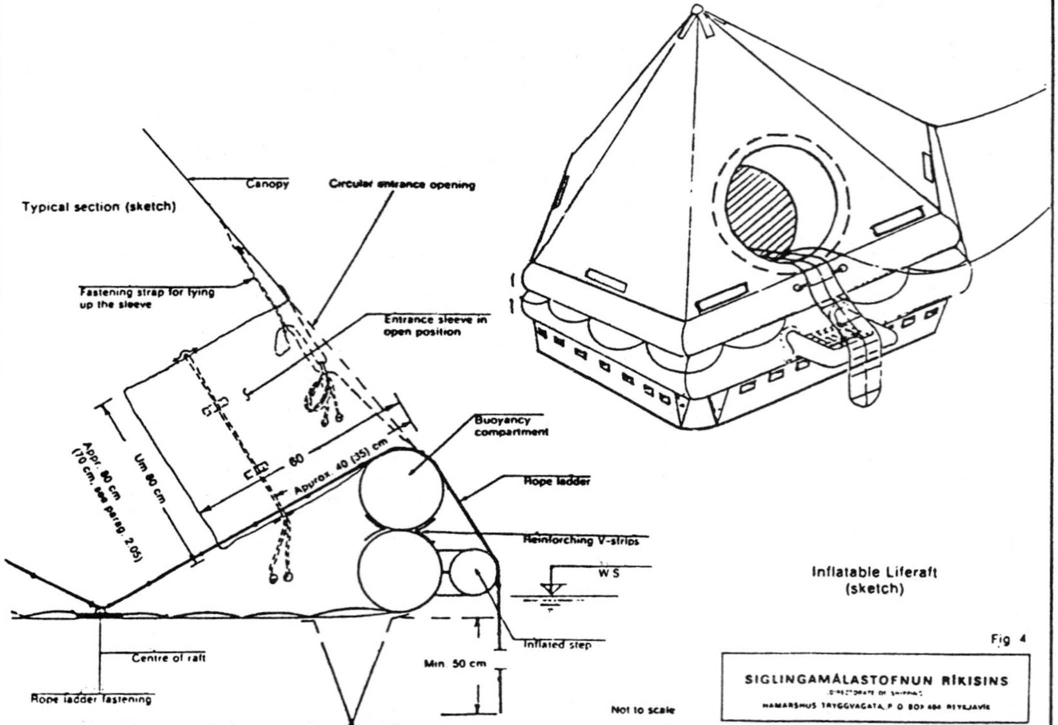
Typical section (sketch)
Canopy
Circular entrance opening
Fastening strap for tying up the sleeve
Entrance sleeve in open position
Buoyancy compartment
60
Approx. 40 (35) cm
Um 80 cm
Appr. 80 cm (70 cm, see parag. 2.05)
Rope ladder
Reinforching V-strips
W S
Inflated step
Min. 50 cm
Centre of raft
Rope ladder fastening
Not to scale
Inflatable Liferaft (sketch)
SIGLINGAMÁLASTOFNUN RÍKISINS
HAMARSHUS TRYGGVAGATA, P O BOX 484 REYKJAVIK

Fig 4

liferaft intended for less than 9 persons, the bigger opening shall be 40cm. (16 ins.) in diameter, the smaller opening shall be 12cm. (5 ins.) in diameter, and the length 80cm. (32 ins.).

There shall be at least 4 shroud lines made of braided nylon of the same length as the sea anchor.

The sea anchor shall be fastened to the liferaft at 90 degrees to the entrance openings. A catching line (lazy line) connected to the line to the sea anchor shall be at hand at an entrance opening. The tensile strength of the line to the sea anchor for a liferaft intended for 9 persons or more, shall be 1,200kp, but for a liferaft of less than 9 persons the tensile strength shall be 800kp.

2.08. A sea-water pumping device shall be fitted in every liferaft and it shall be equally operational with the canopy open or closed and capable of pumping empty a full raft within 15 minutes.

There shall be two ventilating openings on every raft, not less than 50mm (20 ins.) in diameter, so designed that they will stay open, but the opening can be regulated.

NEW ICELANDIC RELEASE DEVICE FOR INFLATABLE LIFERAFTS

This device is designed by Mr. Sigmund Johannsson. While still in the machine shop, this release device had been inspected in February 1981 by an inspector from the Icelandic Directorate of Shipping.

In principle the idea behind the design of this release device for inflatable liferafts is to facilitate the liferaft launching in emergency situations. Thus the device makes it possible to release the liferaft manually either from the wheelhouse or from the boat deck and automatically via an automatic activator. The automatic action is initiated by a water quick-dissolving pellet, which releases compressed gas from a pressure-bottle, and thus disengages mechanically the launching device.

The function of the release device is such that when activated either manually or automatically the inflatable liferaft is released.

The release device may also be connected to a special launching arm in which the inflatable liferaft rests. In this case a balloon behind the launching arm inflates and moves the arm outwards. It may then be adjusted to conditions in which position of the arm, the liferaft starts to inflate.

The control box of the release device, with its pipes and wires for the remote control, manual or automatic, is located in the wheel house.

IMO. Requirements for Rigid Liferafts, on ships built after 1st. July 1986

1. Rigid liferafts shall comply with the general requirements for all liferafts and, in addition, shall comply with the following requirements.

2. **Construction of rigid liferafts**

(a) The buoyancy of the liferaft shall be provided by approved inherently buoyant material placed as near as possible to the periphery of the liferaft. The buoyant material shall be fire retardant or be protected by fire retardant covering.
(b) The floor of the liferaft shall prevent the ingress of water and shall effectively support the occupants out of the water and insulate them from the cold.

3. **Carrying capacity of rigid liferafts**

The number of persons which a liferaft shall be permitted to accommodate shall be equal to the lesser of:
(i) The greatest whole number obtained by dividing by 0.096 the volume measured in M_3 of the buoyancy material multiplied by a factor of 1 minus the specific gravity of that material; or
(ii) the greatest whole number obtained by dividing by 0.372 the horizontal cross-section area of the floor of the liferaft measured in M_2; or
(iii) the number of persons having an average mass of 75kg. (165lbs), all wearing lifejackets, that can be seated with sufficient comfort and headroom without interfering with the operation of any of the liferaft's equipment.

4. **Access into rigid liferafts**

(a) At least one entrance shall be fitted with a rigid boarding ramp to enable persons to board the liferaft from the sea. In the case of a davit-launched liferaft, having more than one entrance the boarding ramp shall be fitted at the entrance opposite to the bowsing and embarkation facilities.
(b) Entrances not provided with a boarding ramp shall have a boarding ladder, the lowest step of which shall be situated not less than 0.4m (16ins) below the liferaft's light waterline.
(c) There shall be means inside the liferaft to assist persons to pull themselves aboard the liferaft from the ladder.

5. **Stability of rigid liferafts**

(a) Unless the liferaft is capable of operating safely whichever way up it is floating, its strength and stability shall be such that it is either self-righting or can be easily righted in a sea-way and in calm water by one person.
(b) The stability of a liferaft when loaded with its full complement of persons and equipment shall be such that it can be towed at speeds of up to 3 knots in calm water.

6. **Rigid liferaft fittings**

(a) The liferaft shall be fitted with an efficient painter. The strength of the painter system, including its means of attachment to the liferaft, except the weak link required to be fitted, shall be not less than 10.0kN for liferafts permitted to accommodate nine persons or more, and not less than 7.5kN for any other liferaft.

(b) A manually controlled lamp visible on a dark night with a clear atmosphere at a distance of at least two miles for a period of not less than 12 hours shall be fitted to the top of the liferaft canopy. If the light is a flashing light it shall flash at a rate of not less than 50 flashes per minute for the first two hours of operation of the 12 hour operating period. The lamp shall be powered by a sea- activated cell and shall light automatically when the liferaft canopy is set in place. The cell shall be of a type that does not deteriorate due to damp or humidity in the stowed liferaft.

(c) A manually controlled lamp shall be fitted inside the liferaft, capable of continuous operation for a period of at least 12 hours. it shall light automatically when the canopy is set in place and be of sufficient intensity to enable reading of survival and equipment instructions.

7. **Markings on rigid liferafts**

The liferaft shall be marked with:

(a) name and port of registry of the ship to which it belongs;
(b) maker's name or trade mark;
(c) its serial number;
(d) name of approving authority;
(e) number of persons it is permitted to accommodate over each entrance in characters of a colour contrasting with that of the liferaft not less than 100mm (4in) in height;
(f) SOLAS;
(g) type of emergency pack enclosed;
(h) length of painter;
(i) maximum permitted height of stowage above waterline (drop test height);
(j) launching instructions.

8. **Davit launched rigid liferafts**

In addition to the above requirements, a rigid liferaft for use with an approved launching appliance shall, when suspended from its lifting hook or bridle, withstand a load of 4 times the mass of its full complement of persons and equipment.

The "Floating Igloo", a Norwegian invention, can be manufactured as either a square or circular raft. Expanded plastic foam is used for the buoyancy, which when covered with a strong coated nylon cloth gives the liferaft exceptional strength combined with elasticity.

To ensure that the raft is ready for immediate use on being launched overside, it is identical on both top and bottom, each side being provided with a self erecting canopy, operated by a tug on the painter or by separate release lines. The equipment, stowed in a drum embedded in the floor of the raft, is equally obtainable from either side. Th permanent buoyancy cannot be affected by puncturing or leakage. I never needs to be righted and the lower canopy filled with water, give the raft tremendous stability.

Surveys , inspections and repairs can all be carried out on board th ship. This is a tangible liferaft with which the crew can become familia whilst carrying out lifeboat drills. It can also be placed in the water fo drill purposes and then restowed by the crew, when it is immediatel available for any emergency, provided always that the batteries hav been removed before the drill and are replaced afterwards.

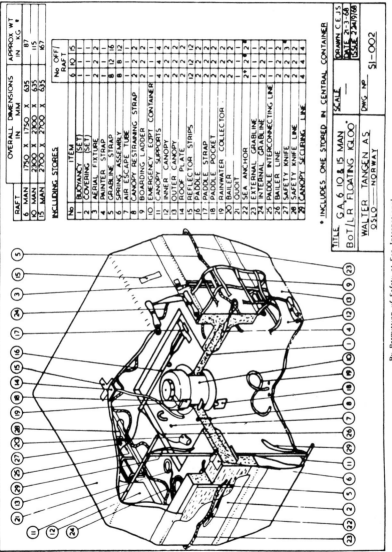

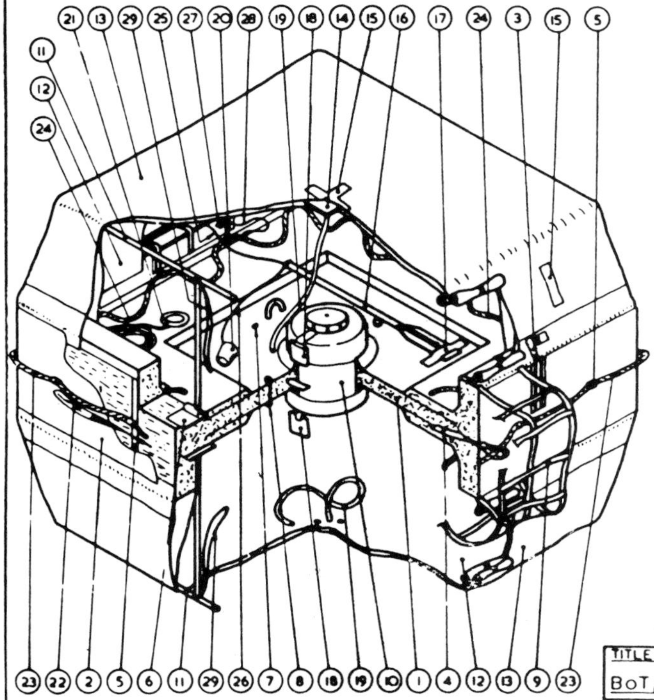

RAFT	OVERALL DIMENSIONS IN MM	APPROX WT IN KG *
6 MAN	1750 X 1750 X 635	87
10 MAN	2300 X 2300 X 635	115
15 MAN	2700 X 2700 X 635	167

INCLUDING STORES

No	ITEM	No OFF/ RAFT 6	10	15
1	BUOYANCY (SET)	1	1	1
2	COVERING (SET)	1	1	1
3	AERIAL FIXTURE	2	2	2
4	PAINTER STRAP	1	1	1
5	GRABLINE STRAP	8	12	16
6	SPRING ASSEMBLY	8	8	12
7	AIR ESCAPE TUBE	1	1	1
8	CANOPY RESTRAINING STRAP	2	2	2
9	BOARDING LADDER	2	2	2
10	EMERGENCY EQPT CONTAINER	1	1	1
11	CANOPY SUPPORTS	4	4	4
12	INNER CANOPY	2	2	2
13	OUTER CANOPY	2	2	2
14	ROOF PLATE	2	2	2
15	REFLECTOR STRIPS	12	12	12
16	PADDLE	2	2	2
17	PADDLE STRAP	2	2	2
18	PADDLE POCKET	2	2	2
19	RAINWATER COLLECTOR	2	2	2
20	BAILER	2	2	2
21	QUOIT	1	1	1
22	SEA ANCHOR	2*	2*	2*
23	EXTERNAL GRABLINE	1	1	1
24	INTERNAL GRABLINE	2	2	2
25	PADDLE INTERCONNECTING LINE	1	1	1
26	BAILER LINE	2	2	2
27	SAFETY KNIFE	2	2	3*
28	SAFETY KNIFE LINE	2	2	2
29	CANOPY SECURING LINE	4	4	4

* INCLUDES ONE STORED IN CENTRAL CONTAINER

TITLE	SCALE	
GA 6 10 & 15 MAN BoT/LR 'FLOATING IGLOO'	—	DRAWN C E J S DATE 21-3-68 ISSUE 2 24/9/68
WALTER TANGEN AS OSLO — NORWAY	DWG No SI-002	

By Permission of Safety at Sea International

THE FLOATING IGLOO

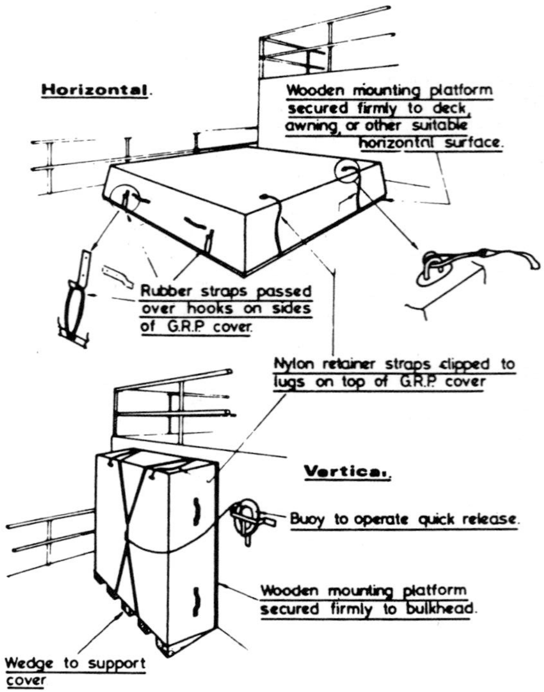
Horizontal.
Wooden mounting platform secured firmly to deck, awning, or other suitable horizontal surface.
Rubber straps passed over hooks on sides of G.R.P. cover.
Nylon retainer straps clipped to lugs on top of G.R.P. cover
Vertical.
Buoy to operate quick release.
Wooden mounting platform secured firmly to bulkhead.
Wedge to support cover

STOWAGE SUGGESTIONS.

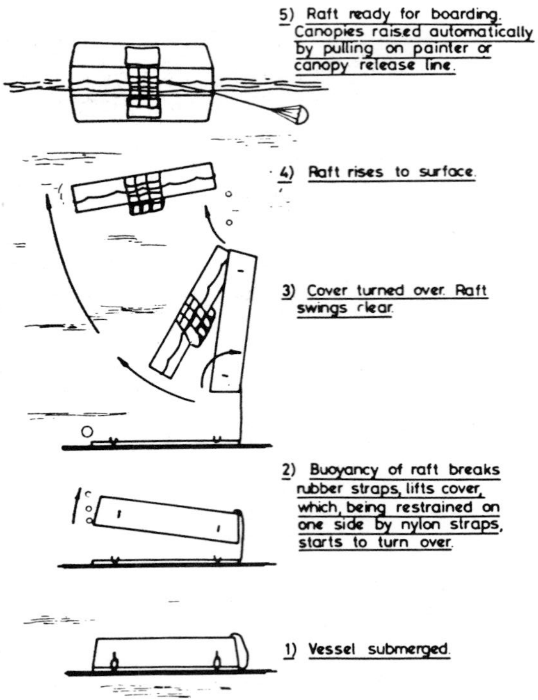
5) Raft ready for boarding. Canopies raised automatically by pulling on painter or canopy release line.
4) Raft rises to surface.
3) Cover turned over. Raft swings clear.
2) Buoyancy of raft breaks rubber straps, lifts cover, which, being restrained on one side by nylon straps, starts to turn over.
1) Vessel submerged.

AUTOMATIC RELEASE SYSTEM.

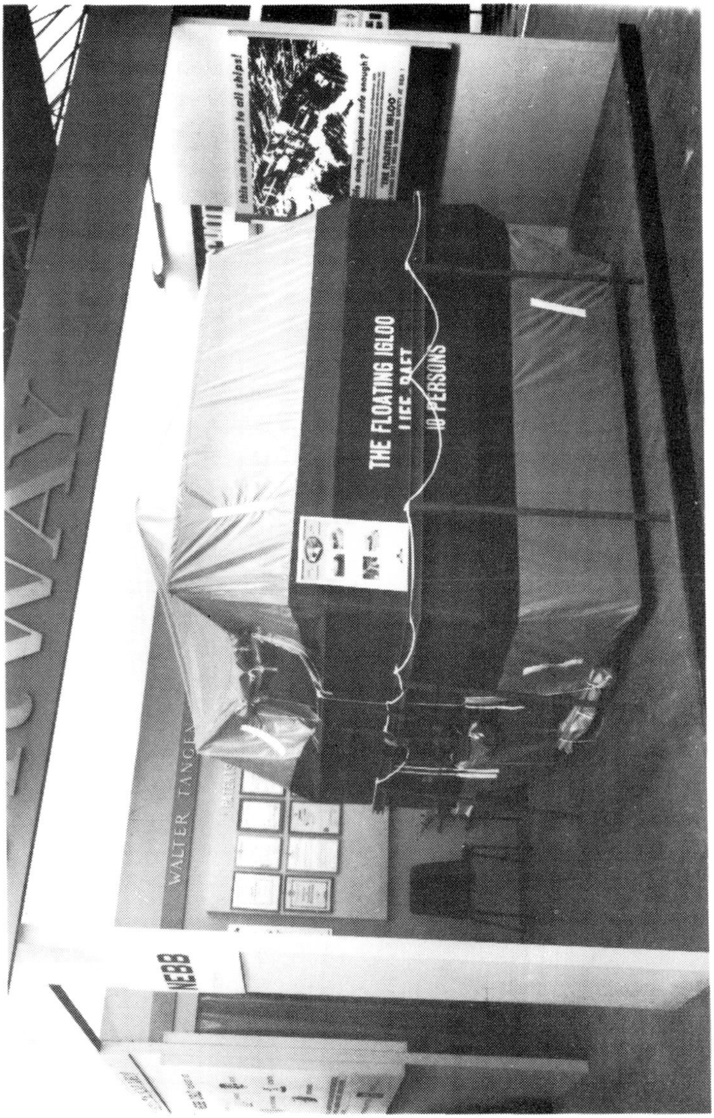
THE FLOATING IGLOO

The "Floating Igloo" afloat with survivors.

BUOYANT APPARATUS. SOLAS 1974 REGULATIONS

Buoyant apparatus is required to be carried on passenger vessels constructed prior to 1st. July 1986, unless they already comply with the 1986 Regulations. Buoyant apparatus is also required to be carried on ferries and similar vessels.

Buoyant apparatus is defined as flotation equipment (other than lifebuoys and lifejackets) designed to support persons who are in the water. It must retain its shape when on board ship and when in the water and must not require adjustment prior to use. Retro-reflective tapes are to be attached.

On Foreign going passenger steamers engaged on long international voyages, it must be capable of withstanding a drop test of not less than 60 feet (18m.)

Made of either glass reinforced plastic containing expanded plastic or a hardwood framework containing air cases, there are to be no projections on the outside and approved grablines are to be fitted all round the apparatus and each loop is required to have a cork or light wood float. The attachments are required to be strong enough to permit the apparatus being lifted by the grablines.

Buoyant apparatus must not exceed 400 lbs. (180kg.) in weight unless a means are provided to enable it to be launched without lifting by hand.

A side or end exposed to the view of the passengers must be marked as follows:-

"Buoyant apparatus", "Number of persons it is permitted to support",

"Height from which it may be dropped", "The surveyor's initials",

"The date of manufacture" and "The Administration stamp".

A painter is to be provided of 2½ inch (20mm) approved cordage, strong enough to lower the apparatus into the water and long enough to reach from the deck at the ship's lightest sea-going draught, to the water, plus 6 feet (1.88m.)

Buoyant apparatus is to be distributed in suitable positions about the ship. It is not to be stowed on the top of deck houses or in isolated positions or beneath any deck but must be capable of being readily launched or float-free. Lashings are to be of a type that can be easily slipped. It may be stowed in tiers of not more than 5 units.

Mashford Buoyant Apparatus

THE SALTER GLASS FIBRE BUOYANT APPARATUS

This 20 person Buoyant Apparatus is approved by the British Department of Trade for use on all classes of vessels, where this type of safety equipment is specified. Home Trade and Foreign Going ships, including passenger craft, ferries, fishing vessels and pleasure boats.

Only best quality glass fibre and resins are used in its manufacture. Each unit is conveniently moulded to stack in tiers on the boat deck, ready for immediate use.

A strong grabline is fastened completely around the unit, and a point is provided for the attachment of a painter at one end, of an approved length in accordance with the statutory requirements.

The glass fibre casing of the apparatus is foam filled and is extremely strong and durable, needing the minimum of maintenance.

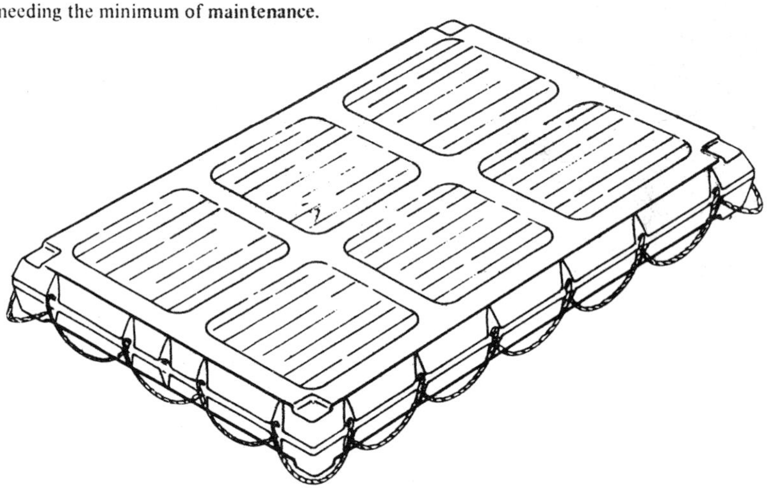

Approved standard model:

Length approx. 6ft.
Depth approx. 7½in.
Standard colour ORANGE

Width approx. 4ft.
Weight approx. 90lbs.

CHAPTER V. ANCILLARY EQUIPMENT

IMO REGULATIONS FOR SURVIVAL CRAFT RADIO INSTALLATIONS ON VESSELS BUILT AFTER 1st JULY 1986

Radiotelegraph Installation for fitting in motor lifeboats

1. The radiotelegraph installation required to be installed in motor lifeboats on passenger ships and those installed in motor lifeboats on cargo ships, shall include a transmitter, a receiver and a source of energy. It shall be so designed that it can be used in an emergency by an unskilled person.
2. The transmitter shall be capable of transmitting on the radiotelegraph distress frequency.
3. In addition to a key for manual transmissions. The transmitter shall be fitted with an automatic keying device for the transmission of the radiotelegraph alarm and distress signals.
4. On the radio distress frequency the transmitter shall have a normal range of 25 miles using the fixed antenna.
5. The receiver shall be capable of receiving the radiotelegraph distress frequency.
6. The source of energy shall consist of an accumulator battery with sufficient capacity to supply the transmitter for four hours continuously under normal working conditions. If the battery is of a type that requires charging, means shall be available for charging it from the ship's power supply. In addition there shall be a means of charging it after the lifeboat has been launched.
7. When the power for the radiotelegraph installations and the searchlight are drawn from the same battery, it shall have sufficient capacity to provide for the additional load of the searchlight.
8. A fixed-type antenna will be provided together with means for supporting it at the maximum practicable height. In addition an antenna supported by a kite or balloon shall be provided if practicable.
9. At sea a radio officer shall at weekly intervals test the transmitter using a suitable artificial antenna, and shall bring the battery up to full charge if it is of a type which requires charging.

Portable radio apparatus for survival craft

1. The apparatus shall include a transmitter, a receiver, an antenna and a source of energy. It shall be so designed that it can be used in an emergency by an unskilled person.
2. The apparatus shall be readily portable, watertight, capable of floating in sea-water and capable of being dropped into the sea without damage. New equipment shall be as light-weight and compact as practicable and shall preferably be capable of use in both lifeboats and liferafts.
3. The transmitter shall be capable of transmitting on the radiotelegraph distress frequency. However. the Administration may permit the transmitter to be capable of transmitting on the radiotelephone distress frequency, as an alternative or in addition to transmission on the radiotelegraph frequency assigned for survival craft.
4. In addition to a key for manual transmissions, the transmitter shall be fitted with an automatic keying device for the transmission of

the radiotelegraph alarm and distress signals. If the transmitter is capable of transmitting on the radiotelephone distress frequency, it shall be fitted with an automatic device, for transmitting the radiotelephone alarm signal.

5. The receiver shall be capable of receiving the radiotelegraph distress frequency. If the transmitter is capable of transmitting on the radiotelephone distress frequency the receiver shall also be capable of receiving that frequency.

6. The antenna shall be either self-supporting or capable of being supported by the mast of a lifeboat at the maximum practicable height. In addition it is desirable that an antenna supported by a kite or balloon shall be provided if practicable.

7. The transmitter shall supply an adequate radio frequency power to the antenna and shall preferably derive its supply from a hand generator. If operated froma battery, the battery shall comply with conditions laid down by the Administration to ensure that it is of a durable type and is of adequate capacity.

8. At sea a radio officer or a radiotelephone operator, as appropriate, shall at weekly intervals test the transmitter, using a suitable artificial antenna and shall bring the battery up to full charge if it is of a type which requires recharging.

9. For the purpose of this Regulation, new equipment means equipment supplied to a ship after the date of entry into force of the 1974 Convention for the Safety of life at sea.

Survival craft emergency position—indicating radio beacons (EPIRB's).

1. Survival craft emergency position-indicating radio beacons required to be carried in survival craft shall provide transmissions to enable aircraft to locate the survival craft and may also provide transmissions for alerting purposes.
2. Survival craft EPIRB's shall, at least, be capable of transmitting alternately or simultaneously signals on the frequencies 121.5MHz and 243.0MHz.

**Frequencies will be changed when satellite operations become available.*

3. Survival craft EPIRB's shall:

(a) be of a highly visible colour, designed so that they can be used by an unskilled person and constructed so that they may be easily tested and maintained. Batteries shall not require replacement at intervals of less than 12 months, taking into account testing arrangements;

(b) be watertight, capable of floating and being dropped into the water without damage from a height of at least 20m. (65ft).

(c) be capable only of manual activation and de-activation;

(d) be portable, lightweight and compact;

(e) be provided with an indication that signals are being emitted;

(f) derive their energy supply from a battery forming an integral part of the device and having sufficient capacity to operate the apparatus for a period of 48 hours. The transmissions may be intermittent. Determination of the duty cycle should take into account the probability of homing being properly carried out, the need to avoid congestion on the frequencies and the need to comply with the requirements of the International Civil Aviation Organization (ICAO); and

(g) be tested and, if necessary, have their source of energy replaced at intervals not exceeding 12 months.

Periodic inspection and testing of EPIRB's

Emergency position-indicating radio beacons shall at intervals not exceeding 12 months be inspected and, if necessary, have their source of energy replaced. However in cases where it appears proper and reasonable, the Administration may extend this period to 17 months.

Two-way radiotelephone apparatus for survival craft

1. The apparatus shall be so designed that it can be used in an emergency by an unskilled person.
2. The apparatus shall be portable and capable of being used for on-board communications.
3. The apparatus shall conform to the requirements laid down in the relevant Radio Regulations for equipment used in the maritime mobile service for on-board communications and shall be capable of operation on those channels specified by the Radio Regulations and as required by the Administration. If the apparatus is operating in the VHF band, precautions shall be taken to prevent the inadvertant selection of VHF Channel 16 on equipment capable of being operated on that frequency.
4. The apparatus shall be operated from a battery of adequate capacity to ensure 4 hours operation with a duty cycle of 1 : 9.
5. While at sea, the equipment shall be maintained in satisfactory condition, and, whenever necessary, the battery shall be brought to the fully charged condition or replaced.

The Marconi Salvare 4 motor lifeboat radio is a complete radio tation for installation in ships' lifeboats and is designed to meet the equirements for use in compulsory radio equipped motor lifeboats.

The equipment is housed in two separate cabinets. One cabinet ontains the 500kHz and 2182kHz transmitter with an integral adiotelephone alarm, and the antenna switching unit, the other ontains the 8364kHz transmitter, the receiver, automatic keying levice, a charging unit and stowage for the headset and morse key. Vhen not in use the cabinets are fitted with quick-release detachable ront covers to prevent the ingress of water.

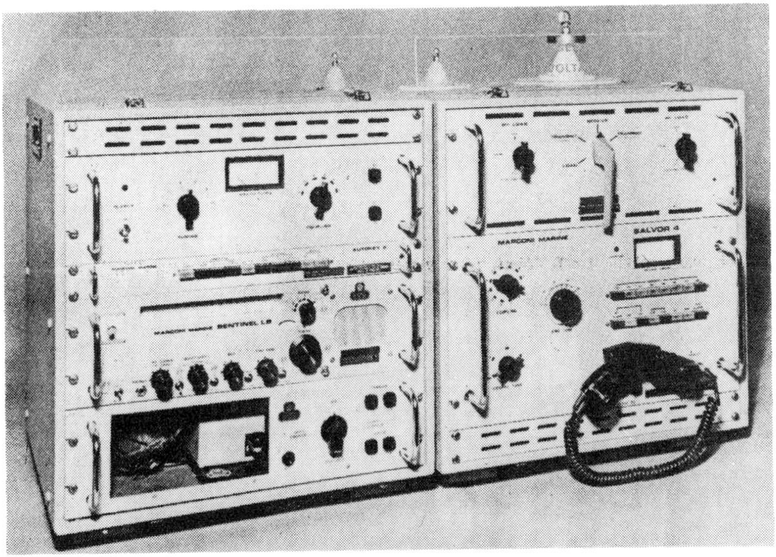

The Marconi Salvare 4 motor lifeboat radio

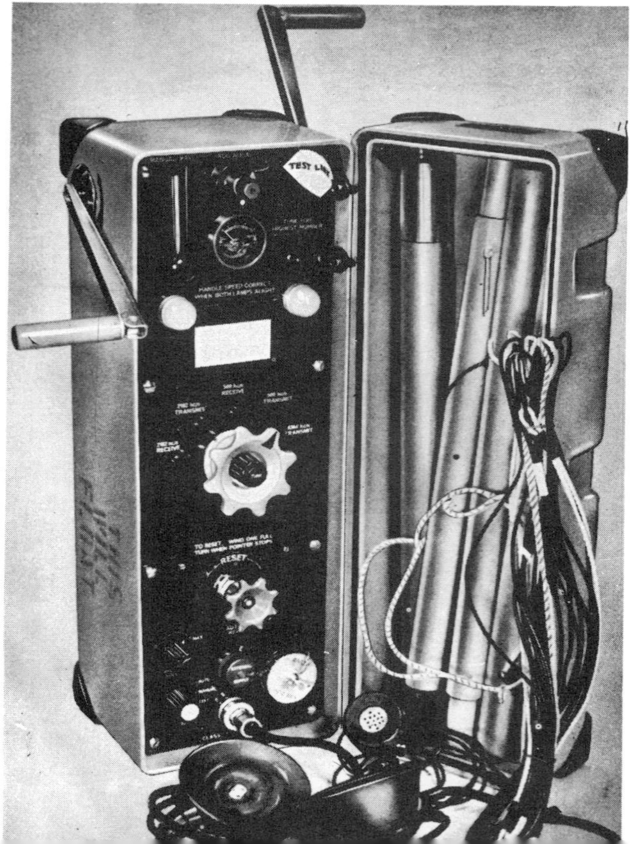

Marconi Marine "Survivor 3" survival craft emergency radio equipment

Contained in a compact vivid yellow case, will float and is designed to withstand drops into the water of 20m. (65 ft.)

INSTALLATION IN A LIFEBOAT

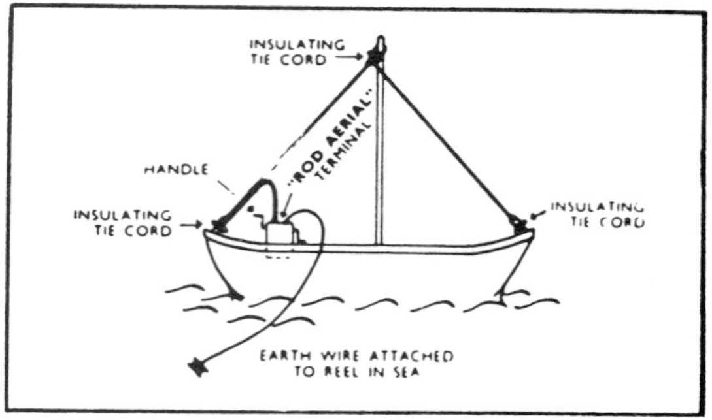

1. Remove short wire aerial from its bag within lid bag.
2. Lash set near stern using webbing straps.
3. Rig aerial using lifeboat mast (see diagram).
4. Ensure wire aerial does not touch mast, sails, etc.
5. Connect wire from aerial to "Rod Aerial" terminal on set.
6. Remove earth reel, unwind wire and drop reel into sea.
7. Insert handle(s) into socket(s) on box side(s).

SURVIVAL CRAFT EMERGENCY RADIO EQUIPMENT by The Marconi International Marine Co. Ltd.

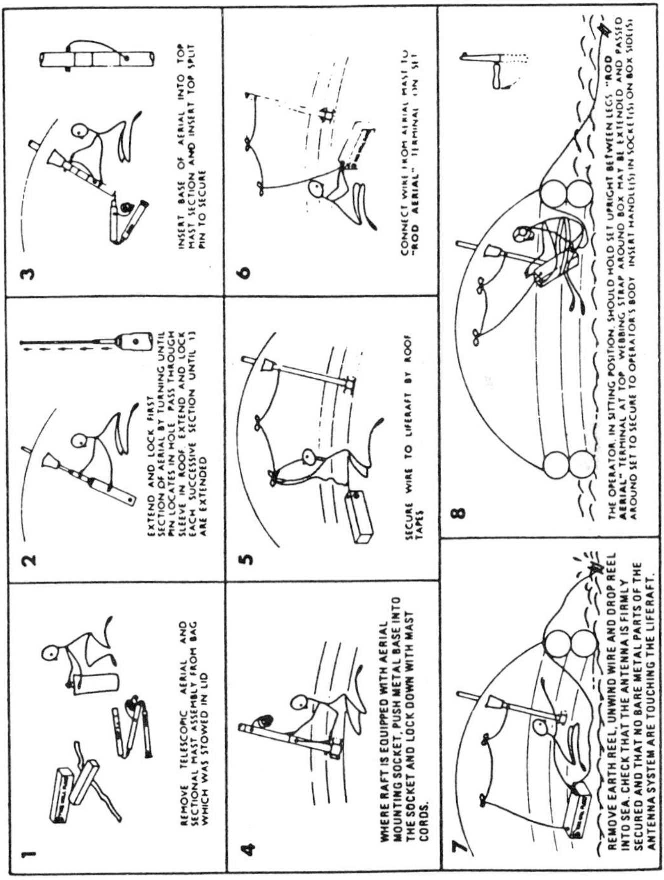
1
REMOVE TELESCOPIC AERIAL AND SECTIONAL MAST ASSEMBLY FROM BAG WHICH WAS STOWED IN LID
2
EXTEND AND LOCK FIRST SECTION OF AERIAL BY TURNING UNTIL PIN LOCATES IN HOLE. PASS THROUGH SLEEVE IN ROOF. EXTEND AND LOCK EACH SUCCESSIVE SECTION UNTIL 13 ARE EXTENDED
3
INSERT BASE OF AERIAL INTO TOP MAST SECTION AND INSERT TOP SPLIT PIN TO SECURE
4
WHERE RAFT IS EQUIPPED WITH AERIAL MOUNTING SOCKET, PUSH METAL BASE INTO THE SOCKET AND LOCK DOWN WITH MAST CORDS.
5
SECURE WIRE TO LIFERAFT BY ROOF TAPES
6
CONNECT WIRE FROM AERIAL MAST TO "ROD AERIAL" TERMINAL ON SET
7
REMOVE EARTH REEL, UNWIND WIRE AND DROP REEL INTO SEA. CHECK THAT THE ANTENNA IS FIRMLY SECURED AND THAT NO BARE METAL PARTS OF THE ANTENNA SYSTEM ARE TOUCHING THE LIFERAFT.
8
THE OPERATOR, IN SITTING POSITION, SHOULD HOLD SET UPRIGHT BETWEEN LEGS "ROD AERIAL" TERMINAL AT TOP. WEBBING STRAP AROUND BOX MAY BE EXTENDED AND PASSED AROUND SET TO SECURE TO OPERATOR'S BODY. INSERT HANDLE(S) IN SOCKET(S) ON BOX SIDE(S)

OPERATING INSTRUCTIONS FOR "SURVIVOR 3 EMERGENCY RADIO

The following instructions will be found on the weatherproo operating instructions card in the lid of the equipment.

1. After installation, turn one or both handles to keep both lights o during sending, receiving and testing. All transmissions can be heard i the earphones.

2. **Automatic distress signal**

a) Turn yellow knob to 500kHz TRANSMIT or 8364kHz transmi as required.

b) Turn blue knob to AUTO KEY and press AUTO KEY STAR button. The automatic distress signal (alarm signal), (S.O.S three times and two long dashes) is sent out automatically, takin about two minutes.
The AUTO KEY light flashes when the AUTO KEY is working

c) Adjust red knob for maximum deflection on meter. The bes adjustment may be obtained when the AUTO KEY light i indicating two long dashes.

d) The AUTO KEY START button must be pressed to start eac automatic transmission.

e) The AUTOMATIC DISTRESS SIGNAL should be sen particularly during silence periods at a quarter to and a quarte past each hour G.M.T.

3. **Manual morse**

a) Set yellow knob to 500kHz TRANSMIT or 8364kH TRANSMIT as required.

b) Set blue knob to MORSE KEY and send message by operatin MORSE KEY.

4. **Receive 500kHz**

Turn yellow knob to 500kHz RECEIVE and set green knob t maximum until reply is received.

5. **Telephone alarm**

a) Turn yellow knob to 2182kHz TRANSMIT.

b) Turn blue knob to 2-TONE ALARM.

c) Adjust red knob for maximum deflection on meter.

d) Alarm signal should be sent for 30 seconds to 60 second particularly during SILENCE PERIODS at the hour and hal hour G.M.T. This should alternate with speech.

6. **Telephony (Speech)**

a) Turn yellow knob to 2182kHz TRANSMIT.

b) Turn blue knob to SPEECH and check that red knob is adjuste for maximum deflection on meter.

c) Transmit distress call as follows:-

MAYDAY. MAYDAY. MAYDAY. THIS IS........
(IDENTIFICATION OF VESSEL OR CRAFT).

Repeat call three times then add your position, nature of distress and assistance required.

. **Receive 2182kHz**

Turn yellow knob to 2182kHz RECEIVE and set green knob to maximum until reply is received.

. **Receive 8364 kHz**

Turn the yellow knob to HF RECEIVE and set the green knob to maximum.
Tune the black knob for 8364kHz reception. Reduce the green knob setting when a reply is received. It may be necessary to listen on another HF RECEIVE frequency after contact has been established. In this case, return the green knob to maximum and tune the black knob until the transmission on the new frequency is received. Reduce the green knob setting if necessary.

ELECTRONICS MARINE LTD.

LTD 40 Emergency Position-Indicating Radio Beacon. (EPIRB)

Frequency:- 121.5 and 243 MHz Battery storage life. 5 years.

Weight 2kgs. (4.5lbs.)

The Electronics Marine LTD 40 when mounted shall be so positioned as to be capable of floating free and is automatically activated in the event of the ship sinking. An optional heater unit for the mounting can be provided for low temperature operation.

TOP VIEW LDT 40 EPIRB
— DOME REMOVED

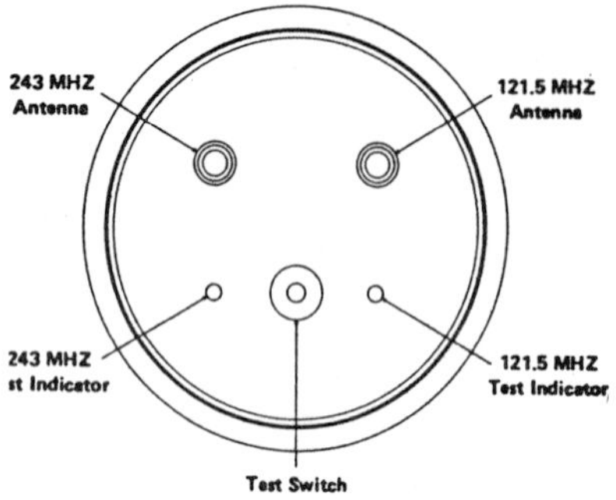

FRONT VIEW LDT 40 IN AUTORELEASE

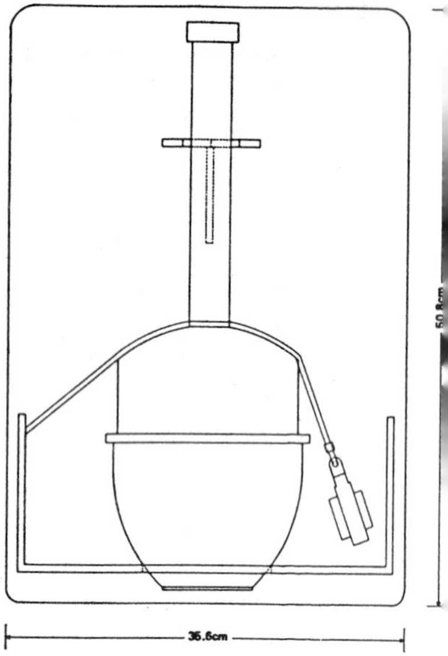

SIDE VIEW
LDT 40 IN AUTORELEASE

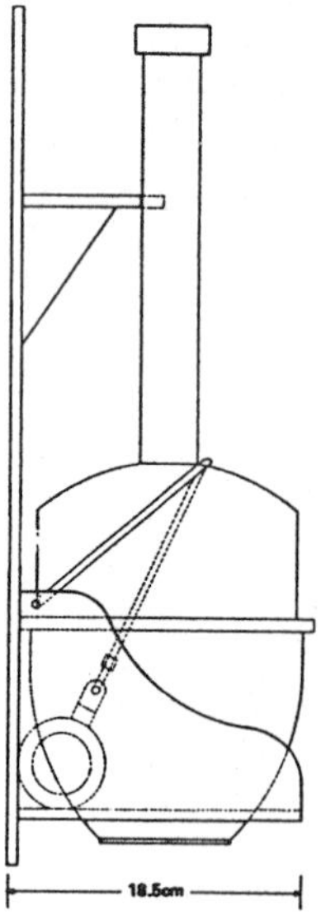

LDT 40
EPIRB

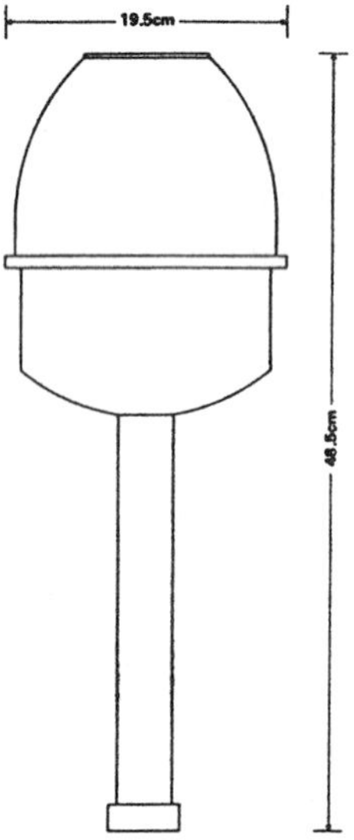

Can be modified for satellite transmissions when the system becomes available.

JOTRON ELEKTRONIC

Manually operated emergency position-indicating radio beacon

Fig. 1. The beacon stored in container for mounting in wheelhouse and lifeboats.

Fig. 2. The beacon stored in container for mounting on board rafts.

Stored in the container the main switch of the beacon is «off». This is made possible by means of a permanent magnet. Consequently no voltage is applied to the electronic unit during transport and storage. The beacon should always be stored in the container.

The mounting bracket may be omitted e.g. onboard rafts or dinghies where the space for installation is limited. The container should be fastened to the bulk-head or the raft-structure.

The TRON -1C is corrosion-proof.

To Start the beacon:

Take the beacon out of the container and pull out the locking pin. TRON-1C then transmits automatically and may be held in the hand or the lanyard tied to the lifeboat and the beacon thrown into the sea.
Change of battery is easily carried out by unscrewing the screwring giving access to the battery.

Can be modified for satellite transmissions when the systems become available.

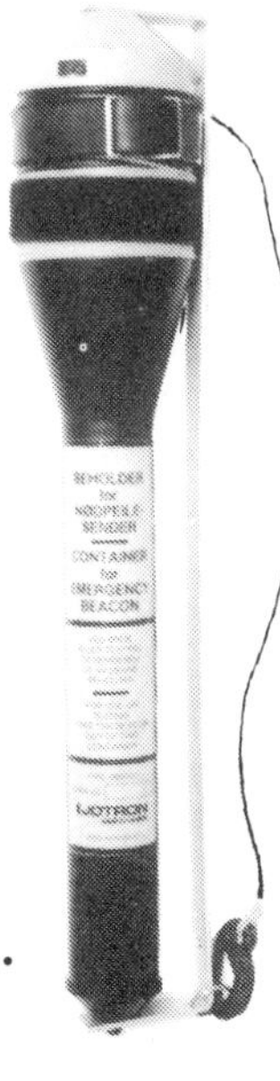

Fig. 1.

Fig. 2.

USE OF EMERGENCY POSITION INDICATING RADIO BEACONS (EPIRBs) AND HAND-HELD EPIRBs ON FREQUENCIES 121·5 MHz AND 243 MHz

Notice to Owners, Masters and Officers of Merchant Ships, Owners and Skippers of Fishing Vessels and Owners of Yachts

This Notice cancels M.863

1. Merchant Shipping Notice No. M.863 was issued in November 197 and set out the Department of Trade's policy towards the carriage o EPIRBs and hand-held EPIRBs designed to operate on the frequencie 121·5 MHz and 243 MHz, which are used primarily for aeronautica purposes and which are an integral part of the aeronautical emergenc system. The Department has reviewed this policy in the light of experienc and information gained over the past two and a half years, and the polic contained in this present notice supersedes that contained in M Notice No 863.

USE OF EPIRBs AND HAND-HELD EPIRBs AS HOMING AIDS

2. Small, lightweight EPIRBs and hand-held EPIRBs which operate o these frequencies and which can be carried on board ships and surviva craft can be useful aids to the maritime Search and Rescue (SAR) services primarily to assist SAR aircraft to locate units or persons in distress. Man of the UK SAR aircraft—and many of those operated by other nationa SAR services—are fitted with equipment which can identify signals fron these beacons at distances of up to 100 miles and then home-in on them The chances of these aircraft locating units or persons in distress wi therefore be greatly improved where these beacons are used. The aircraf will not be constrained either by the prevailing conditions or by the siz of the units in distress, as is the case with visual searches. The effectivenes of the homing equipment in the SAR aircraft may be seriously reduced however, if there are several transmissions on the same frequency emanatin from a limited sea area. Such a situation would have occurred if beacon had been activated by all those boats which were in distress in the 197 Fastnet Race.

LIMITED ALERTING CAPABILITY

3. The beacons also have an alerting capability, but it is severely restricted In most distress incidents, rapid assistance can best be provided by near by shipping or, if this is not possible, by SAR units organised through the SAR services. However, neither ships nor the maritime watchkeepin facilities provided by HM Coastguard and the Post Office—the latte through their Coast Radio Stations—monitor the frequencies used by the beacons. Therefore unless a distress incident occurs within a few miles o an aeronautical shore station monitoring 121·5 MHz and/or 243 MHz a beacon operating on these frequencies will need to rely on an overflyin aircraft to receive the alert. The chances of that, however, are not goo because of the pattern of aircraft routeing, the regularity of flights and the speed at which aircraft transit an area. Moreover, in the UK domesti

ight information regions and on the North Sea and European routes, the ilots of most commercial aircraft monitor transmissions essential for the afe conduct of the flight and do not monitor 121·5 MHz unless specifically equested to do so by Air Traffic Control. Pilots of all aircraft on trans-tlantic flights do watch this frequency in the main Oceanic area, but the nain purpose is to pick up an emergency call by another aircraft within ange and to relay the message to the responsible Oceanic Control Centre n V.H.F.

RECOMMENDATION

. The Department of Trade therefore considers that EPIRBs and and-held EPIRBs transmitting radio signals on the frequencies 121·5 MHz nd 243 MHz are very useful aids to SAR aircraft searching for units or ersons in distress, and that they are a useful supplement to conventional narine radio equipment operating on the international distress frequencies 00 kHz and 2182 kHz and on VHF Channel 16 (156·8 MHz). This onventional marine radio equipment is designed to ensure that the all-mportant distress alert will be received and can be acted upon quickly by hose best in a position to help or arrange assistance. UK sea-going ships of 300 tons and above and fishing vessels of 12 metres or more in length egistered in the United Kingdom are required to carry marine radio quipment by the Merchant Shipping (Radio Installations) Regulations 1980 and by the Merchant Shipping (Radio) (Fishing Vessels) Rules 1974 espectively. The Merchant Shipping (Life-Saving Appliances) Regulations 1980 and the Fishing Vessel (Safety Provisions) Regulations 1975 cover the arriage of radio equipment in survival craft. The Department of Trade, herefore, has no objections to these ships and vessels carrying types of eacons which the Home Office have found to be technically suitable for ransmitting on the frequencies 121·5 MHz and 243 MHz (see paragraph 7 (below) *in addition to* the statutory requirements for marine radio quipment. Similarly, the Department has no objections to such beacons eing carried on small craft not covered by the Merchant Shipping and Fishing Vessel Regulations, but *strongly recommends* that such vessels carry conventional maritime radiotelephone equipment capable of operating on the international maritime distress frequencies 156·8 MHz (VHF Channel 16) or 2182 kHz *as well.*

THE IMPORTANCE OF CORRECT USE AND STOWAGE

5. It is very important that owners and potential users of these beacons are aware of the possible consequences of their misuse. First, they should remember that the frequency 121·5 MHz is the frequency used by civil aircraft in an emergency and that it is possible that misuse or accidental activation of a beacon could mask a genuine alert by an aircraft in trouble. Secondly, they should remember that the SAR services have no way of telling whether or not the alert signal is genuine. Once these services are made aware of an alert they will respond but the resources they are able to call upon are expensive and may need to be diverted from genuine distress situations elsewhere. Repeated false alarms could easily bring these beacons into disrepute, which in view of their undoubted usefulness as homing aids would be most unfortunate.

6. In distress situations therefore users should, wherever possible, first attempt to obtain assistance using conventional maritime radio equipment

and procedures, and only activate their beacons if they are unable to obtain assistance by conventional means or when the SAR services request that they be activated in order to help SAR aircraft to locate them. Once activated, the beacons should not be turned off until the emergency is over. It is recommended that the beacons are stowed in the vicinity of the bridge in the case of merchant ships, in the vicinity of the wheelhouse in the case of fishing vessels and near to the helm in the case of yachts.

LICENSING

7. In order to comply with Section 1 of the Wireless Telegraphy Act 1949 owners of EPIRBs and hand-held EPIRBs capable of operating on the aeronautical emergency frequencies must have them licensed by the Home Office. The Home Office will grant licences only in respect of those beacons which they consider to be technically suitable for transmitting on these frequencies and which meet the technical and performance standards laid down in the relevant Home Office specification. The beacons will be licensed only as voluntary equipment for carriage on board individual ships. Where these ships are required to fit conventional maritime radio equipment the beacons will be licensed as an addition to and not a replacement for any of this compulsorily fitted equipment. Licensing will be effected by attaching a form of authorisation to the ship's licence. The use or installation of a beacon without a licence or otherwise than in accordance with the terms of the licence could lead to prosecution. The maximum penalties in the case of these radio beacons are a fine of £400 or up to three months imprisonment or both on summary conviction.

8. Enquiries and applications concerning licensing and type-testing should be addressed to the Radio Regulatory Department, Home Office, Waterloo Bridge House, Waterloo Road, London SE1 8UA. A list of the beacons which have been found to be technically suitable for operating on the aeronautical frequencies 121·5 MHz and 243 MHz and which meet the Home Office technical and performance standards is available from the Home Office Radio Regulatory Department and from the Department's Marine Survey Offices.

USE OVERSEAS

9. It should be recognised that the UK may apply search and rescue arrangements different to some other countries. Therefore foreign vessels in UK waters and UK vessels in foreign waters may find that any locating devices they carry voluntarily for safety purposes may not be the most suitable.

CODE OF PRACTICE

10. A summary of the main points of this Notice is contained in the Annex.

Department of Trade
Marine Division
London WC1V 6LP
August 1981.

ANNEX TO MERCHANT SHIPPING NOTICE No. M.982

EMERGENCY POSITION INDICATING RADIO BEACONS (EPIRBs) AND HAND-HELD EPIRBs OPERATING ON 121·5 MHz AND 243 MHz

CODE OF PRACTICE

1. The main function of these devices is as a homing aid in conjunction with SAR aircraft fitted with direction-finding equipment operating on these frequencies.

2. The devices have an alerting capability but it is limited and should not be relied upon. Commercial aircraft flying over the approaches to the United Kingdom within about 150 miles of the coast and over the North Sea do not normally keep watch on these frequencies. They do keep a watch on 121·5 MHz in Oceanic areas on trans-atlantic flights, although this watch is primarily for aircraft in distress.

3. Conventional maritime radio equipment is the best means of alerting near-by shipping and the maritime SAR services to a distress situation. Either of those is normally in the best position to help. Although the EPIRBs and hand-held EPIRBs operating on the aeronautical emergency frequencies are useful supplements to this equipment, they are not substitutes.

4. The carriage of EPIRBs and hand-held EPIRBs needs to be licensed. Only those devices which are technically suitable for operation on the aeronautical emergency frequencies will be licensed.

5. In order to protect the primary use of the frequencies—for civil aviation emergency purposes—and to avoid the misuse of the maritime SAR organisation, it is especially important that the devices are:

(i) handled competently and neither dropped nor knocked;

(ii) stowed appropriately (readily available should they be required but out of reach of unsupervised persons and so placed as to prevent accidental operation) and stored safely when in harbour.

6. Once activated, the devices should not be switched off until the emergency is over.

7. It should be noted that SAR aircraft will need to make special adjustments to their homing equipment when responding to transmissions on the same frequency from more than one vessel in a limited sea area.

"AMEECO" "HIGGS" SEARCH INITIATOR BUOY. MODEL 3.

Life-saving Emergency Position-Indicating Radio Buoys are regarded as an improvement of the original EPIRB which can also be modified for satellite operation.

They are the invention of a Canadian, Captain William York Higgs. Basically they consist of a buoy made of orange coloured glass reinforced plastic and foam filled to ensure buoyancy even if damaged. The buoy contains two radio transmitters which give omni-directional radio signals on both 121.5 and 243MHz, the international military and civil distress frequencies. The buoy is topped with a high intensity xenon strobe light, which flashes at a rate of 50-60 flashs per minute and has a very high visibility range. There are also canisters embodied in the base which will automatically release a marker dye and calming oil over a prolonged period both to aid daylight location and provide maximum calming effect for survival craft attached to the buoy. In addition there are lifelines attached to four grab handles. A 15m. (49ft) floating liferaft mooring line provides a static marshalling point for survivors and to which survival craft may be tethered.

The buoy is carried in a glass reinforced plastic cradle, in which heating elements have been inserted to ensure that the buoy is not trapped by frost, ice or snow, the power for the heating elements being supplied by the ship's electrical supply. The buoy is attached to the ship by means of a steel mooring wire contained within the base of the buoy. This wire, which is 914m. (2970 ft.) long, is of sufficient length to ensure that the buoy remains attached to the ship in coastal and continental shelf waters. It has a controlled release system to ensure clear running and restrained drift to enable swimmers and survival craft to connect with it. A weak link on the ship end of the wire, ensures automatic release should the ship sink to a greater depth than 914m. (2970ft). The wire will then act as a drogue to reduce leeway and help to keep any life support systems head to wind and sea, thereby reducing the chance of capsize.

The beacon acts as a homing beacon for search and rescue operations, and automatically provides an alarm, when, in the event of sudden accident, no S.O.S. or MAYDAY signal has been made. It also marks the sunken vessel in shallow waters so that it can be permanently marked and charted. Maintenance is minimal and can be carried out and tested on board by the simple exchange of modules and requires no technical ability.

The buoys will normally be stowed in cradles on the upper deck and in a position where direct seas will not dislodge them situated as high as possible above the waterline, in a situation where they will be reasonably free from disturbance. Lashings are not normally provided as the cradle is considered to be high enough to prevent accidental dislodgement of the buoy. If however, lashings should be necessary, then they should be secured by means of a pre-tested monofilament line or by a hydrostatic release system with a manual release facility.

The Canadian Authorities have made it mandatory for most tugs flying the Canadian flag to carry a modified L.E.P.I.R.B.

These buoys can be modified for satellite transmissions when the systems become available.

AMEECO HIGGS LIFE-SAVING EMERGENCY POSITION INDICATING BUOY. MODEL 3

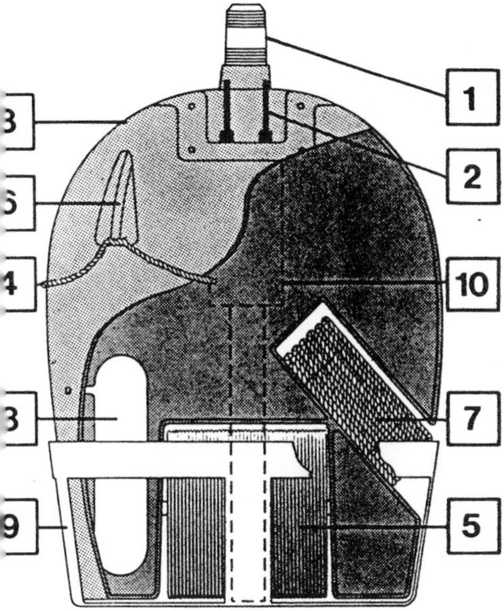

1 Xenon strobe light
2 Two channel radio distress beacon transmitter
3 Buoy body, Orange coloured and foam filled
4 Lifelines fitted to grab handles
5 Controlled release mooring wire
6 Stainless steel grab handles moulded in
7 15m of floating lifeline with light attached
8 Dye or oil canister
9 Cradle
10 Electronics module. Replaced when servicing

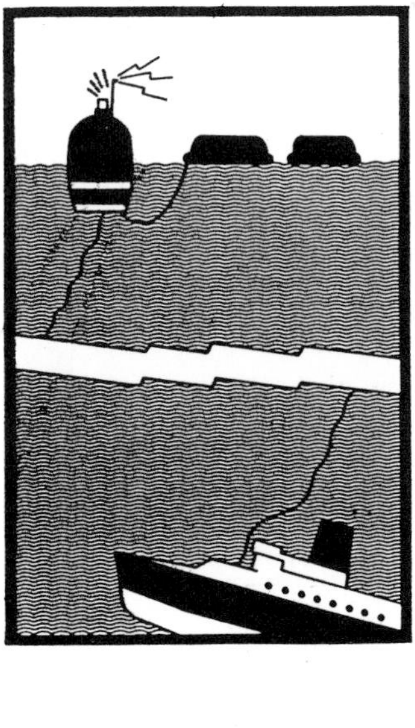

THE "MARCONI" MARINER 16 PORTABLE DISTRES RADIOTELEPHONE

The Mariner 16 is a high power portable distress radiotelephon which provides two-way communication on the international distre frequency 2182kHz. The equipment has been designed to me the performance specification for radiotelephone equipmer compulsorily fitted in fishing vessel survival equipment.

Operation is simple. Just release the earth bobbin and drop it in th sea, then extend the 2m (6.5ft) long telescopic antenna and turn th volume control clockwise to the receiver volume required. To transmi press the speaker switch and speak. The two-tone alarm signal can b transmitted by turning the alarm switch clockwise for a period of 3 seconds to one minute. To receive, release microphone switch an adjust the volume as required. When transmitting speech, keep th microphone about 6 ins (150mm) from the mouth.

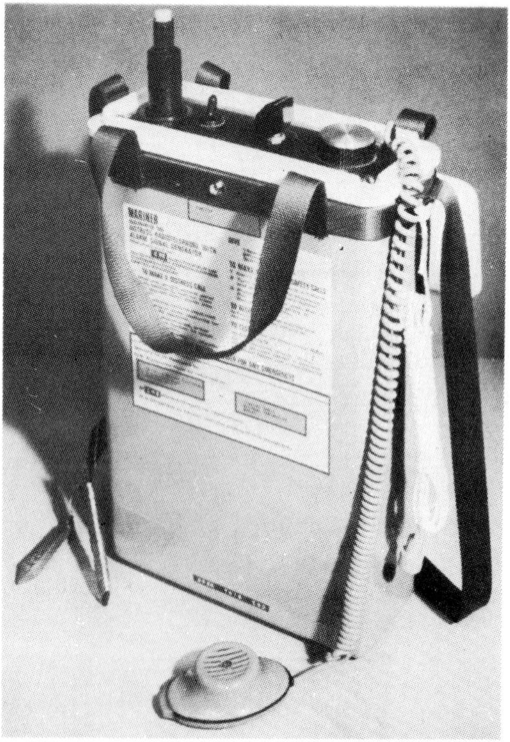

"JOTRON" TRON 2L EMERGENCY PORTABLE RADIOTELEPHONE

To Operate: To receive; Select the distress frequency 121.5MHz (upper switch). Transmit by pressing lower switch and talk into the microphone.

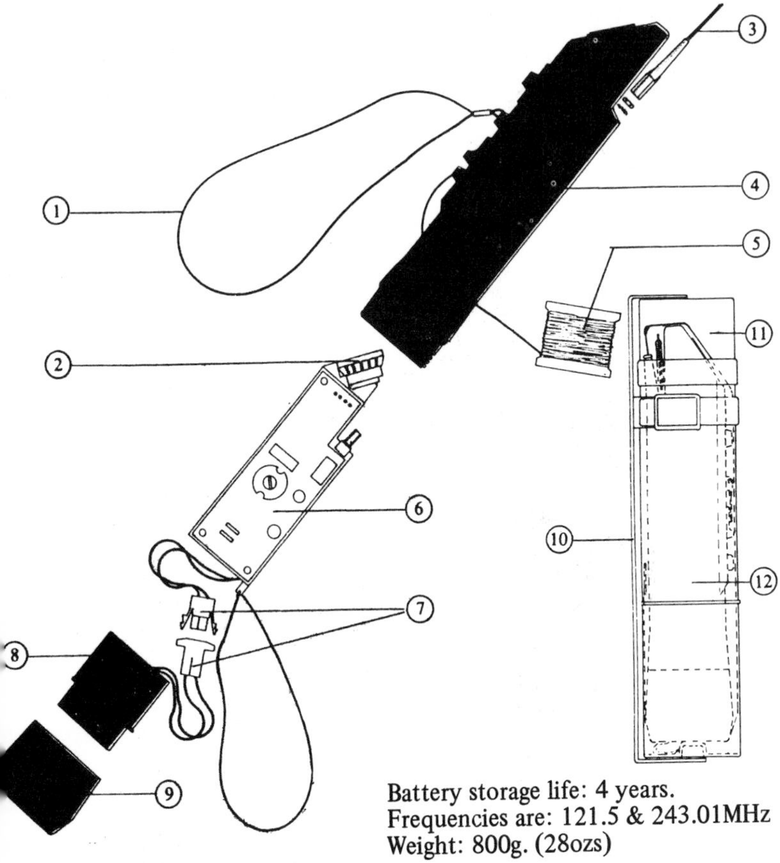

Battery storage life: 4 years.
Frequencies are: 121.5 & 243.01MHz
Weight: 800g. (28ozs)

1. Neck-strap (nylon)
2. Microphone - loudspeaker
3. Antenna (laminated, flexible steel)
4. Housing (ABS)
5. Anchoring-line (25 m - nylon)
6. Electronic-unit (with noryl-frame)
7. Battery-plugs
8. Battery-unit (with lithium-cells)
9. Battery-compartment (ABS)
10. Mounting-bracket (aluminium)
11. Container-lid (transparent-plastic)
12. Container (ABS).

DEPARTMENT OF TRANSPORT MERCHANT SHIPPING NOTICE
No. **M.1056**

USE AND FITTING OF RETRO-REFECTIVE MATERIAL ON LIFE-SAVING APPLIANCES

Notice to Owners of Merchant Ships and Fishing Vessels, Shipbuilders and Repairers, Masters, Skippers, Officers and Seamen

This notice supersedes Notices M. 696

(1) The International Maritime Organisation Resolution A.27· (VIII) of 20 November 1973 recommends, inter alia, that Contractin Governments should encourage owners of all vessels under their flag to fit retro-reflective material to life-saving appliances as an additiona aid to search and rescue operations.

(2) The Department fully supports the resolution and in 197· issued Merchant Shipping Notice No. 696 recommending that · phased programme of fitting retro-reflective material to life-savin appliances should commence with the hope that at the end of tw years all life-saving appliances included in the Resolution would hav been fitted with retro-reflective material.

(3) The Department hereby repeats its recommendation that th life-saving appliances mentioned in this Notice not yet fitted wit retro-reflective material be so fitted at the earliest opportunity. Nev applications of material should be of the highest intensity type and of type which has been approved by the Department.

(4) The I.M.O. guide lines for the siting of retro-reflective materia are set out in the appendix to this Notice, but the following additiona comments should be noted:

(a) *Liferafts*
Variation in design will require differing quantities an location of retro-reflective material in order to achiev optimum results; at the owner's request, manufacturers wil undertake to apply retro-reflective material to new raft before delivery, and to existing rafts at approved servic stations during servicing.

(b) *Buoyant apparatus*
On buoyant apparatus which may float either way up, th retro-reflective material must be visible in either attitude.

(c) *Lifejackets*
Variations in design will again require differing quantitie and location of retro-reflective material. Leaflets showin the most effective arrangements for various lifejacket (based on Department of Trade tests) are available a Marine Offices.

Inflatable boats

(5) Although not dealt with in the I.M.O. Recommendations, th Department recommends that inflatable boats should also be fitte with retro-reflective material. They are subject to the same comment as for liferafts in paragraph 4(*a*) above.

(6) The Department will welcome reports on the performance o

etro-reflective material and on its success or otherwise in aiding earch and rescue operations. Such reports should be addressed to the Deputy Surveyor General (C), Department of Trade, 90 High Holborn, London WC1V 6LP.

April 1983

APPENDIX

Guidelines for the Use and Fitting of Retro-reflective Material on Life-saving Appliances

. *Lifeboats*

Retro-reflective tapes should be fitted on top of the gunwale as well s on the outside of the boat as near the gunwale as possible. The tapes hould be sufficiently wide and long (approximately 5×30 cm) and hould be spaced at suitable intervals (approximately 50 cm). If a anopy is fitted it should not be allowed to obscure the tapes fitted on he outside of the boat and the top of the canopy should be fitted with etro-reflective tapes similar to those mentioned above but shaped in he form of a cross and spaced at suitable intervals (approximately 50 m).

. *Liferafts*

Retro-reflective tapes should be fitted around the canopy of the raft t suitable intervals (approximately 50 cm) and at a suitable height bove the waterline. On inflatable liferafts retro-reflective tapes should lso be fitted on the underside of the floor (four tapes fitted at equal ntervals around the outer edges on the bottom of the liferafts). The apes should be sufficiently wide and long (approximately 5×30 cm). A uitable cross-shaped marking of two such tapes should also be pplied to the top of the canopy. On rafts which are not equipped with anopies at least four such tapes should be attached to and evenly paced on the buoyancy chamber in such a manner that they are isible both from the air and sea.

. *Lifebuoys*

Retro-reflective tape of a sufficient width (approximately 5 cm) hould be applied around or on both sides of the body of the buoy at our evenly-spaced points.

. *Buoyant apparatus*

Buoyant apparatus should be fitted with retro-reflective tapes in the ame manner as rafts without canopies, always depending on the size nd shape of the object. The reflectors should be visible from both the ir and sea.

. *Lifejackets*

Unless its cover material is retro-reflective, a lifejacket should be itted with retro-reflective tapes sufficiently wide and long (approximately 5×10 cm). These tapes should be placed as high up on the acket as possible in at least six places on the outside of the jacket and n at least as many places on the inside of the same by virtue of the fact hat lifejackets are reversible.

IMO PERSONAL LIFE-SAVING APPLIANCE (SPECIFICATION) FOR SHIPS BUILT AFTER 1st. JULY 1986

Lifebuoys

1. **Lifebuoy specification**

Every lifebuoy shall:

(a) have an outer diameter of not more than 800mm (31½ ins and an inner diameter of not less than 400mm (16¼ ins.);
(b) be constructed of inherently buoyant material; it shall nc depend upon rushes, cork shavings or granulated cork, any othe loose granulated material or any air compartment which depend on inflation for buoyancy;
(c) be capable of supporting not less than 14.5kg (321lbs) of iro in fresh water for a period of 24 hours;
(d) have a mass of not less than 2.5kg (5¾ lbs.)
(e) not sustain burning or continue melting after being totall enveloped in a fire for a period of 2 seconds;
(f) be constructed to withstand a drop into the water from th height at which it is stowed above the waterline in the lightest sea going condition or 30m (97.5ft), whichever is the greater, withou inpairing either its operating capability or that of its attache components;
(g) if it is intended to operate the quick release arrangemen provided for the self-activated smoke signals and self-ignitin lights, have a mass sufficient to operate the quick releas arrangement or 4kg. (9 lbs.), whichever is the greater;
(h) be fitted with a grabline not less than 9.5mm in diameter (1 inch rope) and not less than 4 times the outside diameter of th body of the buoy in length. The grabline shall be secured at fou equidistant points around the circumference of the buoy to forr 4 equal loops.

2. **Lifebuoy self-igniting lights**

Self igniting lights shall:

(a) be such that they cannot be extinguished by water;
(b) be capable of either burning continuously with a luminou intensity of not less than 2 cd in all directions of the uppe hemisphere or flashing (discharge flashing) at a rate of not les than 50 flashes per minute with at least the correspondin effective luminous intensity;
(c) be provided with a source of energy capable of meeting th requirements of sub-paragraph (b) above for a period of at leas two hours;
(d) be capable of withstanding the drop test required by sub paragraph 1 (f) above.

3. **Lifebuoy self-activating smoke signals**

Self-activating smoke signals shall:

(a) emit smoke of a highly visible colour at a uniform rate for a period of at least 15 minutes when floating in calm water;

(b) not ignite explosively or emit any flame during the entire smoke emission time of the signal;

(c) not be swamped in a sea-way;

(d) continue to emit smoke when fully submerged in water for a period of at least 10 seconds;

(e) be capable of withstanding the drop test required by sub-paragraph 1. (f) above.

4. Buoyant lifelines attached to lifebuoys shall:

(a) be non-kinking;

(b) have a diameter of not less than 8mm (one inch rope);

(c) have a breaking strength of not less than 5kn.

Manoverboard

An emergency light and smoke marker

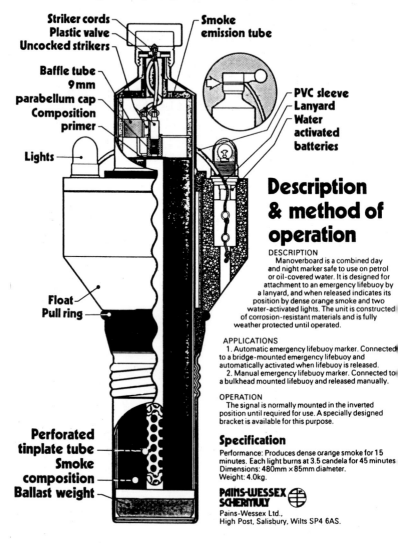

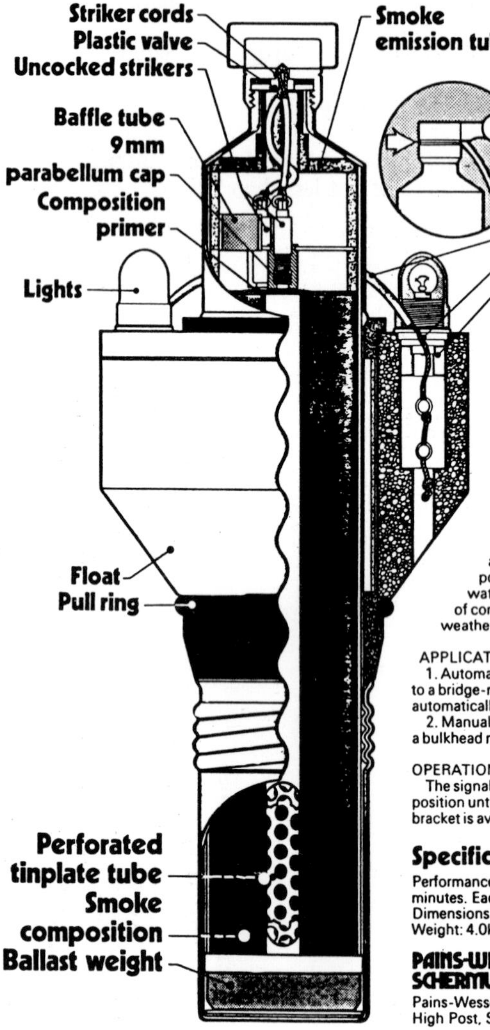

Description & method of operation

DESCRIPTION

Manoverboard is a combined day and night marker safe to use on petrol or oil-covered water. It is designed for attachment to an emergency lifebuoy by a lanyard, and when released indicates its position by dense orange smoke and two water-activated lights. The unit is constructed of corrosion-resistant materials and is fully weather protected until operated.

APPLICATIONS

1. Automatic emergency lifebuoy marker. Connected to a bridge-mounted emergency lifebuoy and automatically activated when lifebuoy is released.

2. Manual emergency lifebuoy marker. Connected to a bulkhead mounted lifebuoy and released manually.

OPERATION

The signal is normally mounted in the inverted position until required for use. A specially designed bracket is available for this purpose.

Specification

Performance: Produces dense orange smoke for 15 minutes. Each light burns at 3.5 candela for 45 minutes
Dimensions: 480mm × 85mm diameter.
Weight: 4.0kg.

PAINS-WESSEX SCHERMULY

Pains-Wessex Ltd.,
High Post, Salisbury, Wilts SP4 6AS.

LIFEBUOY MARKERS

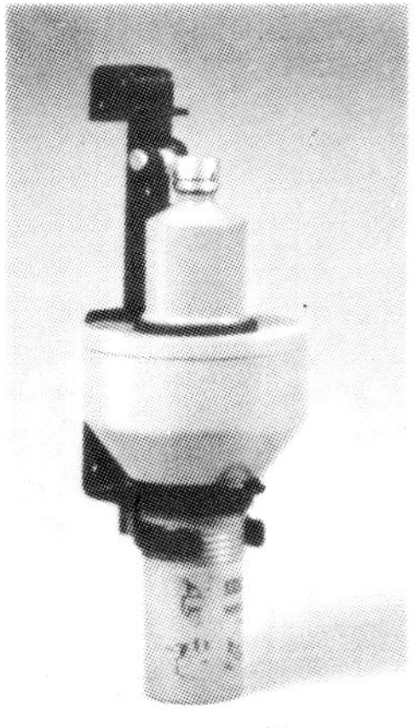

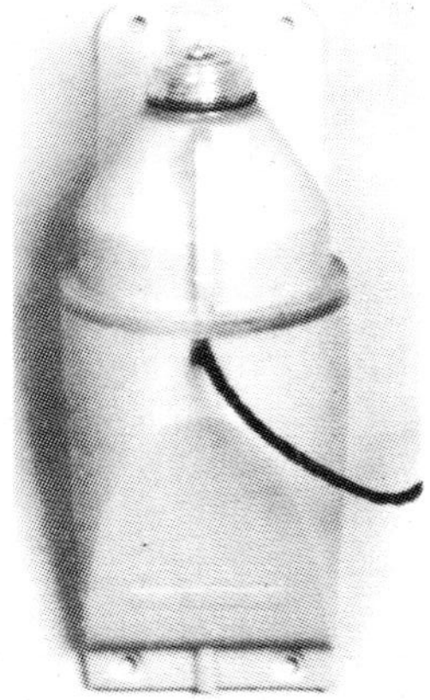

Manoverboard and Buoysmoke

Manoverboard — combined day and night marker safe to use on petrol or oil covered water. Connects to bridge or bulkhead mounted lifebuoy and can be automatically activated when lifebuoy is released. Produces dense orange smoke for 15 minutes, and two lights operate at 3.5 candela for 45 minutes.

• Manoverboard 1652. • Buoysmoke — as Manoverboard without lights 1651. • Mounting bracket enabling signal to be stowed in inverted position, necessary for release with lifebuoy 1661.

Lifebuoy Lights

Continuous output McMurdo Aqualite with lanyard for attachment to emergency lifebuoy. On contact with water, light functions automatically and operates at 2 candela for 45 minutes.
Dry-battery operated McMurdo Apollo 2 light with lanyard for attachment to emergency lifebuoy. Flashes for 12 hours minimum at 0.6 candela.

• Aqualite 3501. • Apollo 2 3511.

Lifejackets

1. IMO General requirements for lifejackets

(a) A lifejacket shall not sustain burning or continue meltin after being totally enveloped in a fire for a period of 2 seconds.

(b) A lifejacket shall be so constructed that:

(i) after demonstration, a person can correctly don within a period of one minute without assistance;

(ii) it is capable of being worn inside-out or is clearl capable of being worn in only one way and, so far as possibl cannot be donned incorrectly;

(iii) it is comfortable to wear;

(iv) it allows the wearer to jump from a height of at lea 4.5m (14ft 6in) into the water without injury and withou dislodging or damaging the lifejacket.

(c) A lifejacket shall have sufficient buoyancy and stabilit in calm fresh water to:

(i) lift the mouth of an exhausted or unconscious perso not less than 120mm (4¾ in) clear of the water with the bod inclined backwards at an angle of not less than 20 degrees an not more than 50 degrees from the vertical position;

(ii) turn the body of an unconscious person in the wat from any position to one where the mouth is clear of th water in not more than 5 seconds.

(d) A lifejacket shall have buoyancy which is not reduced b more than 5% after 24 hours submersion in fresh water.

(e) A lifejacket shall allow the person wearing it to swim a sho distance and to board a survival craft.

(f) Each lifejacket shall be fitted with a whistle firmly secure by a cord. *(pea whistles are unacceptable)*

2. **Inflatable lifejackets**

A lifejacket which depends on inflation for buoyancy sha have not less than two separate compartments and comply with th requirements of paragraph 1 above and shall:

(a) inflate automatically on immersion, be provided with a devic to permit inflation by a single manual motion and be capable c being inflated by mouth;

(b) in the event of loss of buoyancy in any one compartmer be capable of complying with the requirements of sub-paragrapl 1(b), (c) and (e) above;

(c) Comply with the requirements of sub-paragraph 1 (d) aft inflation by means of the automatic mechanism.

3. **Lifejacket lights**

(a) Each lifejacket light shall:

(i) have a luminous intensity of not less than 0.75 cd;

(ii) have a source of energy capable of providing a luminou intensity of 0.75 cd for a period of at least 8 hours;

(iii) be visible over as great a segment of the upper hemisphe as is practicable when attached to a lifejacket.

(b) If the light referred to in paragraph (a) above is a flashin light it shall in addition:

(i) be provided with a manually operated switch;
(ii) not be fitted with a lens or curved reflector to concentrate the beam;
(iii) flash at a rate of not less than 50 flashes per minute, with an effective luminous intensity of at least 0.75 cd.

The U.K. Merchant Shipping (Life-Saving Appliances) Regulations 1980 require all approved lifejackets to be marked on both sides "Person of 32kg" or "Person under 32 kg" they must also be marked on one side "Accepted by Dept. of Transport" and to provide a minimum of 155 newtons buoyancy in fresh water for 24 hours. They must not be adversely affected by oil or oil products and shall be of a highly visible colour. They shall be fitted with a ring or loop of adequate strength to facilitate rescue. The fastening tapes are to be rot-proof. Metal fastenings when used shall be of a size and strength consistant with the fastening tapes and of corrosion resistant material. When buoyancy is provided by kapok, the kapok must be protected from oil and is normally contained in plastic bags. Inflatable lifejackets are only to be supplied to cargo ships and are to be marked "Crew only". Inflatable lifejackets are unsuitable for tankers.

Because a lifejacket invariably turns its wearer into the wind and because a considerable amount of body heat is lost through the head, some manufacturers now produce lifejackets that are fitted with a stole containing a hood, sometimes with an attached transparent visor, both to reduce heat loss from the head and to keep wind and spray out of the survivors face. Even in a slightly choppy sea, wind and spray can cause considerable discomfort to the wearer of a lifejacket in the water and may shorten the possible survival period considerably. However, although hoods and visors are supplied to many navies throughout the world, they are not an IMO requirement.

The Perry Lifejacket for persons of 32kg or more

1 Place lifejacket over head ensuring that arms pass over side tapes.

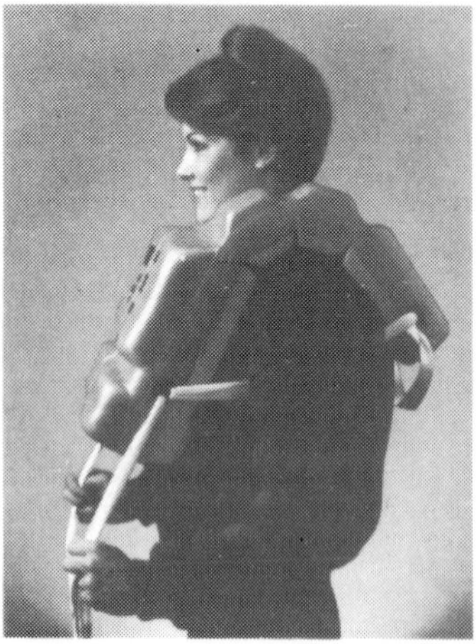

2 Draw tape end downwards ensuring back pad is firmly pulled on to shoulders.

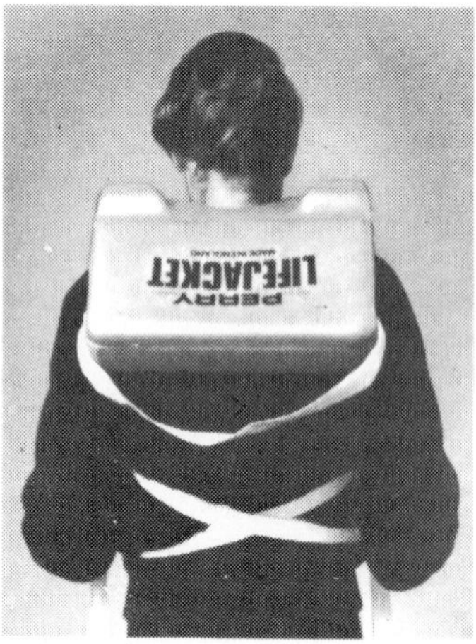

3 Cross tapes behind back and return to front.

4 Tie tapes securely in recess on front.

NOTE

1. Loop of tape at back is to be left free for recovery purposes
2. Whistle to aid recovery is in pocket at side of shoulder.

DUNLOP

General Purpose Naval Lifejackets

Designed for general issue in all parts of the world, the General Purpose Naval Lifejacket is hard wearing, ozone resistant and rotproof. Comfortable to wear, easy to use and positive in action, it consists of an inflatable stole, housed in a fabric pouch, which in turn is attached to a strong webbing waist belt. It is positively self righting when fully inflated and possesses correct flotation characteristics.

The inflatable stole with head aperture is made from well proven two-ply butyl proofed synthetic fabric, coloured bright yellow to make the wearer easy to spot in an emergency. The rear panel of the lifejacket is fitted with a broad fabric loop retained in the pouch by a webbing waist belt that has a webbing lifting harness attached.

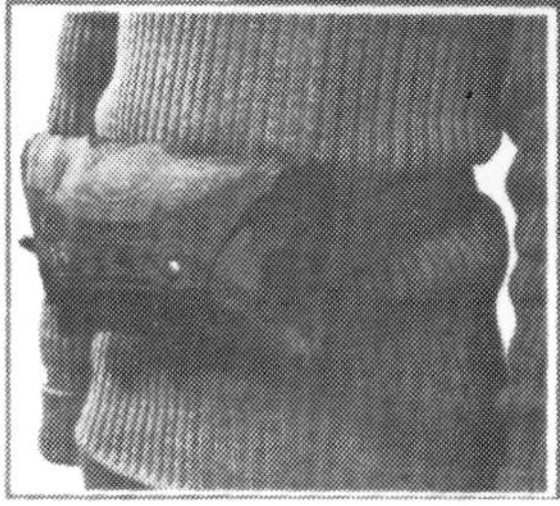

The pouch, made from hard wearing synthetic fabric, contains the folded lifejacket compactly. It is internally reinforced at points of stress and has a flap, closed by two press fasteners.

An adjustable waistbelt made from 45 mm (1¾ in) wide webbing passes through side openings in the pouch and the broad loop on the stole so that the folded lifejacket can be housed easily within the pouch. There is an adjustable nylon fastener for the waist belt. The pouch can be moved around the waist belt so that it is in the position most convenient to the wearer.

A 25 mm (1 in) wide harness of synthetic webbing is attached to the waist belt to enable a survivor to be lifted easily from the water. When not in use this harness is neatly retained to the stole and the waist belt by press fasteners and webbing loops, minimising any risk of snagging.

A plastic whistle is provided, to attract the attention of rescuers and housed in the pouch are a lifeline and toggle.

A battery and light assembly, to aid location when it is dark, are available as an optional extra.

DUNLOP

DUNLOP LIMITED MARINE SAFETY PRODUCTS
GRG DIVISION
(A subsidiary of Dunlop Holdings Limited)
Atherton Road, Hindley Green, Wigan, Lancs. England
Tel: (0942) 57181 Telex: 67171 Cables: Dunlop Hindley

Department of Transport Merchant Shipping Notice
No. **M.1212**

LIFEJACKET LIGHTS

Notice to Owners and Managers of Merchant Ships and Fishin Vessels, Masters, Skippers, Officers, Seamen and Yachtsmen

(1) The usefulness of lifejacket lights has long been recognised i search and rescue situations when it had proved difficult during th hours of darkness to locate survivors who have been unable to mak use of a survival craft. Two recent Formal Investigation Reports o Court* have respectfully drawn attention to, and recommended th provision of, lifejacket lights in advance of such provision becoming statutory requirement.

(2) The revision of Chapter III of the 1974 SOLAS Conventio will introduce a requirement for the fitting of lights on lifejacket carried on passenger ships engaged on long international voyages, an on cargo ships of 500 grt and over engaged on international voyages This requirement will apply in the first instance to ships whose keel are laid on or after 1 July 1986. Ships whose keels are laid before tha date will be required to be provided with lifejacket lights by 1 Jul 1991.

(3) The Department is of the view that lifejacket lights are a important addition to life-saving equipment currently carried an strongly recommends that Owners and Managers initiate a pro gramme of fitting lights to lifejackets carried on the ships referred to i paragraph 2 in advance of the statutory dates for fitting.

(4) Although the revision of Chapter III only applies to ships o Class I, to ships of Classes VII, VII (A), VII(T), VIII and VIII(T) o 500 grt and over and to ships of Class IX of 500 grt engaged o international voyages, it is recommended that Owners and Manager also give consideration to providing lifejacket lights for all on boar ships of Classes VII, VII (A), VII(T), VIII and VIII(T) of less than 50 grt and for all on board ships of Classes VIII(A), VIII (A)(T), IX, X and XI, and pleasure craft (other than passenger ships) which procee to sea.

(5) Lights fitted should be of a type which has been approved b the Department as complying with the revision of Chapter III of th 1974 SOLAS Convention.

*Report of Court No. S501, published August 1979 (mvf Boston Sea Ranger) an report of Court No. 8071, published March 1984 (mv Grainville). These Reports ar available from HMSO.

Department of Transport
Marine Directorate
London WC1V 6LP
January 1986

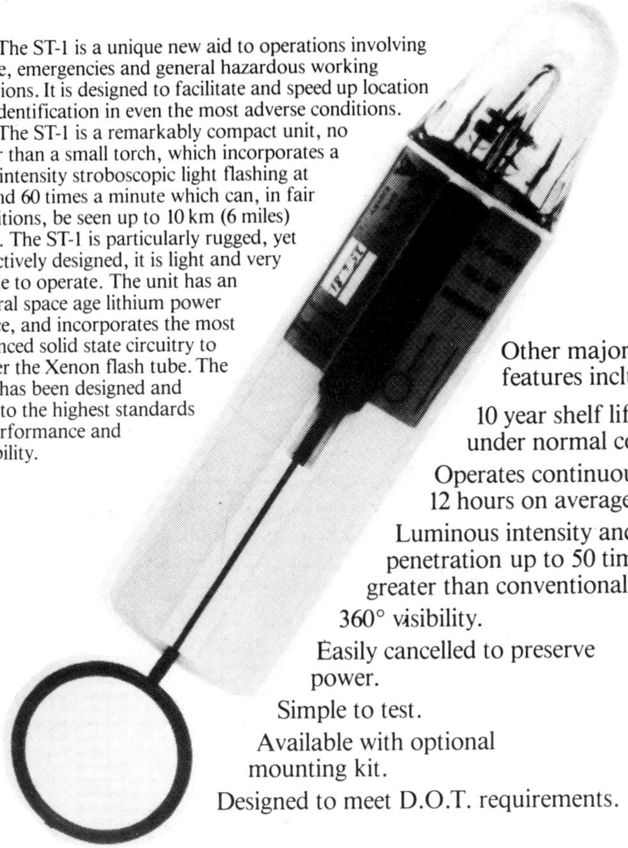

Immersion suits

1. IMO General requirements for immersion suits

(a) The immersion suit shall be constructed with waterproo materials such that:

(i) it can be unpacked and donned without any assistanc within 2 minutes, taking into account any associated clothin and a lifejacket if the immersion suit is to be worn i conjunction with a lifejacket;

(ii) it will not sustain burning or continue melting afte being totally enveloped in a fire for a period of 2 seconds;

(iii) it will cover the whole body with the exception of th face. Hands shall also be covered unless permanently attache gloves are provided;

(iv) it is provided with arrangements to minimize or reduc free air in the legs of the suit;

(v) following a jump from a height of not less than 4.5m (14.9 ft) into the water there is no ingress of water into th suit.

(b) An immersion suit which also complies with the requirements for a lifejacket, may be classified as a lifejacket.

(c) An immersion suit shall permit the person wearing it, and also wearing a lifejacket, if the immersion suit is to be worn with a lifejacket to;

(i) climb up and down a vertical ladder at least 5m (16.25ft) in length;

(ii) perform normal duties during abandonment;

(iii) jump from a height of not less than 4.5m (14.9 ft) into the water without damaging or dislodging the immersion suit, or being injured; and

(iv) swim a short distance through the water and board a survival craft.

(d) An immersion suit which has buoyancy and is designed to be worn without a lifejacket shall be fitted with a light and a whistle.

(e) if the immersion suit is to be worn in conjunction with a lifejacket, the lifejacket shall be worn over the immersion suit A person wearing such an immersion suit shall be able to don a lifejacket without assistance.

2. **Thermal performance requirements for immersion suits**

(a) An immersion suit made of material which has no inherent insulation shall be:

(i) marked with instructions that it must be worn in conjunction with warm clothing;

(ii) so constructed that when worn in conjunction with warm clothing, and with a lifejacket if the immersion suit is to be worn with a lifejacket, the immersion suit continues to provide sufficient thermal protection, following one jump by the wearer into the water fron a height of 4.5m (14.9ft), to ensure that when it is worn for a period of 1 hour in calm circulating water at a temperature of 5 degrees C (42 deg.F), the wearer's body core temperature does not fall more than 2 degrees C. (3.5 deg.F)

(b) An immersion suit made of material with inherent insulation, when worn either on its own or with a lifejacket, if the immersion suit is to be worn in conjunction with a lifejacket, shall provide the wearer with sufficient thermal insulation, following one jump into the water from a height of 4.5m (14.9ft) to ensure that the wearer's body core temperature does not fall more than 2 degrees C (3.5 deg.F) after a period of 6 hours immersion in calm circulating water at a temperature of between 0 degrees and 2 degrees C (32 and 35.5 deg,F).

(c) The immersion suit shall permit the person wearing it with hands covered to pick up a pencil and write after being immersed in water at 5 degrees C (41 deg.F) for a period of 1 hour.

3. **Buoyancy requirements**

(a) A person in fresh water wearing an immersion suit and, if the immersion suit is to be worn in conjunction with a lifejacket, a lifejacket, shall:

(i) be able to turn from a face-down to a face-up position in not more than 5 seconds;

(ii) float in a stable face-up position acceptable to the Administration, with the mouth not less than 120mm (4¾ ins) clear of the water.

Thermal Protective Aids

1. A thermal protective aid shall be of waterproof material having a thermal conductivity of not more than 0.25 W/(mK) and shall be so constructed that, when used to enclose a person, it shall reduce both the convective and evaporative heat loss from the wearer's body.

2. The thermal protective aids shall:

(a) cover the whole body of a person wearing a lifejacket with the exception of the face. Hands shall also be covered unless permanently attached gloves are provided;

(b) be capable of being unpacked and easily donned without assistance in a survival craft or rescue boat;

(c) permit the wearer to remove it in the water in not more than 2 minutes, if it impairs the ability to swim.

3. The thermal protective aid shall function properly throughout an air temperature range of minus 30 degrees C to plus 20 degrees C (-22 to 63 deg. F)

OFFSHORE SURVIVAL SUIT

The Beaufort Offshore Survival Suit (BOSS) has been developed to meet the special survival requirements of workers on oil-rigs and platforms.

Wearing the suit over only light underwear it aims to give:– Ability to be afloat in water temperatures of 0 degrees C without losing more than 1 degree C body temperature in the first hour, thus providing a minimum survival time of 4 hours at 0 degrees C. At sea temperatures of 6 to 8 degrees C, winter survival time in the North Sea, for instance, is expected to be 12 to 15 hours. Protection in a temperature of 800 degrees C for up to three minutes without substantial damage to the suit. Ease of evacuation from difficult situations such as a helicopter which has ditched in the sea.

The Beaufort design is a one-piece suit with integral hood, boots and gloves, manufactured from neoprene-proofed nylon with head- and oil-resistant properties. It is completely waterproof and gives good thermal protection. A full-length waterproof zip from crotch to chin permits quick and easy donning. A quilted inner lining for warmth, includes closed cells foam inserts for flotation and can be detached for cleaning. Zips are fitted to the legs so that when the wearer is in water with the zips closed air is excluded, resulting in a good flotation angle which will keep the wearer's face above the surface of the water.

The suit provides a buoyancy of 7 kg (15½ lb). Zips and Velcro fastenings at the wrists permit the gloves to be rolled and stowed out of the way when working with the hands through the wrist openings, which also contribute to ventilation. When required the gloves can be quickly released, the hands inserted into the gloves and the zips closed to give complete immersion protection. A breast pocket is provided and includes a window area for displaying the identification of the wearer and a whistle stowage.

The outer hood is manufactured from fluorescent-coloured material and includes stripes of retroreflective tape, to aid search parties in poor light or darkness.

The complete suit is packed into a valise which is designed to be held or carried as a shoulder bag. A small repair kit will be included in the valise which when packed measures 49 cm x 34 cm x 13 cm and weighs 5½ kg (12¼ lb). The suit is manufactured in four sizes, small, medium, large and extra large.

AUTOMATIC OFFSHORE LIFEJACKET MK.1

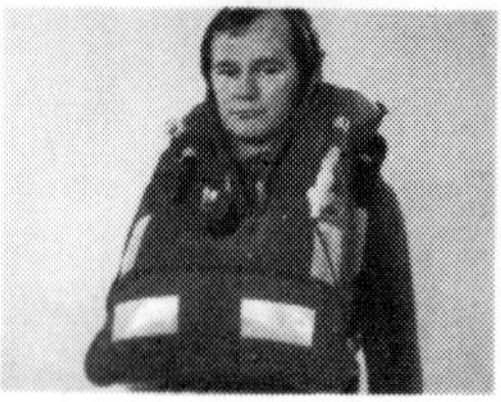

D.T.I. Approved for all Offshore Installations

Introduction

The Automatic Offshore Lifejacket is designed for persons who require immediately available buoyancy in an emergency, but who are unable to wear a restrictive, bulky lifejacket.

Description

The lifejacket is manufactured from top quality synthetic materials. The inflatable stole is divided into two independent chambers – front and rear – each of which is fitted with both an automatic/manual gas inflation system and an oral inflation system. The design of the lifejacket is such that, should one chamber sustain damage, the wearer will be supported by the remaining chamber.

The front panel of the lifejacket is fitted with the following:– (a) Whistle and lanyard, (b) Reflective strips, (c) Light (automatically activated when jacket inflates in water), (d) Lifting becket (for lifting wearer out of water by hand or by helicopter winch). The rear panel of the lifejacket is fitted with the following:– (a) Water activated battery to power the light, (b) Battery plug ejection system. Three alternative valises are available, all of which protect the stole from damage and prevent rain or spray from penetrating the operating heads and causing an inadvertent inflation. The three alternatives are identified as follows:–

Mk. 1 – standard lifejacket
Mk. 1T – training lifejacket
Mk. 1W – welders lifejacket

Operation

In an emergency, the wearer can fall or jump into the water and the jacket will inflate automatically. It will then turn the wearer – conscious or unconscious – face uppermost with the nose and mouth held clear of the water. Should the automatic jacket fail, the wearer can inflate the lifejacket manually simply by pulling two readily accessible knobs.

INSTRUCTIONS

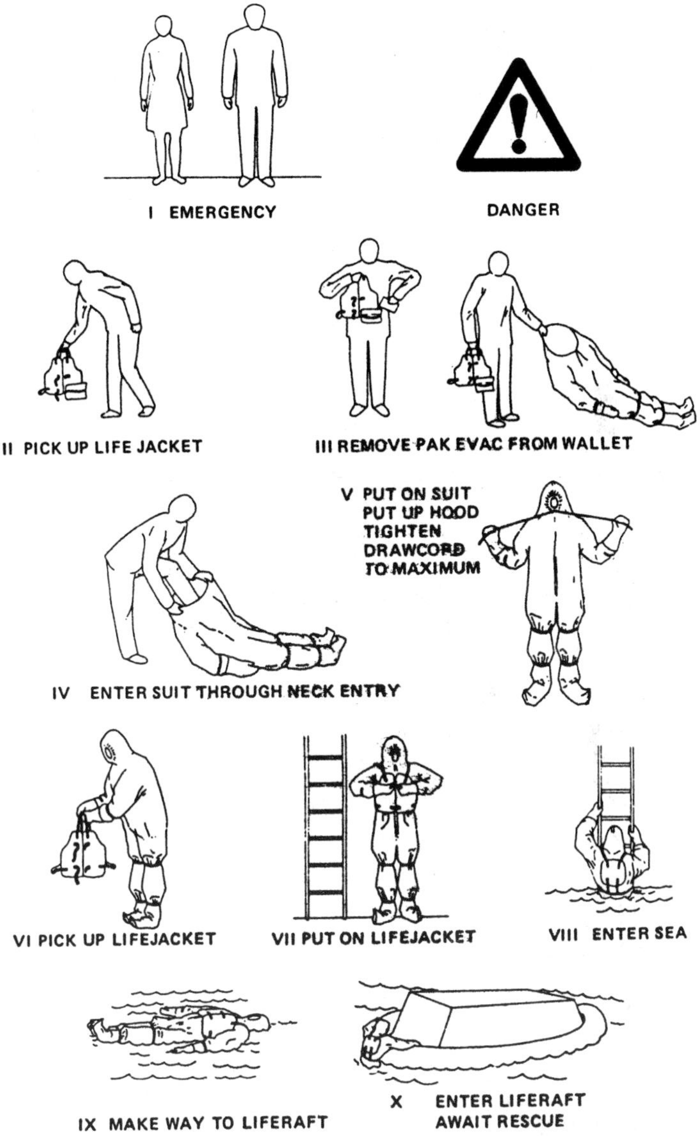
I EMERGENCY
DANGER
II PICK UP LIFE JACKET
III REMOVE PAK EVAC FROM WALLET
V PUT ON SUIT
PUT UP HOOD
TIGHTEN
DRAWCORD
TO MAXIMUM
IV ENTER SUIT THROUGH NECK ENTRY
VI PICK UP LIFEJACKET
VII PUT ON LIFEJACKET
VIII ENTER SEA
IX MAKE WAY TO LIFERAFT
X ENTER LIFERAFT
AWAIT RESCUE

PACKED EVACUATION SUIT FOR AN ABANDON SHIP SITUATION

PAK EVAC© a premeditated evacuation suit, also termed, 'once-only suit'.

G. R. Woodford "PAKEVAC" Once-only Suit
A Thermal Protective Aid

In the event of an accident at sea this survival suit, with neck entry, should be donned, a life jacket should be donned over the suit, the individual should enter the sea with as little impact as possible in order to prevent water entering the reduced facial aperture, and the individual should make his way to a life-raft.

The PAK EVAC suit is a fully waterproof garment, so designed to keep the wearer dry whilst in a life-raft awaiting rescue, therefore eliminating sitting in cold water in ordinary clothing, offsetting the risk of hypothermia through the core of the body.

Diving into the sea should at all times be avoided — easing oneself into the sea is recommended at all times.

Should it be necessary to jump into the sea, use both hands to cover the face aperture to minimise entry of water into the suit through the facial aperture.

Should water, of any volume, enter the facial aperture due to not drawing the drawcord to it's maximum closure, (or not using the hands to minimise water entering the suit through the facial aperture), should jumping be necessary, then it is recommended that once in a life-raft one should attempt a posture to get the water into the feet of the garment and the wearer should draw his legs up the suit out of contact with the water remaining trapped in the feet of the suit.

COLOUR: Orange SIZE: One-size only

TECHNICAL DATA

CONSTRUCTED OF: Lock-knit Polyamide, plasticized coated, all seams stitched and welded.

COLD RESISTANCE Does not crack when folded rapidly at −30°C. (−22°F.)

WATER RESISTANCE The material is waterproof

Oil & Flame Resistant
Ministry of Defence approved
M.O.C.R.A. approved
A.Y.R.S. approved
Numerous approvals pending.

G. R. WOODFORD LTD.,

Military Apparel and Offshore Survival Wear

This garment is a premeditated evacuation suit, also termed, 'once-only suit', which is supplied in a wallet. The wallet is so designed to be attatched to a life jacket. The wallet dimensions are 36cm x 23cm.

In the event of emergency evacuation, the garment would be removed from the wallet, the garment would be donned and the hood would be raised, the drawcord would be drawn to the face, a life jacket donned and the wearer would then enter the sea to await location or upon entry to the sea make his way to a life support raft. This garment is designed for all sea-going vessels in view of the ability of this garment to increase the survival time of an individual in cold water, and we recommend that a PAK EVAC suit be attached to every life jacket on board.

DE-SALTING APPARATUS

Contents of the "Permutit" sea water desalting kit. (approved by D.o.T.).

1 cardboard container;
1 storage bag of rubberised fabric with securing cord;
1 storage bag of rubberised fabric with filter pad, drinking tube, plug and lanyard;
chemical charges, each containing 4 cubes.

After removing the contents of the pack, a metal plug is inserted in the purifier bag outlet tube. The bag is then filled with sea water to the level indicated and one chemical charge of 4 cubes is added. The contents are kneaded for 5 minutes and then shaken occasionally during a period of 30 minutes. The reaction between the chemical charge and the dissolved salts in the water is then complete, and clear drinking water can be squeezed through the outlet tube into the mouth or into a container. Residual solids and salts are retained by a filter pad in the purifier bag; these deposits are rinsed from the bag before the next de-salting operation. Full instructions are printed on the storage bag.

Solar still

While solar stills are not approved de-salting apparatus, there is no reason why they should not be included in any survival craft. They are approved for use in liferafts carried by aircraft and various navies throughout the world.

Light in weight, they will continue to produce fresh drinking water day after day. The hotter the day, the more water they produce, so that the supply increases with the survivors' need. On a hot day in the tropics a single still may produce as much as one-and-a-half pints (¾ litre) of water. On the other hand, with wintry conditions and the hours of sunshine greatly reduced, the productivity of the still tends to be reduced accordingly.

To operate:–

Inflate the still, place sea water in the base chamber and tow the still; water vapour condenses on the inside surface of the still as small droplets of fresh water. The droplets enlarge and the rocking action of the still sends them rolling down inside the still and into the fresh water collecting ring. From the collecting ring, the fresh water drains into the fresh water container beneath the still. Do not attempt to hurry the still, give it time to work. These stills are very fragile and should not be lifted out of the water except by their cord loops.

To help alleviate thirst, suck a large button or something similar.

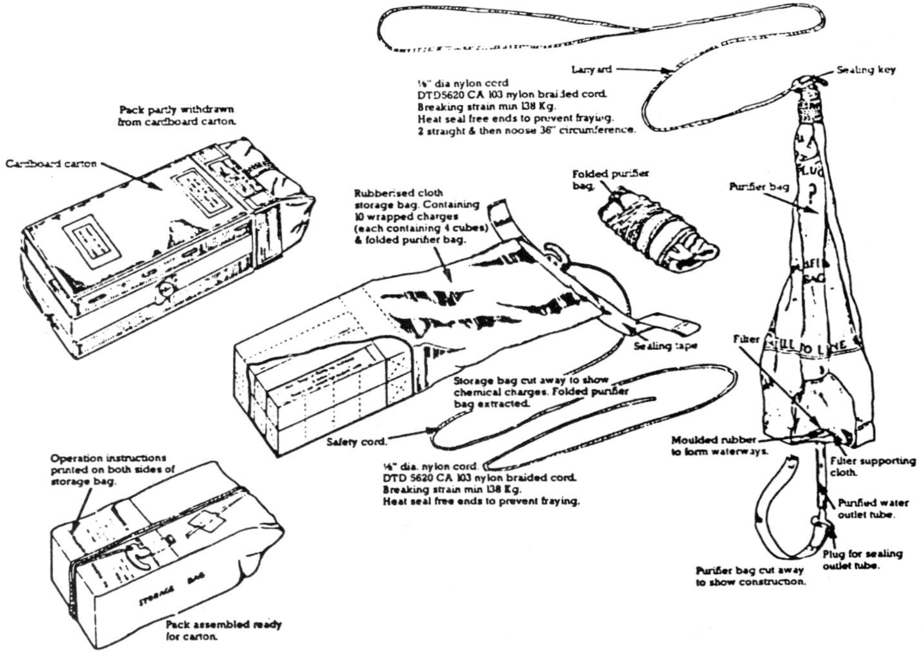

Permutit Sea-Water Desalting Apparatus

Six Pint Kit

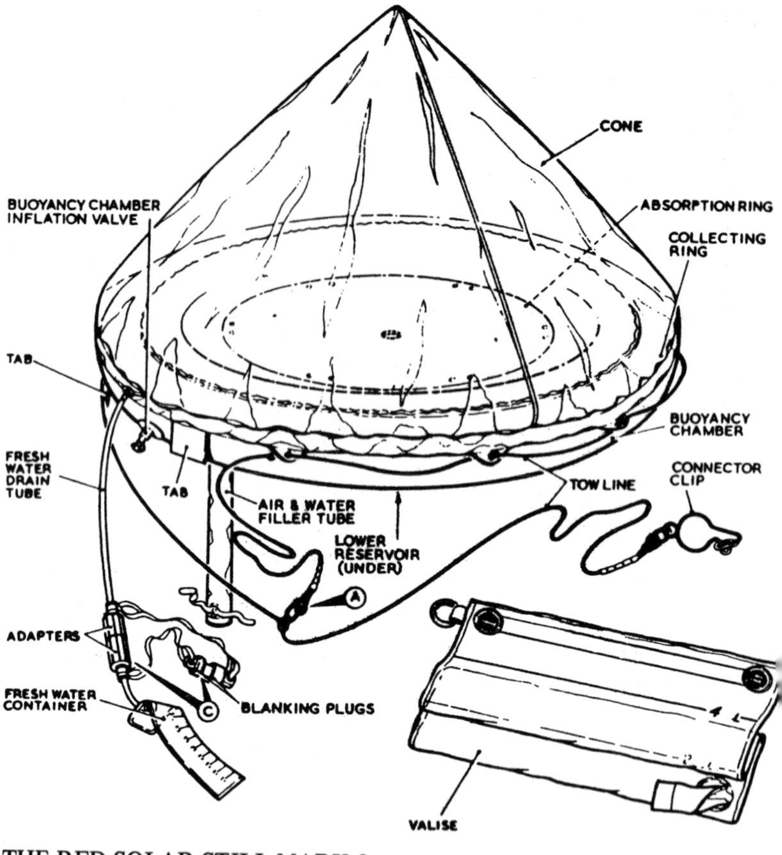

THE RFD SOLAR STILL MARK 3.

Operation commences when the buoyancy chamber and still cone are inflated. Sea-water, contaminated water or urine is placed in the base chamber sump. Distillation will then commence as the still is exposed to the sun. Always ensure exposure is away from the shaded areas to obtain maximum yield, although the still will continue to produce drinkable water even in overcast weather. The still is capable of achieving 175cc drinking water per hour at peak performance.

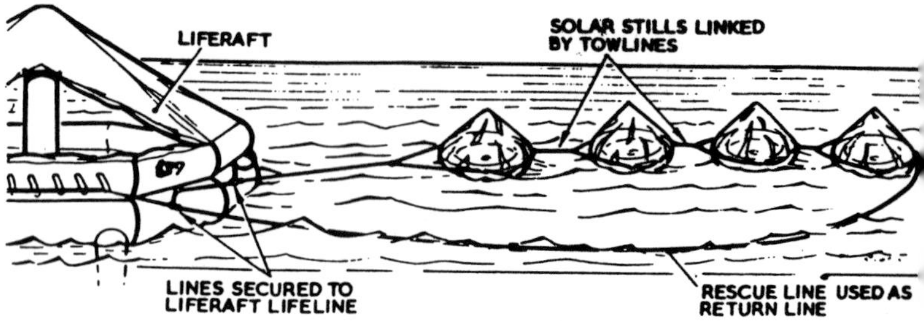

RADAR REFLECTORS

A radar scanner is only as effective as the reflector which returns its signal. Therefore, for a reflector to return a signal, the reflector must be visible to the scanner.

Large high-sided ships such as VLCC's, particularly when they are proceeding light, will have a large area ahead which is screened from the scanner, by the bows. Moreover, they will probably be proceeding ahead at a fast speed.. Obviously then, unless an object is seen by the scanner well ahead of the ship, it may not be seen at all. By the time it is sighted by the look-out man (if indeed it is sighted) it may be too late for the ship to alter course, should it be on a collision course with a small boat or liferaft.

It is therefore essential that a radar reflector is mounted as high as possible. A minimum height of 4m (13ft) above sea level will give an effective coverage over a minimum radius of 5 miles. The reflector should, if at all possible, never be mounted any lower than this. It should be mounted in as near a vertical position as possible and should not be masked by any metal part of the superstructure (metal masts will not create enough "shadow" to cause concern).

Some liferafts now include radar reflective strips in their canopies.

On liferafts, lifeboats and sailing yachts which are not provided with a radar reflector, the only alternative is to keep the sails or canopy wet with sea water. However, the actual effectiveness of a wet sail or canopy is very dubious, while the continued effort required would mean that it could only be carried out when a vessel was sighted.

THE FIRDELL BLIPPER RADAR REFLECTOR

Fully encased to prevent corrosion, to guard against rigging and sail chafe, and without projecting metal attachment points that could cause damage to decks and crew, Firdell radar reflectors offer less than one-seventh of the wind resistance of the long discredited 'standard octahedral' reflectors.

A consistent echo is the main requirement of any radar reflector, particularly now that modern radar scanners are programmed to eliminate 'erratic' echoes from the screen. The optimised arrays of reflective corners which make up Firdell reflectors give a consistently large echo through 360° of azimuth and to some ± 30° of heel.

All three reflectors give a *consistent* echo, so your choice of Firdell reflector will often depend on the size, type, stowage space, rigging, or use made of the boat it is to protect. Yachtsmen and powerboat owners with boats up to 40/50' LOA and more, prefer the popular Blipper, compact, smart, and integrally moulded in a polythene case. Blippers can also be seen on offshore beacons, for example around the coast of Guernsey in the Channel Islands.

Firdell's newest reflector, the Blipper 300, in its special high impact resistant ABS derivative casing, has been designed to meet the special requirements of commercial, civil and military workboats, survival craft, buoyage and beacon systems etc., but like Firdell's original reflector, the Pentland, is also used on larger leisure craft.

reports on Firdell's reflectors:

* 'The company aims to maintain the highest possible standards of safety for its seagoing staff and we are confident that the Pentland measures up to those requirements.'—*Capt. Richardson, Chief Marine Superintendent, CP [Canadian Pacific] Ships. All CP Ships lifeboats are equipped with Firdell reflectors*

* 'The quality of [Firdell's] product is confirmed by the news that [Blippers] are now being fitted to many buoys and beacons round our coasts.'

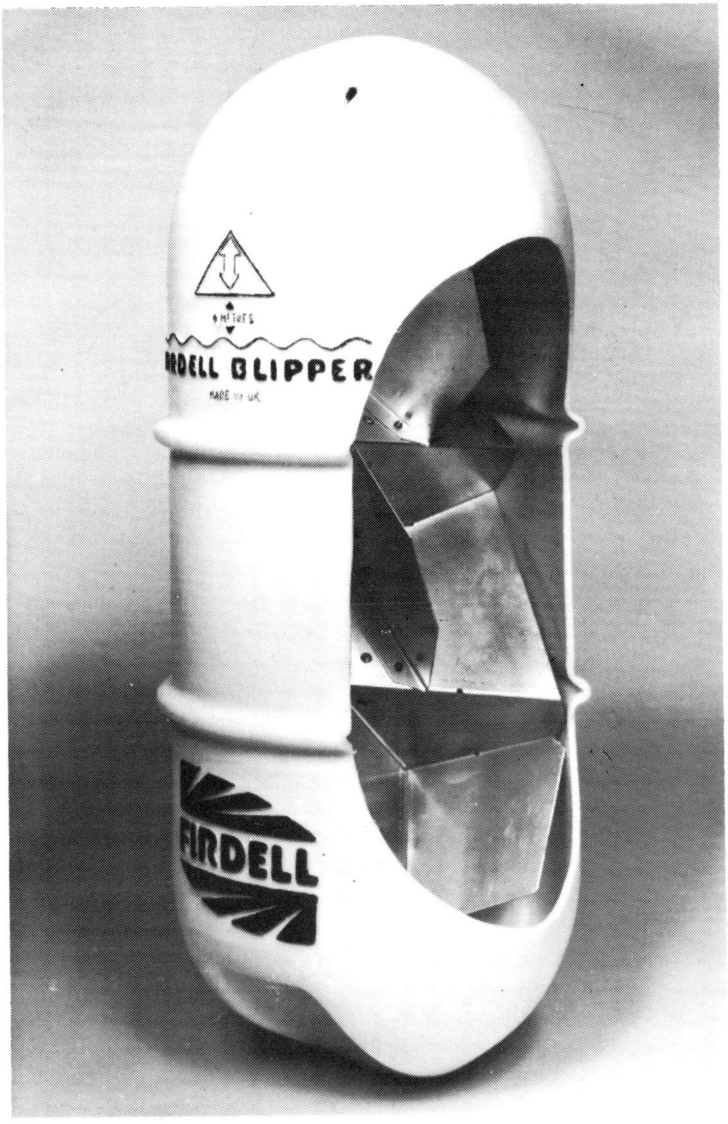

The "Firdell Blipper" (Cut away)

"Firdell Blipper" mounted on the stern of a United States high speed semi-rigid inflatable boarding and rescue craft.

Replies from life-saving stations or maritime rescue units to distress signals made by a ship or person.

SIGNAL

Day signals—Orange moke signal or ombined light and ound signal (thunder-ight) consisting of three ingle signals which are ired at intervals of pproximately one ninute.

Night signals—White tar rocket consisting of 3 Single signals which are fired art intervals of pproximately one ninute.

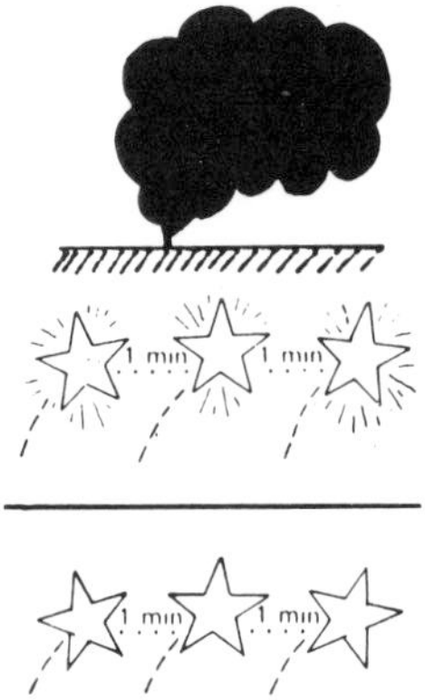

SIGNIFICATION

You are seen—assistance will be given as soon as possible.

(Repetition of such signal shall have the same meaning).

If necessary the day signal may be given at night or the night signals by day.

Landing signals for the guidance of small boats with crews or persons in distress.

SIGNAL

Day and night signals—Vertical motion of a white flag, arms or white light or flare or firing green star signal or code letter K given by light or sound signal apparatus.

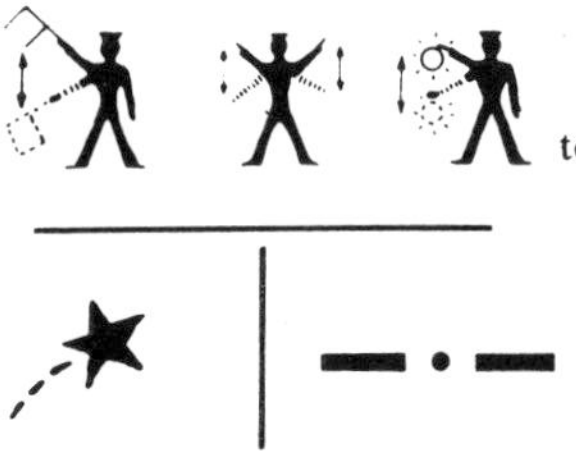

SIGNIFICATION

This is the best place to land.

A range (indication of direction) may be given by placing a steady white light or flare at a lower level and in line with the observer. If a second steady light, or flare is shown at a lower level: Line the two up and come in on this approach.

SIGNAL

Day and night signals—Horizontal motion of a white flag, or arms extended horizontally—or white light or flare or firing of a red star signal or code letter S given by light or sound signal apparatus.

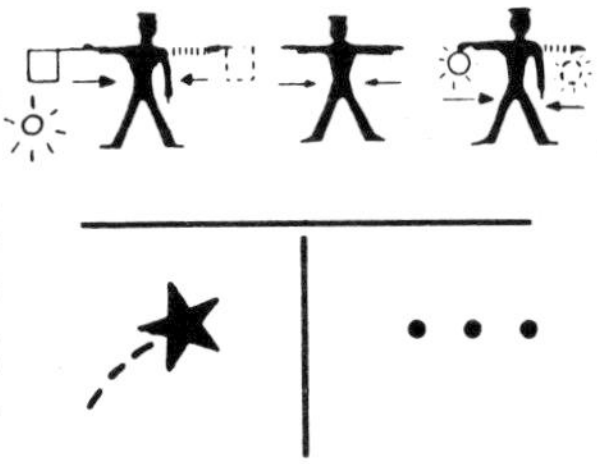

SIGNIFICATION

Landing here highly dangerous.

SIGNAL		SIGNIFICATION
Day and night signals—1. Horizontal motion of a white flag, light or flare followed by 2. the placing of the white flag, light or flare on the ground and 3. by the carrying of another white flag light or flare in the direction to be indicated.	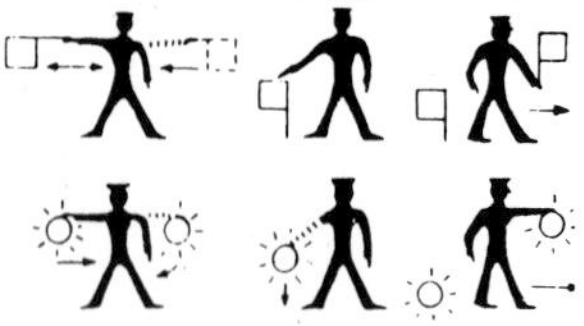	
The firing of a red star or the Morse code letter S (· · ·) given by light or sound.		Landing here highly dangerous.
		A more favourable location for landing is in the direction indicated.
1. or firing of a red star signal vertically and 2. a white star signal in the direction towards the better landing place.	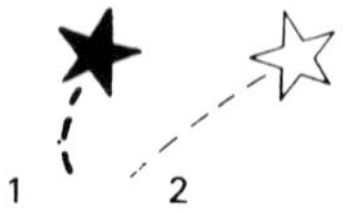	
1. or signalling the code letter S (· · ·) followed by the code letter R (· - - ·) if a better landing place for the craft in distress is located more to the **right** in the direction of approach. 2. or signalling the code letter S (· · ·) followed by the code letter L (· - · ·) if a better landing place for the craft in distress is located more to the **left** in the direction of the approach.		

Surface to air visual signals.

Communication from surface craft or survivors to an aircraft.

Use the following surface-to-air visual symbols by displaying the appropriate symbol on the deck or on the ground.

Message	ICAO visual Symbol	
— Require assistance	V	
— Require medical assistance	X	
— No or negative	N	
— Yes or affirmative	Y	
— Proceeding in this direction	↑	

Reply from an aircraft observing the above signals from a surface craft or survivors. Signification

 Drop a message or	 Rock the wings (during the daylight) or	 Flash the landing lights or navigation lights on and off twice (during hours of darkness) or	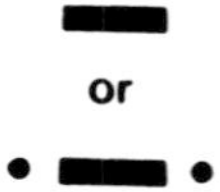 Flash Morse Code signal 'T' or 'R' by light or	Use any other suitable signal	Message understood
 Fly straight and level without rocking wings or	 Flash Morse code signal 'RPT' by light or	Use any other suitable signal			Message not understood (repeat)

THE USE OF DISTRESS SIGNALS

Like most man-made devices there is a right way and a wrong way to use distress signals.

The following hints are intended to assist you to obtain the maximum benefit from your distress signals should you be in the unfortunate position of having to use them.

SIGNALLING BY NIGHT.

The maximum ranges at which distress signals can be seen at night in good visibility conditions are as follows:–

Rocket Parachute Signals	25–35 miles
Hand Flares	5–10 miles

NOTE-Remember that these ranges will be reduced considerably in poor visibility conditions such as usually exist when distress signals are required to be used.

In view of the above it is obviously a waste of valuable signals to display them unless the following conditions exist:–

(a) PARACHUTE SIGNALS. You have good reasons to believe that a possible rescue ship, or an aircraft, or an inhabited shore, is within the estimated visibility range of your signals.

(b) HAND FLARES. The lights of a ship or aircraft or lights on shore are visible to you.

TO SAVE YOUR SIGNALS MAY SAVE YOUR LIFE.

SIGNALLING BY DAY.

The official daylight signal is the Buoyant Orange Smoke Signal. These, however, have a very limited visibility range especially in a stiff breeze.

Many tests have shown that in normal daylight conditions a hand flare or a rocket parachute signal can be seen at a greater range than can an orange smoke signal. The ideal combination, therefore, is smoke and flare.

RADAR REFLECTIVE ROCKET SIGNALS.

These signals, not widely used as yet, eject a cloud of radar reflective material which produces a distinctive echo on a ship or aircraft radar for a period of 10 to 20 minutes and from a considerable range.

They can therefore be used both by day and night when it is considered that a search for survivors is in progress.

READ THE INSTRUCTIONS

Whatever the signals, always carefully read the instructions affixed to them before attempting to fire them .... and follow those instructions precisely.

ATTRACTING ATTENTION.

INTERNATIONAL DISTRESS SIGNALS.

A gun or other explosive signal fired at intervals of about one minute.

A continuous sounding with any fog signal apparatus.

Rockets or shells, throwing red stars fired one at a time at short intervals.

The signal S.O.S. . . . — — — . . . sent in morse code.

A signal sent by radiotelephony consisting of the spoken word "MAYDAY".

The international code signal of distress indicated by N.C.

A signal consisting of a square flag having above it or below it a ball.

Flames on the vessel.

A rocket parachute or a hand flare, showing a red light.

A smoke signal giving off a volume of orange coloured smoke.

Slowly and repeatedly raising and lowering arms outstretched to each side.

The radiotelephone alarm signal consists of two tones transmitted alternately over a period of from 30 seconds to one minute on a frequency of 2182 kc/s.

The automatic radio alarm signal consists of a series of twelve dashes, sent in one minute, each dash having a duration of four seconds, with an interval of one second between every two consecutive dashes on a frequency of 500 kc/s.

A piece of orange coloured canvas with either a black square and circle or other appropriate symbol (for identification from the air).

A dye marker.

Although not laid down as international distress signals, articles of clothing or an ensign flown upside-down at the masthead, are generally recognised signals of distress.

ON NO ACCOUNT ARE ANY OF THE ABOVE SIGNALS TO BE MADE EXCEPT IN EMERGENCY.

ISTRESS SIGNALS.

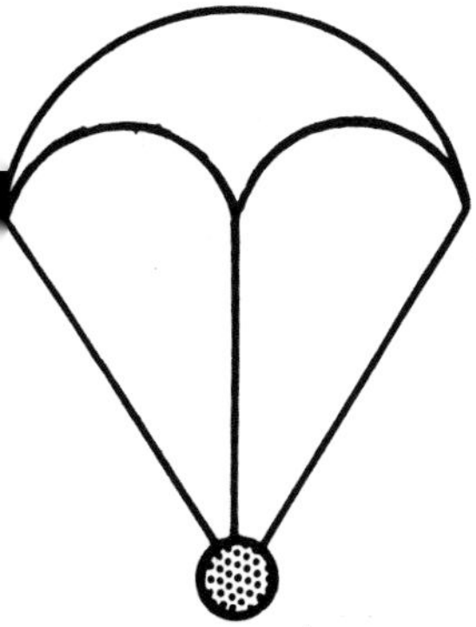

A red parachute flare.

Flames.

Red flare.

Yellow smoke.

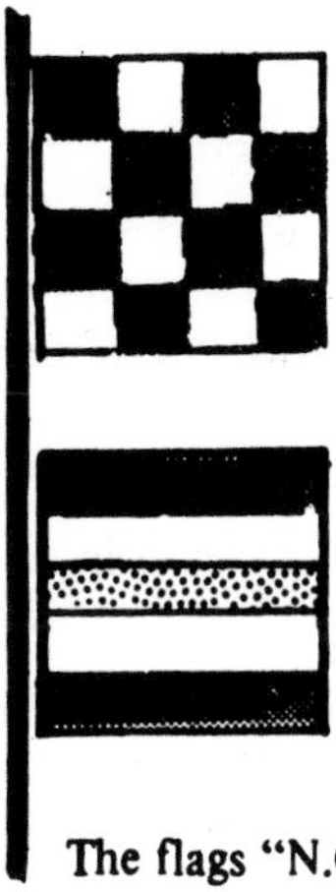

The flags "N.C."

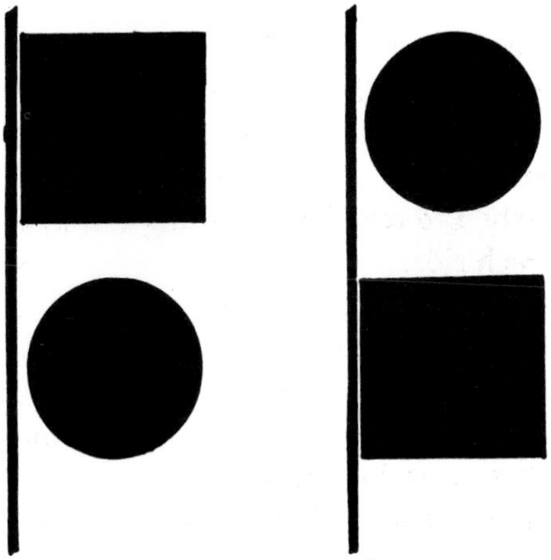

A square flag having above or below it a ball.

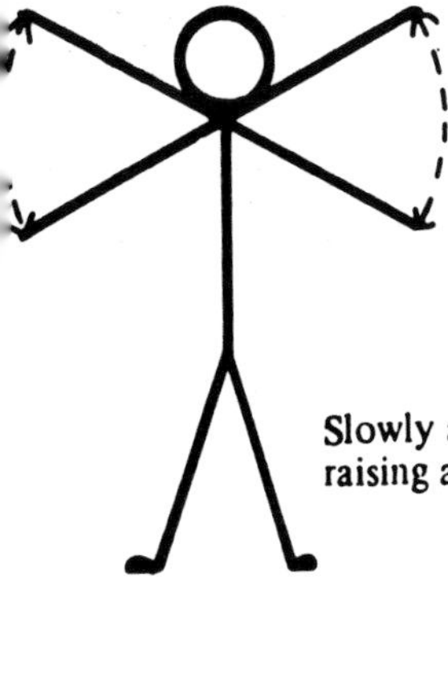

S.O.S. by light or sound.

Slowly and repeatedly
raising and lowering the arms - outstretched.

ROCKETS, FLARES AND SMOKE FLOATS

Rocket parachute flares for ships and survival craft

1. The rocket parachute flare shall:
   - (a) be retained in a water resistant casing;
   - (b) have brief instructions or diagrams clearly illustrating the use of the rocket parachute flare printed on its casing;
   - (c) have integral means of ignition;
   - (d) be so designed as not to cause discomfort to the person holding the casing when used in accordance with the manufacturer's operating instructions.
2. The rocket shall, when fired vertically, reach an altitude of not less than 300m (975ft). At or near the top of its trajectory, the rocket shall eject a parachute flare, which shall:
   - (i) burn with a bright red colour;
   - (ii) burn uniformly with an average luminous intensity of not less than 30,000 cd;
   - (iii) have a burning period of not less than 40 seconds;
   - (iv) have a rate of descent of not more than 5 m/s;
   - (v) not damage its parachute or attachments while burning.

   When firing a rocket, it should always be fired slightly to leeward because a rocket has a tendency to climb to windward. It will not normally gain its full height if fired vertically. In cloudy weather fire the rocket at an angle of about 45 degrees, to keep it below the cloud. For use by day or night.

Hand flares

1. The hand flare shall:
   - (a) be contained in a water resistant casing;
   - (b) have brief instructions or diagrams clearly illustrating the use of the hand flare printed on its casing;
   - (c) have a self-contained means of ignition.
   - (d) be so designed as not to cause discomfort to the person holding the casing and not endanger the survival craft by burning or glowing residues when used in accordance with the manufacturer's operating instructions.
2. The hand flare shall:
   - (a) burn with a bright red colour;
   - (b) burn uniformly with an average luminous intensity of not less than 15,000 cd;
   - (c) have a burning period of not less than 1 minute;
   - (d) continue to burn after having been immersed for a period of 10 seconds under 100mm (4 ins) of water.

CAUTION: **when firing a hand flare, if you have open sleeves, roll them up.** A little burning residue falling down your sleeve will cause considerable injury from burning. If you have a glove, wear it.

Flares may be used by day or night and are provided to pin-point your position to a rescue craft or aircraft that is in sight.

In addition to the above, U.K. Regulations demand that all rockets and flares shall be fitted with an integral means of firing that is easy to operate with wet, cold or gloved hands in adverse conditions without external aid and requiring the minimum of preparation. Sealing shall not depend on adhesive tapes.

The signal shall be so constructed that the end from which the rocket or flare is emitted can be positively identified by day or night.

The signal shall be capable of functioning after immersion for 2 hours under 1m (3.25ft) of water. In the ready to fire condition, the signal shall function after immersion for one minute under 10mm (4 ins) of water.

Buoyant smoke signals

. The buoyant smoke signal shall:
(a) be contained in a water resistant casing;
(b) not ignite explosively when used in accordance with the manufacturer's operating instructions;
(c) have brief instructions or diagrams clearly illustrating the use of the buoyant smoke signal printed on its casing.

. The buoyant smoke signal shall:
(a) emit smoke of a highly visible colour, at a uniform rate for a period of not less than 3 minutes when floating in calm water;
(b) not emit any flame during the entire smoke emission time;
(c) not be swamped in a sea-way;
(d) continue to emit smoke when submerged in water for a period of 10 seconds under 100mm (4ins) of water.

Buoyant smoke signals are provided to attract aircraft that can be seen in daylight. Owing to wind dispersal, they are not readily noticeable from the bridge of a ship. Ignite the signal and cast it into the sea.

In addition to the above, U.K. Regulations demand that all smoke signals shall be fitted with an integral means of ignition, easy to operate with wet, cold or gloved hands in adverse conditions without external aid and be so designed as to enable the signal to be released from a survival craft without harm to the occupants.

The signal shall be capable of functioning after immersion for two hours under 1m (3.25ft) of water.

The signal shall be safe to operate in oil covered waters.

On all rockets, flares and smoke signals, the date of manufacture and the date of expiry shall be marked indelibly on the signal. All components, composition and ingredients shall be of such a character and quality as to enable them to burn evenly and maintain their serviceability under good average storage conditions in the marine environment for a period of at least three years.

CAUTION: all rockets, flares and smoke signals are to be fired on the lee side of the survival craft. **Always read the instructions before firing.** *Calcium flares are unsuitable as smoke signals and are not safe on oil covered waters).*

Radaflare

Radaflare, manufactured by "Pains-Wessex" is for use when visibility is poor and the conventional parachute distress signal rocket is of limited use. Specially designed for poor visibility, "Radaflare" carries 210,000 radar reflective dipoles and two free-falling red stars, to a height of 1250 fe (380m) by means of the "Icarus" rocket. The echo is clearly detected b aircraft radar at 20 miles and ship's radar at 10 miles for a period of up t 15 minutes.

Miniflare packs (Personal survival for yachtsmen)

Miniflare packs are a complete signalling set consisting of eight single-st cartridges and a pen-sized firing projector packed into a compact, wate proof plastic pouch which can easily be kept in a pocket or attached by lanyard to a lifejacket. Should a yachtsman fall into the water unnotice these flares can be fired from the water. Each star ejects to 80m (130 f and burns for five seconds at 5,000 candels.

Line-throwing appliances

1. Every line-throwing appliance shall:
   (a) be capable of throwing a line with reasonable accuracy;
   (b) include not less than four projectiles each capable of carryi the line at least 230m (750 ft) in calm weather;
   (c) include not less than four lines each having a breaking stra of not less than 2,000 N;
   (d) have brief instructions or diagrams clearly illustrating the u of the line-throwing appliance.
2. The rocket, in the case of a pistol fired rocket, or the assembly, the case of an integral rocket and line, shall be retained in a wat resistant casing. In addition, in the case of a pistol fired rocket, t line and rockets together with the means of ignition shall be stow in a container which provides protection from the weather.

CAUTION: always wear gloves and shield the eyes when firing a rocket li from a pistol, in case there should be a flash-back.

Always fire a line throwing rocket slightly to leeward of the target, allow for the rocket climbing into the wind.

In addition to the above, U.K. Regulations demand that all line throwi appliances shall be so constructed that the end from which the rocket ejected can be positively identified by day or night.

All components, compositions and ingredients of the rockets and t means of igniting them shall be of such character and quality as to enab them to maintain their serviceability under good average storage conditio in the marine environment for a period of at least three years.

The rocket, or the assembly shall function after immersion for one minute under 10cm (4 ins) of water. The date of manufacture and the date of expiry shall be marked indelibly on the rockets and cartridges.

Clear and concise directions for use, in the English language shall form an integral part of all rockets, flares, smoke signals and line-throwing apparatus.

Para Red Mk 3

A hand-held distress signal rocket

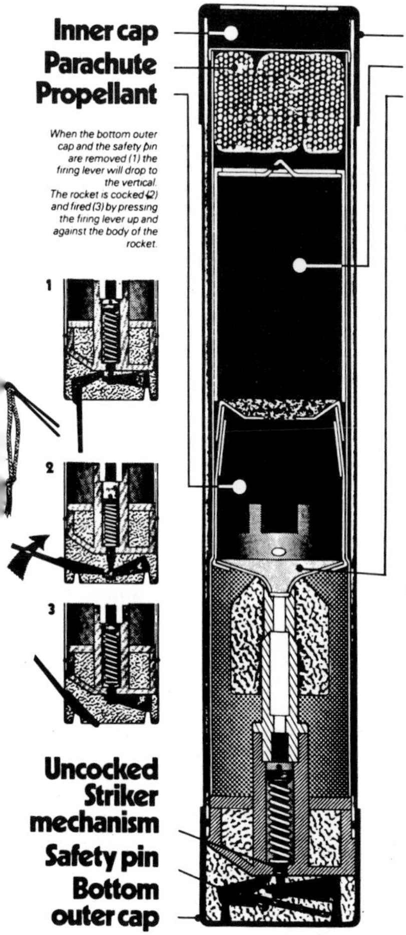

Description and method of operation

DESCRIPTION
The Para Red Mk. 3 is a hand-held distress rocket, ejecting a parachute suspended red flare at 300m altitude. Although based on the long established and well proven Mk. 1 version, the Mk. 3 incorporates several major improvements.

(a) Waterproofing tapes have been eliminated hence quicker and easier use. These tapes have now been replaced by 'O' ring seals at the top and bottom of the signal to give greater environmental protection.
(b) The well proven trigger-lever operation is still used, but with the addition of an uncocked striker and a safety pin to prevent inadvertent ignition.
(c) Improved waterproofing in the "ready-to-fire" condition to ensure greatest possible protection against water immersion in bad weather.

The rocket casing is of a tough plastic to ensure optimum protection and reliability in severe marine weather conditions.

APPLICATIONS
Long range distress signalling.
This signal is approved by International Authorities for use on ships of all types and in ship's lifeboats and liferafts.

OPERATION
Although the signal is normally fired vertically to provide maximum range of visibility, in low cloud conditions (below 300m), it is advisable to fire the rocket at an angle of 45°.

1. Remove top end cap.
2. Remove bottom end cap and safety pin.
3. Hold signal firmly.
4. Squeeze trigger lever. Fire vertically, slightly downwind.

Specification

Deployment height: 300m when fired vertically, 200m at 45° angle.
Burning time: 40 seconds.
Light output: 40,000 candela.
Dimensions: 266mm×47mm diameter.
Weight: 360g.
Explosive content: 143g.

PAINS-WESSEX SCHERMULY

Pains-Wessex Limited
High Post Salisbury Wiltshire SP4 6AS

Pinpoint Mk 6

A hand-held red distress flare

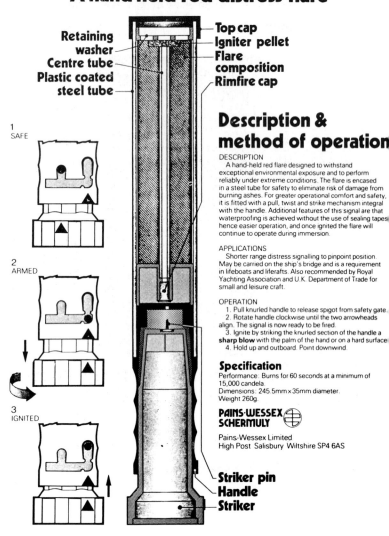

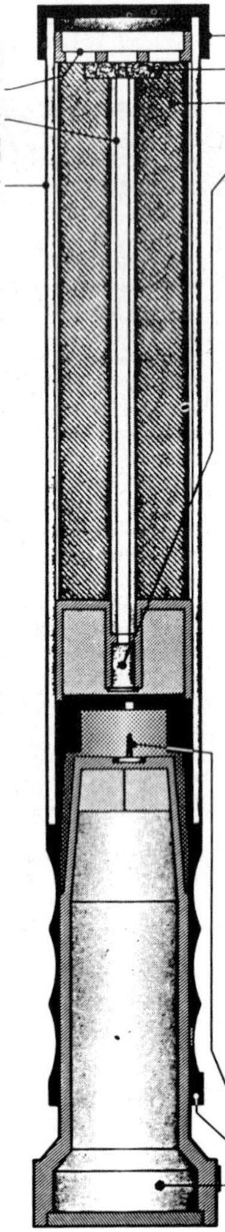

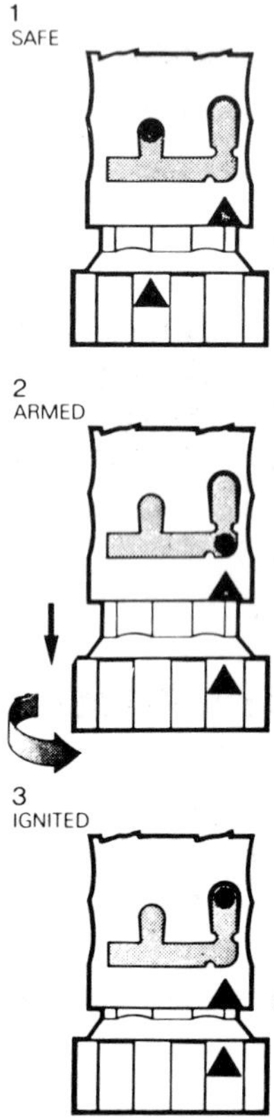

Description & method of operation

DESCRIPTION

A hand-held red flare designed to withstand exceptional environmental exposure and to perform reliably under extreme conditions. The flare is encased in a steel tube for safety to eliminate risk of damage from burning ashes. For greater operational comfort and safety, it is fitted with a pull, twist and strike mechanism integral with the handle. Additional features of this signal are that waterproofing is achieved without the use of sealing tapes hence easier operation, and once ignited the flare will continue to operate during immersion.

APPLICATIONS

Shorter range distress signalling to pinpoint position. May be carried on the ship's bridge and is a requirement in lifeboats and liferafts. Also recommended by Royal Yachting Association and U.K. Department of Trade for small and leisure craft.

OPERATION

1. Pull knurled handle to release spigot from safety gate.
2. Rotate handle clockwise until the two arrowheads align. The signal is now ready to be fired.
3. Ignite by striking the knurled section of the handle a **sharp blow** with the palm of the hand or on a hard surface.
4. Hold up and outboard. Point downwind.

Specification

Performance: Burns for 60 seconds at a minimum of 15,000 candela.
Dimensions: 245.5mm x 35mm diameter.
Weight 260g.

PAINS-WESSEX SCHERMULY

Pains-Wessex Limited
High Post Salisbury Wiltshire SP4 6AS

Lifesmoke Mk 3

A buoyant orange smoke signal

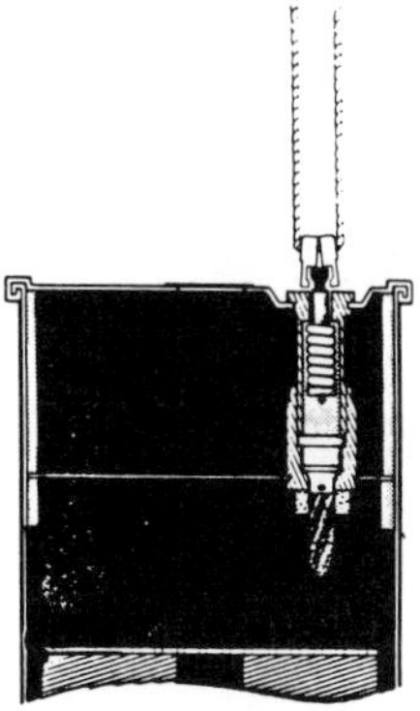

1. Remove plastic lid.
2. Pull cord firmly.
3. Plastic grip separates and releases striker.
4. Throw signal overboard DOWNWIND.

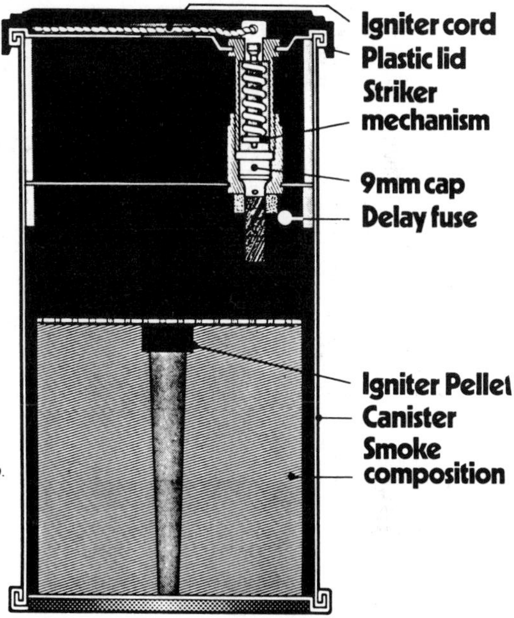

Description and method of operation

DESCRIPTION

A buoyant orange smoke signal safe to operate on petrol or oil covered water. The signal consists of a metal case containing smoke composition and is fitted with a simple pull-cord ignition.

APPLICATION

Daylight distress signalling. Required in ship's lifeboats and suitable for use in other commercial and pleasure craft.

OPERATION

The unit has been designed for maximum ease of operation:

1. Remove plastic top cap.
2. Grasp cord firmly and pull vertically.
3. Throw signal overboard downwind.

Specification

Performance: Produces dense orange smoke for 3 minutes. Dimensions: 160mm×85mm diameter. Weight: 459g.
Note: A 4-minute smoke duration version is available to special order.

Speedline

A self-contained linethrowing unit

Description & method of operation

DESCRIPTION

A complete, self-contained linethrowing unit with two important advantages over the older pistol-type apparatus.

1. The set of 4 units normally carried can be dispersed in strategic positions throughout the vessel.

2. Each unit can be fired independently as required.

Each unit consists of a weatherproof plastic body/launcher incorporating the handle and trigger assembly, and containing the rocket, igniter and 275m of ready-flaked line.

APPLICATIONS

1. Ship-to-ship, ship-to-shore or shore-to-ship linethrowing.

2. Rescue of swimmers in distress using the optional buoyant head to keep the rocket afloat.

OPERATION

Speedline has been designed for maximum ease of operation. Full pictorial instructions are printed on both sides of the unit and can be read by either right or left-handed users.

Specification

Performance: Projects a line to at least 230m in calm weather. Dimensions: 330mm x 190mm diameter. Weight: 4.kg.

Pains-Wessex Ltd., and Schermuly Ltd.,
High Post, Salisbury, Wilts SP4 6AS.

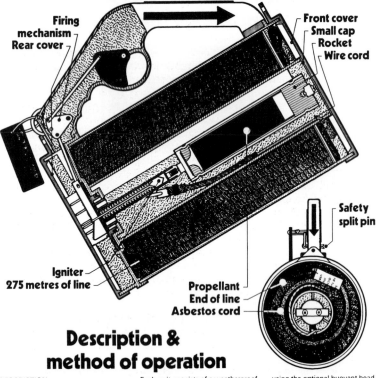

CHAPTER VII FIRST AID

1. Hypothermia (loss of body heat)

The majority of deaths occuring during and after shipwreck are caused by hypothermia.

Symptoms

- Slowing of physical and mental responses.
- Irritability or unreasonable behaviour.
- Cramps or shivering.
- Unsteadiness.
- Difficulties with speech or vision.

The colder the climate, the quicker the body heat evaporates. In order to retain body heat, warm and preferably wind and waterproof clothing is essential. If the patient has been immersed in a cold sea and is taken aboard a survival craft, his clothing should be kept on. At most, remove only the top layer, wring it out and put it on again as quickly as possible; this is to help retain the body heat that still remains in the body. Tight clothing should be loosened. If available, dress the patient in a thermal protective aid, alternatively, add blankets or additional clothing, cover the head and face as much as possible, allowing him to breathe; 40% of heat loss is through the head and face. Place the patient in the most sheltered position available, keeping both your own and the patient's clothing on, cuddle together to impart body heat to the patient.

Do not massage, do encourage warmth and rest, give a little water if thirsty. **Stimulants should not be given.** In order to avoid loss of blood pressure it is essential that the return of body heat is gradual.

If the patient is ice cold, stiff and unconscious, attempts at cardiac massage and artificial respiration are more likely to do harm until the body's temperature has risen. By this time there may be evidence of spontaneous heart action and/or respiration.

2. Unconscious persons

An unconscious person should, if possible, always be laid on their side in a prone position with the forearm on which they are lying behind their back; bend the knee of the same side slightly, place the other (upper) limb in a convenient position, bending both hip and knee to nearly a right angle. This position will prevent the tongue obstructing the air passage and will also prevent the patient from being asphyxiated by their own vomit. Loosen the clothing, take out false teeth and keep warm.

Fits, Epilepsy

Keep the patient horizontal and try to prevent injury. Place something hard between the teeth at the side of the mouth if possible, do not use force to restrain them, loosen the clothing. After the attack, lay them on their side in a prone position. Encourage rest, warmth and sleep.

3. Shock

Encourage warmth, rest and sleep, give a little water if thirsty, loosen tight clothing. Raise the legs. Stimulants should not be given.

4. Control of bleeding

Apply pressure in the form of a cotton wool pad, securely bandaged. Tourniquets should be avoided, they can do much harm by restricting the return flow of blood from the limb. **If used, they must be loosened every 15 minutes** to restore circulation. Treat for shock. **Raise the bleeding part.**

5. **Fractures**

In the absence of anything suitable to use as a splint, bind a broken leg to the sound leg. For a broken arm, shoulder or ribs, support the arm in a sling and strap to the chest. Keep the patient as still as possible and treat for shock. Fit survivors may be used as chocks to prevent undue movement in a sea-way.

Symptoms

- Pain
- Swelling
- Deformity
- Loss of power
- Abnormal movement

When in doubt, always treat the condition as a fracture. If there is excessive swelling, do not bandage too tightly as this could stop circulation to the rest of the limb. If a hand or foot below the fracture becomes cold, blue and swollen – loosen the bandage. If possible, keep the limb raised.

6. **Sun burn, burns and scalds**

Place hands or feet in plastic bags obtained from the equipment and tie the neck to exclude air. Failing this, immerse in sea water for at least 10 minutes (but this will be painful). If the first-aid kit contains a suitable burn ointment, spread as much of the ointment as may be necessary on gauze and place it on the burn. Do not attempt to rub ointment into burnt flesh. Do not bandage. Do not prick or burst any blisters. Do not attempt to remove clothing adhering to the burn. Treat for shock.

Sun burn is best avoided by remaining in the shade if possible and retaining clothing on the body.

7. **Heat stroke (hyperpyrexia)**

Causes

A high atmospheric temperature with a hot dry wind, high humidity and lack of air movement.

Symptoms

- Restlessness
- Passing urine frequently
- Rapid and stertorous breathing
- High temperature
- Headache, dizziness and feeling hot
- Confused, stuporous, or in a coma

Strip the patient, wrap in wet, cold clothing, reduce the temperature if possible by evaporation from the skin. Give water if thirsty, encourage rest and sleep.

8. **Frost bite**

Frost bite may occur in very cold weather; the ears, toes, nose, chin and fingers are the most frequently affected. The affected part will feel cold, painful and stiff , feeling and power of movement are either diminished or lost. It may be recognised by blanching and numbness of the part. **Unless immediate action is taken, gangrene and death may be the result.** Remove constricting clothing, rings, boots etc. Cover an affected part on the face with a dry gloved hand and scarves until colour and sensation are restored. Put an affected hand under the clothing in the armpit. Wrap up affected feet with a blanket or woollen scarf.

Do not massage. It is important to thaw the affected part gradually and restore the circulation.

9. **Immersion foot**

Immersion foot causes swelling, blistering, ulceration and numbness. Remove footwear and rest legs horizontally. The patient should be encouraged to move his feet and toes. **Do not massage.**

10. Foreign bodies in the eye

A foreign body beneath the lower lid or on the white of the eye may be removed with the moistened corner of of a hankerchief. Do not remove anything that adheres to the clear part of the eye. A foreign body under the upper lid can often be removed by pulling the upper lid down over the lower lid when the lashes may sweep the foreign body out of the eye. Where the eyes have been in contact with a chemical substance, wash them well with sea water.

11. Cramp

When suffering from cramp, try to stretch the shortened muscle. Keeping the knee straight, stretch your leg and pull the toes towards you. Forcibly straighten fingers and toes.

12. Recuscitation

WHEN BREATHING STOPS, WHATEVER THE CAUSE, START ARTIFICIAL RESPIRATION AT ONCE. THE ONLY EXCEPTION IS IF A BODY FEELS ICE COLD WITH STIFF MUSCLES AND WIDE PUPILS –IN THIS CASE START REWARMING EVEN BEFORE ARTIFICIAL RESPIRATION – SEE HYPOTHERMIA.

DROWNING – URGENT TREATMENT IS VITAL – SECONDS COUNT

Start artificial respiration by mouth to mouth method with patient flat if possible. There is no time to remove dentures, loosen clothing or to try and find out if the heart is still beating.

MOUTH TO MOUTH RESUSCITATION

1. lay the patient on his back and, if on a slope, have the stomach slightly lower than the chest.
2. Make a brief inspection of the mouth and throat to ensure that they are clear of obvious obstruction.
3. Give the patient's head the maximum backwards tilt so that the chin is prominent, the mouth closed and the neck stretched to give a clear airway. – Fig A.
4. Open your mouth wide, make an air-tight seal over the nose of the patient and blow. The operator's cheek or the hand supporting the chin can be used to seal the patient's lips – Fig B. Or if the nose is blocked, open the patient's mouth using the hand supporting the chin; open you mouth wide and make an airtight seal over his mouth and blow – Fig C. This may also be used as an alternative to the mouth to nose technique.
5. After exhaling, turn you head to watch for chest movement whilst inhaling deeply in readyness for blowing again – Fig. D.

A

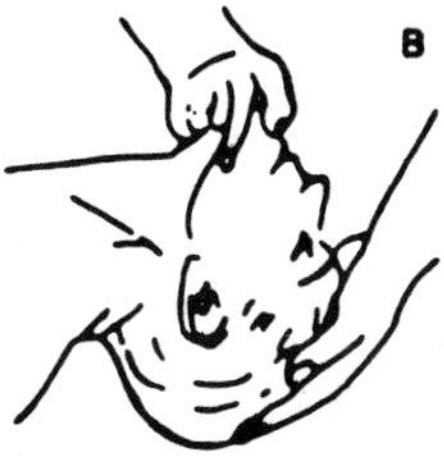

B

C

6. If the chest does not rise, check that the patient's mouth and throat are free of obstruction and the head is tilted backwards as far as possible. Blow again.

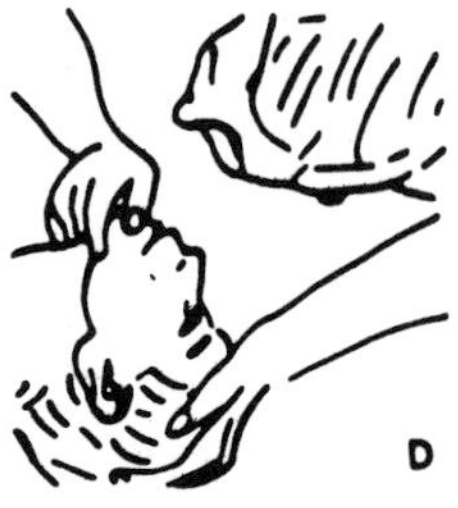

D

In general. To help to alleviate thirst, suck a large button or something similar. To help the circulation, gently exercise separate parts of the body, one at a time, fingers, toes, lips, ears, eyelids, face wrinkles etc. To help retain stamina, keep the survival craft and yourself dry, eat all your rations in the first four days, while you still have water. Keep warm in the cold and try to avoid perspiration in the heat, wear some clothing in the sun to avoid sun burn. Evacuate the bowels as soon as possible and then forget about them. Never drink salt water or urine. Do not take alcohol, dump any that is available Alcohol aids hypothermia to set in and helps cause dehydration.

Warmth and sleep are nature's finest cures and should be encouraged Beware of massage, massage can often do more harm than good. The first aid kit will have clear instructions provided as to the proper use of the contents. **Read those instructions.**

Will power is probably the most important asset a castaway possesses. Never allow yourself to give up hope. Many survivors have kept themselves alive over long voyages by sheer will power.

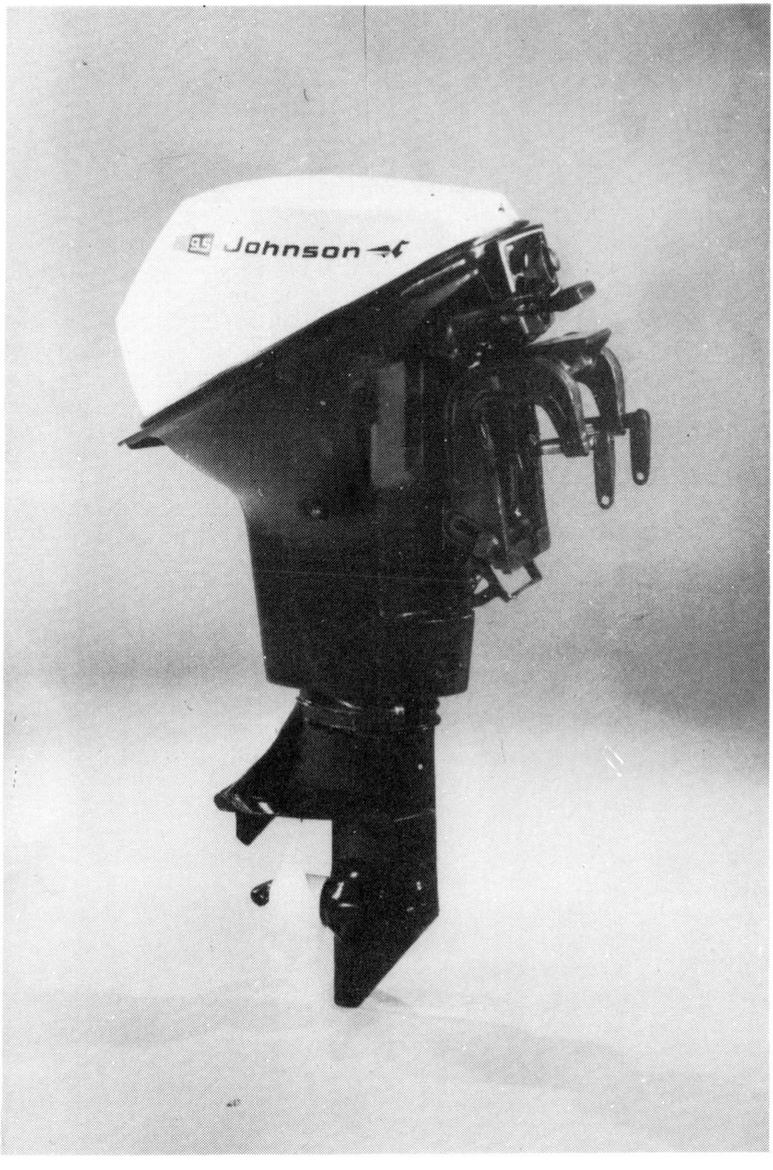

Johnson Outboard Motor
Approved for use with D.O.T.I. Boats. E. P. Barrus (Concession Aires) Ltd.

F.P.T. Industries D.o.T. Inflatable Boat with 8 persons.

Department of Transport Merchant Shipping Notice
No. **M.1204**

HELICOPTER ASSISTANCE AT SEA

Notice to Shipowners, Masters, Shipboard Safety Officers, Shipbuilders and Shiprepairers

The purpose of this Notice is to recommend the provision of suitable facilities to assist helicopter operations on all types of sea-going ships and also to give guidance on:

(a) the selection and preparation of areas suitable for winching operations; and

(b) the contingency plans which should be made and the drills which should be undertaken in anticipation of the need for helicopter assistance.

Guidance on these topics, and others, is contained in the ICS "Guide to Helicopter/Ship Operations (Revised Edition 1982)" which is regarded as one of the recognised reference manuals for this type of operation.

1. GENERAL

A recent passenger ship casualty has once again demonstrated the valuable assistance helicopters can provide to shipping. In that particular instance helicopters evacuated all the passengers and part of the crew, an operation which has been accomplished several times on various types of ships. Additionally, of course, helicopters are now regularly used in emergency to take doctors or specialists to, and to remove sick or injured persons from, ships which are at sea. In the majority of these cases all that was provided was a clear area of deck over which the helicopter could safely hover whilst winching people to or from the ship.

Most of these winching operations have been totally successful due to the skill of the crews of the helicopters but accidents have occurred. Shipowners and seafarers must appreciate the hazard of these operations and be prepared to take all reasonable measures to reduce the risks involved.

2. THE WINCHING AREA

It is recommended that a winching area prepared in accordance with the advice in paragraph 1 of the Annex to this Notice be provided whenever practicable on:

(a) all new sea-going ships over 100 metres in length; and

(b) all other sea-going ships over 100 metres in length which are suitable for routine helicopters operations.

In the case of passenger ships, it is suggested that serious consideration be given to the provision of two winching areas; one forward and one aft.

On ships which are not suited for routine helicopter operations, serious consideration should be given to arranging for emergency helicopter winching operations as described in paragraph 2 of the Annex to this Notice.

3. CONTINGENCY PLANS AND DRILLS

Shipowners should, wherever practicable make any arrangements necessary on the ship to assist winching operations in order to ensure that helicopters may operate safely without risk to persons on board or in the helicopter itself. They should therefore ensure that a sufficient number of crew members have received the necessary training and information.

In addition to the information provided in the ICS Code, further guidance concerning helicopter operations will also be found in MERSAR Chapter 4.2 and Notice No. 4, paragraphs 57 to 68 of the Annual Summary of Admiralty Notices to Mariners.

Department of Transport
Marine Directorate
90 High Holborn
London WC1V 6LP
December 1985

ANNEX

(Information in this Annex is based upon the ICS Publication— "Guide to Helicopter/Ship Operations")

1. HELICOPTER WINCHING AREAS AND ASSOCIATED ARRANGEMENTS

1.1 Size and Marking (figure 1 refers)

A winching area to be used for helicopter operations comprises two zones, ie an inner Clear Zone and an outer Manoeuvring Zone having the following dimensions and markings:

(a) *The Clear Zone*—this is a totally clear circular plated area having a minimum diameter of 5·0 metres. The area selected should be painted yellow and contrast with the adjacent paintwork of the ship.

(b) *The Manoeuvring Zone*—this is a circular area, at least 30·0 metres in diameter, which extends beyond the Clear Zone. The height of obstacles such as companionways, small deck houses, ventilators, etc located in this Zone must not be more than 3 metres above deck level nor encroach into the Clear Zone. All such obstacles should be painted in "day-glo" yellow and black stripes of equal width.

1.2 Location

The winching area should, as far as practicable, be located:

(a) on the port side of the ship, such that a large portion of the

Manoeuvring Zone extends over the ship's side and situated where it will enable the pilot of an attending helicopter to have an obstructed view of the ship;

(b) in a position which will minimise the effect of air turbulance in the Manoeuvring Zone;

(c) clear of areas likely to be affected by flue gases;

(d) where safe means of access are available from at least two widely differing directions;

(e) where there will be an adequate deck area adjacent to the Manoeuvring Zone for people to muster prior to their being transferred to the helicopter. Both the muster area and the deck area in way of the Manoeuvring Zone should have slip resistant deck surfaces;

(f) clear of accommodation spaces.

1.3 **Obstructions and Projections**

The task of lowering or raising persons will be made safer if obstructions and projections within the winching area are kept to a minimum, consequently:

(a) the fitting of aerials, awnings, stanchions, derricks and similar obstructions should be avoided as far as is practicable. Where however the presence of some such items is essential then they should be capable of being easily removed or struck as appropriate;

(b) to reduce the risk of the winching wire becoming fouled, all guard rails and handrails located within the Manoeuvring Zone should be made removable or collapsable;

(c) all moveable objects must be secured or removed from the area before transfer takes place.

1.4 **The Flight Path**

Helicopters will normally approach the winching area and depart from it along a flight path on the port side of the ship. It is therefore necessary to ensure that:

(a) the flight path along the ship's side is kept clear; and

(b) a wind pennant is hoisted in a position where it can be readily seen by the pilot of the helicopter.

1.5 **Operations Conducted at Night**

If winching operations have to be conducted during the hours of darkness then it will be necessary to illuminate the Manoeuvring Zone, the mustering area and the wind pennant. This lighting should:

(a) be supplied from the main and emergency power sources, and

(b) not be directed, or reflected by the ship's structure or at the sea, towards the helicopter.

Significant obstructions, eg a mast, in the vicinity of the Manoeuvring Zone should be illuminated by floodlights or other accepted lighting, supplied from sources as in sub-paragraph (a) above. Guidance regarding the positioning and acceptability of such lighting can be provided by the Department. In some instances formal exemption from the requirements of the International Regulations for Preventing Collisions at Sea, 1972, may be required.

2. FACILITIES FOR EMERGENCY HELICOPTER WINCHING OPERATIONS

2.1 Emergency Winching Area and Wind Pennant

In ships where it is clearly not practicable to incorporate all the arrangements described in paragraph 1 above, it is strongly recommended that facilities for emergency helicopter winching operations should be arranged. Such facilities should incorporate, whenever possible, the zones in sub-paragraph 1.1 and at least the wind pennant in accordance with sub-paragraph 1.4 (b). Emergency winching areas should however *not* be permanently marked.

2.2 Alternative Facilities

Where it is impossible to provide the arrangements described in sub-paragraph 1.1, the highest area clear of obstructions to which a helicopter can safely manoeuvre and over which it may safely hover should be chosen. The clear winch area should be as near to the ships side as possible and there should be no obstruction greater than 3·0 metres in height in the manoeuvring area. Any obstructions such as aerials or stays must be removed or lowered. Because of main deck clutter or perhaps the dangerous nature of the cargo carried, it is most likely that the only areas available will be near the stern of the ship around the bridge or poop. If possible, two positions should be selected—one on each side of the ship to allow for wind direction if the ship is unable to manoeuvre.

Having chosen the most advantageous safe position(s) which could be used for emergency helicopter winch operations, the upper part of any tall obstacles in the vicinity should be painted in a conspicuous colour (see sub-paragraph 1.1).

Where none of these arrangements can be made, operations should be carried out (weather permitting) in a lifeboat positioned to leeward.

3. INSTRUCTIONS FOR WINCHING OPERATIONS

Details of the provisions to facilitate helicopter assistance should be recorded and be readily available to the watchkeepers.

4. EXAMINATION OF ARRANGEMENTS

If required, the Department will arrange for drawings and notes detailing proposed "helicopter assistance arrangements" for any ship to be examined in consultation with the Civil Aviation Authority. These drawings to include, both in plan view and elevation, the location of all obstacles, obstructions and equipment relative to the proposed winching area. Such examinations will be subject to the normal scale of fees.

WINCHING AREA

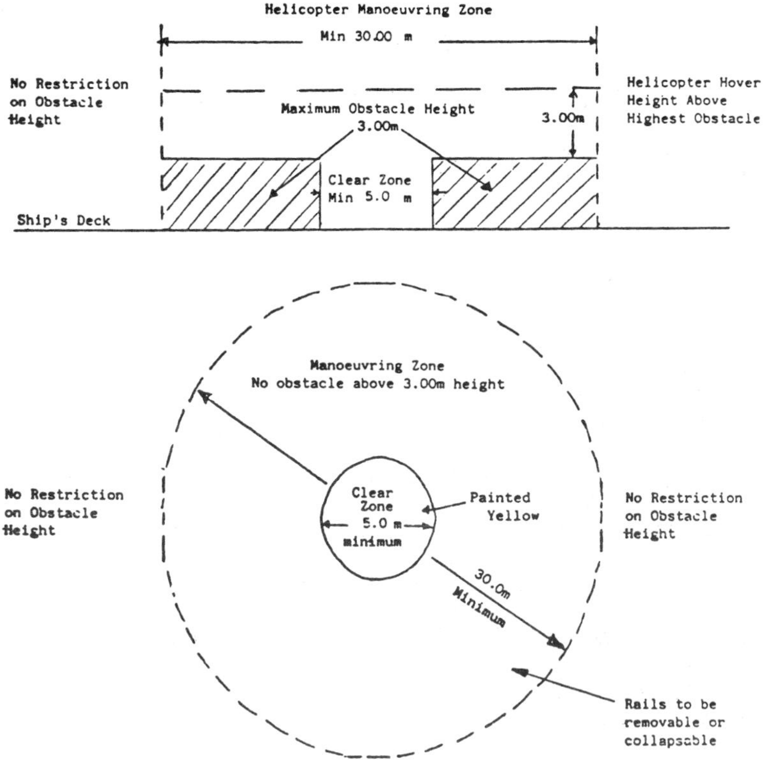

figure 1.

DEPARTMENT OF TRANSPORT MERCHANT SHIPPING NOTICE
No. **M.1066**

SURVEY OF INFLATABLE BOATS

Notice to Owners and Masters of Merchant Ships, Owners and Skippers of Fishing Vessels and Yachtsmen

1. In accordance with a recent Resolution adopted by the Assembly of the International Maritime Organisation the Department requires that inflatable boats carried on British ships are surveyed annually only by authorised competent persons and that information about servicing facilities is readily available to ship owners and masters.

2. Regulation 29(3) of the Merchant Shipping (Life-saving Appliances) Regulations 1980 and Section 12.26.2 of Survey of Fishing Vessels (Instructions for the guidance of surveyors) require inflatable boats on British ships and fishing vessels to be surveyed at intervals of not more than 12 months at a servicing station approved by the Department or at the work's of the manufacturers.

3. The appendix to this notice gives a United Kingdom list of

Lifeguard Equipment Limited	Telephone: 08242 4314
Lon Parcur, Ruthin	
Clwyd LL15 1YU	Telex: 61687

approved servicing stations; prior to the re-survey of an inflatable boat in a place outside the British Isles, however, shipowners and masters are strongly recommended to enquire in good time from the boat's manufacturers, the name of the approved servicing station convenient to the ship's envisaged port of call.

4. The requirement for inflatable boats to be serviced only at service stations approved by this Department and/or the manufacturer of the boat, is because no service station is approved to handle any boat approved by this department, unless those involved in the servicing hold current certificates issued by the manufacturer proving they have been trained in servicing by the manufacturer. Failure to comply with this requirement will render the inflatable boat unacceptable to this Department, and prosecution could be initiated by the Department for carriage of boats which are considered defective.

Department of Transport
Marine Division
London WC1V 6LP
July 1983

APPENDIX

Cosalt Limited, 174 Market Street, Aberdeen.

Seadog Lifesaving Appliances (Scotland) Limited, 4 Constitution Place, Leith Docks, Edinburgh, EH6 7DL.

B. A. Beadle, St. Andrews Dock, Hull, HU3 4PQ.

Cosalt PLC, Battery Green, Lowestoft, Suffolk, NR32 1DN

Seaweather Marine, D22 Fairview Industrial Estate, Motherwell Way, West Thurrock, Essex.

Viking Life-Saving Equipment Limited, Ocean House, Caxton Street North, London, E16 1JL.

R. Perry & Co., Nightingale Grove, Shirley, Southampton.

Salcombe Marine, Island Street, Salcombe, Devon, TQ8 8DP.

R. Perry & Co., 95-101 Whitehouse Lane, Bedminster, Bristol, BS3 4 DN.

R. Perry & Co., Monks Ferry Works, 90 Church Street, Birkenhead, Merseyside.

Cosalt PLC, 196-214 Dock Street, Fleetwood, FY7 6TB.

James Tedford & Co. Ltd., 5 and 9 Donegal Quay, Belfast.

DEPARTMENT OF TRANSPORT MERCHANT SHIPPING NOTICE
No. **M.1207**

MANNING OF SURVIVAL CRAFT

Notice to Shipowners, Masters, Officers and Ratings

This Notice is to be read in conjunction with the Merchant Shipping (Life-Saving Appliances) Regulations 1986 and The Merchant Shipping (Life-Saving Appliances Regulations 1980) (Amendment) Regulations 1986

1. Merchant Shipping (Life-Saving Appliances) Regulations 1986 and the Merchant Shipping (Life-Saving Appliances Regulations 1980) (Amendment) Regulations 1986 which implement the 1983 Amendments to Chapter III of the International Convention for the Safety of Life at Sea 1974 and which come into force on 1 July 1986, include new requirements for the manning of survival craft.

2. Details of the new requirements are set out in the Annex to this Notice. The contents of the Annex are compulsory applicable from 1 July 1986 under Regulation 26(2) of the Merchant Shipping (Life-Saving Appliances) Regulations 1986 and regulation 46(2) of the Merchant Shipping (Life-Saving Appliances) Regulations 1980 as amended by the Merchant Shipping (Life-Saving Appliances Regulations 1980) (Amendment) Regulations 1986 for determining the number of deck officers, certificated persons and other crew members required for the manning of survival craft and handling of launching arrangements.

3. Previous requirements were limited to the provision of a number of certificated lifeboatmen and persons trained in the handling and operation of liferafts on certain Classes of passenger ships. The number of certificated lifeboatmen provided was related to the number of lifeboats carried and the capacity of these lifeboats.

4. The new requirements which apply to cargo ships engaged on international voyages as well as passenger ships of Classes I, II, II(A) and III are related not only to the number of lifeboats carried, but also to their type, and in addition, to the number of liferafts carried and the launching arrangements for the lifeboats and liferafts.

5. The new requirements refer not to a "certificated lifeboatman" but to a "certificated person" who is defined in the Regulations as follows:
A "certificated person" means a member of the crew who holds either
(a) a Certificate of Proficiency in Survival craft under the Merchant Shipping (Certificates of Proficiency in Survival Craft) Regulations 1984 or such certificate issued by or

under the authority of any government outside the United Kingdom which is accepted by the Secretary of State as being the equivalent to a certificate under those Reguations; or

(b) a Certificate of Efficiency as Lifeboatman issued before 28 April 1984 by or under the authority of the Secretary of State or such certificate issued before 28 April 1984 by or under the authority of any government outside the United Kingdom which is accepted by the Secretary of State as being the equivalent of such a certificate issued by or under the authority of the Secretary of State.

. The requirements also refer to a "person practised in the handling nd operation of liferafts". For the purpose of the Regulations such erson is defined as follows:

> A person practised in the handling and operation of liferafts" means a person who has attended a basic sea survival course and has at least two months sea service.

Department of Transport
Marine Directorate
London WC1V 6LP
une 1986

ANNEX

MANNING OF SURVIVAL CRAFT

In ships of Classes I, II, II(A), III, VII, VII(A), VII(T), VII. VIII(T), and ships of Classes IX and XI engaged on internationa voyages the minimum number of deck officers, certificated person and other crew members required to operate the minimum number c survival craft and launching arrangements to be used for aban donment by the total number of persons on board shall be determine as shown in the following table.

	ARRANGEMENT	MINIMUM NUMBER OF DECK OFFICERS OR CERTIFICATED PERSONS
A	Lifeboat boarded at stowed position, which can be lowered from within, and with gripes which can be released from lifeboat.	2 in each lifeboat See Note 1.
B	Lifeboat boarded at stowed position, which can be lowered from within, and with gripes which are released from outside lifeboat.	2 in each lifeboat plus 1 per every 2 lifeboa to release gripes. See Note 1.
C	Lifeboat released and lowered from ship.	2 in each lifeboat plus 1 per lifeboat fo lowering. See Notes 1 and 2.
D	Liferafts.	1 in each lifecraft. See Note 3.
E	Davit launched liferafts.	1 at each davit. See Note 4.
F	Throwover liferafts.	1 at each launching position. See Note 4.

NOTE 1 In the case of a motor lifeboat where the coxswain is unable to operate th engine and steer the lifeboat from one position a crew member capable o operating the engine and carrying out minor adjustments shall also b provided in the lifeboat. Such crew member need not be a certificated person

NOTE 2 Where the person in charge of lowering the lifeboat cannot operate the winch brake and keep the lowering operation in sight a crew member capable o operating the winch brake shall also be provided. Such crew member need not be a certificated person.

NOTE 3 Number required can include deck officers and certificated persons allocated to lowering lifeboats and all but one of such persons allocated to liferaft davi or launch positions. In ships of Classes II, II(A) and III the crew member ir charge of a liferaft need not be a certificated person but, if not, must be a person practised in the handling and operation of liferafts.

NOTE 4 Number required can include up to 3 deck officers or certificated persons on each side in charge of lowering lifeboats or assisting with gripes as required by Arrangement B or C.

In passenger ships of Classes I, II and II(A) where the rescue boat o boats are not lifeboats and are included in the number of boat required to marshall liferafts as prescribed in regulations 5(4)(d) and 6(6)(d) of the Merchant Shipping (Life-Saving Appliances) Regulations 1986, at least 2 deck officers or certificated persons shall be provided for each rescue boat so included.

If the minimum number of deck officers, certificated persons and other crew members as indicated above is insufficient to permit a cargo ship to be abandoned in 10 minutes or a passenger ship in 30 minutes then the number of deck officers, certificated persons and other crew members provided shall be such that these times are not exceeded.

If the survival craft or launching arrangements provided are not of the type referred to in the table the number of deck officers, certificated persons and other crew members shall be sufficient to ensure that the above abandonment times are not exceeded and that there is a deck officer or certificated person in charge at each stage of the abandon ship procedure.

DEPARTMENT OF TRADE MERCHANT SHIPPING NOTICE
No. **M.973**

LAUNCHING CREWS FOR LIFEBOATS, CLASS C BOATS INFLATABLE BOATS AND OTHER BOATS

Notice to Shipowners, Masters, Officers and Seamen of Merchant Ships, and to Owners, Skippers, Mates and Crews of Fishing Vessels

This notice supersedes Notice No. M. 689

1. Merchant Shipping Notice No. M.689 was issued following the loss of a fisherman from a trawler which brought to the Department's attention the possibility of some doubt about the proper number of men for a launching crew of a lifeboat, Class C boat, inflatable boat or other boat on ships which are required to comply with the Department's rules and regulations on life-saving appliances.

2. The purpose of this Notice is to update the Merchant Shipping Notice No. M.689 in accordance with the Merchant Shipping (Life-Saving Appliances) Regulations 1980* and the Fishing Vessels (Safety Provisions) Rules 1975.†

3. In the case where the boat or the means of launching are not of sufficient strength for the boat to be lowered safely into the water when loaded with its full complement of persons and equipment required by the Regulations and Rules the davits or other means of launching shall be conspicuously marked with a RED BAND 150 millimetres wide painted on a white background, in accordance with Regulations 43(19) and Rule 96(17) of the Merchant Shipping (Life-Saving Appliances) Regulations 1980 and Fishing Vessels (Safety Provisions) Rules 1975, respectively.

4. The Department considers that the proper number of men for a launching or recovery crew when the device is "RED-BANDED" is two. Shipowners, Masters, Officers and Seamen of merchant ships, and Owners, Skippers, Mates and Crews of fishing vessels, are therefore asked to note that this number should never be exceeded.

5. A notice should be attached to each relevant set of davits or other device stating "Lower or recover with two man crew only".

6. In the case of a launch/recovery device for an inflatable boat, although in certain circumstances the boat is not required to be fitted with an engine, an allowance is made in all cases for the weight of an engine and its fuel of at least 60 kgs. (132 lbs) in case one is fitted at a later date.

* S.I. 1980 No. 538
† S.I. 1975 No. 330

May 1981

DEPARTMENT OF TRANSPORT MERCHANT SHIPPING NOTICE
No. **M.1218**

UIDELINES FOR TRAINING CREWS FOR THE PURPOSE F LAUNCHING LIFEBOATS AND RESCUE BOATS FROM SHIPS MAKING HEADWAY THROUGH THE WATER

Notice to Shipowners, Masters, Officers and Ratings

his Notice is to be read in conjunction with the Merchant Shipping (Musters and Training) Regulations 1986

. Merchant Shipping (Life-Saving Appliances) Regulations 1986 ome into force on 1 July 1986 and included in these regulations is a equirement that lifeboats carried on cargo ships of 20,000 GRT and pwards, and all rescue boats shall be capable of being launched when ie ship on which they are carried is making headway at speeds up to 5 nots in calm water. These Regulations apply to ships, the keels of hich are laid on or after 1 July 1986.

. There is no requirement that lifeboat and rescue boat launching rills be carried out when a ship is making headway, but regulation (9) of the Merchant Shipping (Musters and Training) Regulations 986 requires that if such drills are conducted, they must be conducted 1 accordance with the guidance specified in this Notice. That uidance which has been developed by the International Maritime)rganization on account of the potential hazards involved, consists of uidlines set out in the Annex to this Notice.

. Such drills should not be carried out with boats other than those escriberd in paragraph 2 of the Annex except where it has been etermined by experience that the boats, release gear and launching rrangements are suitable for the purpose, the boat crews have been ully trained, and the drills are carried out in accordance with well roven procedures.

)epartment of Transport
Aarine Directorate
ondon WC1V 6LP
une 1986

ANNEX

GUIDELINES FOR TRAINING CREWS FOR THE PURPOSE OF LAUNCHING LIFEBOATS AND RESCUE BOATS FROM SHIPS MAKING HEADWAY THROUGH THE WATER

1. There is no requirement in the Merchant Shipping (Musters and Training) Regulations 1986 to carry out training in launching lifeboats and rescue boats from ships making headway through the water. However, these guidelines should be followed if such training is undertaken.

2. These guidelines apply to those launching drills referred to in regulation 6(9) of the above mentioned Regulations undertaken with lifeboats and rescue boats capable of being safely launched with the ship making headway at speeds of up to 5 knots in calm water, as prescribed in regulations 18(6) and 20(3) of the Merchant Shipping (Life-Saving Appliances) Regulations 1986, ie to new cargo ships of 20,000 tons gross tonnage and upwards and other new ships fitted with rescue boats, (and any other ship fitted with a lifeboat or rescue boat or both having on-load release gear adequately protected against accidental or premature use).

3. These guidelineds supplement the procedures to be followed for the particular equipment provided on board a ship and as described in the instructions and information found in the ship's training manual required by the Merchant Shipping (Life-Saving Appliances) Regulations 1986. This will include instructions on launching and recovery, the use of the release gear, clearing the boat from the ship, and where applicable, the use of a painter. The boat's crew should be instructed in the procedures to be followed before the drill commences.

4. Drills should either be carried out on board a ship at anchor or alongside where there is a suitable relative movement between ship and water, or at a suitable shore establishment where similar conditions prevail. Alternatively, at the master's discretion, it may be carried out on board a ship when making headway in sheltered waters. For safety purposes, it is not necessary when training to exercise at the maximum design 5 knot headway launching capability of the equipment. Drills should be carried out with a low relative water speed particularly where inexperienced personnel are involved. When planning the drill consideration should be given to ensuring that, as far as practicable, the relative water speed will be at a minimum when recovering the boat.

5. None of the provisions in these guidelines are intended to inhibit launching drills carried out on ships where such drills are carried out on a frequent and regular basis with fully trained and experienced boat crews.

When planning for and carrying out the launching drills referred to in regulation 6(9) of the Merchant Shipping (Musters and Training) Regulations 1986 the following precautions should be taken:

6.1 Drills should only be carried out under the supervision of an officer experienced in such drills and under calm water and clear conditions.

6.2 Provisions should be made for rendering assistance to the boat to be used in the drill in the event of unforseen circumstances, eg where practicable a second boat should be made ready for launching.

6.3 Where practicable the drill should be carried out when the ship has minimal freeboard.

6.4 Instructions as to procedures should be given to the boat's crew by the officer in charge before the drill commences.

6.5 The minimum number of crew members should be in the boat compatible with the training to be carried out.

6.6 Lifejackets, and where appropriate, immersion suits should be worn.

6.7 Except in the case of totally enclosed boats, head protection should be worn.

6.8 For the purposes of the drill, skates where fitted should be removed unless they are designed to be retained under all launch conditions.

6.9 In the case of totally enclosed boats, all openings should be closed except for the helmsman's hatch which may be open to provide a better view for launching.

6.10 Two-way radiotelephone communications should be established between the officer in charge of lowering, the bridge and the boat before lowering commences, and be maintained throughout the exercise.

6.11 During lowering and recovery and while the boat is close to the ship, steps should be taken to ensure that the ship's propeller is not turning, if practicable.

6.12 Before the boat enters the water the boat's engine should be running.

6.13 The launching and recovery should be followed by a de-briefing session to consolidate the lessons learned.

DEPARTMENT OF TRANSPORT MERCHANT SHIPPING NOTICE
No. **M.1266**

WAVE QUELLING OILS—USE OF OIL IN RESCU OPERATIONS

Notice to Owners, Masters and Skippers

This notice supersedes Notice No. M. 725

1. The Department of Transport advises seafarers against t indiscriminate use of oil to calm the sea by ships coming to the resc of other ships. When survivors are likely to be in the water, th pumping of oil should only be carried out when absolutely necessa and then with the greatest care.

2. Experience has shown that vegetable and animal oils, includin fish oils, are most suitable in such circumstances. If they are n available, lubricating oils should be used. Fuel oils should not be use unless absolutely unavoidable and then only in very limited quantitie Oils of the former types are less harmful to men in the water and a very effective quelling agents. Tests carried out by an independe company have shown that 200 litres of lubricating oil, discharge slowly through a rubber hose with an outlet just above the sea, whi the ship proceeds at slow speed, can be an effective agent for quellin seas over an area of at least 4500 square metres.

Department of Transport
Marine Directorate
London WC1V 6LP
January 1987

DEPARTMENT OF TRANSPORT MERCHANT SHIPPING NOTICE
No. **M.1248**

UTOMATIC RELEASE HOOKS FOR LIFERAFTS AND ISENGAGING GEAR FOR LIFEBOATS AND RESCUE BOATS

Notice to Shipowners, Managers and Masters of Merchant Ships

This notice supersedes Notice No. M. 866

eneral

It is now a requirement for all United Kingdom Ships, the uilding of which commenced on or after the 1 July 1986, to have utomatic release hooks for use with their davit launched liferafts, and isengaging gear—incorporating "on-load" simultaneous release echanisms—fitted in their lifeboats and rescue boats. Such release quipment has already been installed on a number of ships, and in the ase of many others, the lifeboats have been fitted with simultaneous elease mechanisms which are designed to operate in the "off-load" ode.

The primary aim in fitting these particular types of release quipment is to facilitate the safe release of the rafts and boats from e davit falls during an evacuation, especially in heavy weather onditions.

. Whilst this type of release equipment represents a great technical dvance over the traditional hook, which has been used for so many ears, it is also so much more complex. Consequently, to ensure that it ill always function efficiently, it must be properly adjusted, regularly aintained and correctly operated at all times. Of particular nportance is the need to ensure that the correct design of ring or ackle is being used to connect the release equipment to the davit lls. Failure to attend to any of these factors could cause a alfunction of the release equipment, resulting in the raft or boat eing released prematurely or, alternatively, being retained on the falls hen waterborne, with potentially disastrous results in either case.

. Unfortunately several serious accidents involving the use of these elease mechanisms have already been reported to the Department, ey include:—

(a) premature release of lifeboats from the falls; causing death and serious injuries to the occupants and extensive damage to the boats;

(b) retention of a lifeboat on the falls when waterborne; the occupants were uninjured but great difficulties were experienced in releasing the boat;

(c) premature release of davit launched liferafts from the falls; causing loss of confidence to members of the crew witnessing the launches and also damage to the rafts.

Whilst the accidents reported above occurred during drills and test the added seriousness of any such incident occurring during a genuin abandonment will be readily appreciated.

Maintenance, training and testing

5. Investigation of the incidents reported above has shown that th primary cause was due to the equipment being:—

(a) inadequately maintained,
(b) improperly adjusted, or
(c) incorrectly operated.

5.1 The attention of all is therefore drawn to these particula problems and especially the need to ensure that crew members ar given full and regular training in the correct operation of this type c equipment. Furthermore, it is essential for the equipment to b adjusted and maintained in *strict* accordance with the manufacturer' instructions and then tested at least once every three months—a required by the Merchant Shipping (Musters and Training) Regula tions 1986, Regulations 6(7) and (8) refer.

5.2 The testing to be conducted in the following manner:—

(a) the operation of automatic release hooks for liferafts—b suspending an appropriate weight; setting the hook and the lowering the weight to the deck where it should b automatically released;
(b) the operation of "on-load" disengaging gear and "off-load release mechanisms—by lowering the boat into the wate and then pulling on the control lever after which the hook should open. It should be noted that this equipment shoul never be tested whilst the boat is hanging in the davits.

5.3 Regular servicing and proof testing of release equip ment is now also required when it is fitted on any ship th building of which commenced on or after the 1 July 1986 and these procedures are strongly recommended when suc equipment is fitted on any older ship. In the case of:—

(a) automatic release hooks for rafts; such procedures to b conducted at intervals not exceeding 2½ and 5 year respectively;
(b) "on-load" disengaging gear and "off-load" release mechan isms for boats; such procedures to be conducted at interval not exceeding 5 years.

The Merchant Shipping (Life-Saving Appliances) Regulations 1986 Regulation 13 refers.

Offences

6. It is an offence for the life-saving appliances on any ship to be i defective condition, and under the powers contained in Sections 43 and 431 of the Merchant Shipping Act 1894, a ship may be detaine until such defects have been rectified. Additionally, in some cases

·oceedings may be mounted against the owner and master if such ›pliances are not kept at all times fit and ready for use.

xplanatory Note

"Off-load" release mechanisms, fitted in life-boats and the automa-: release hooks used for liferafts, are designed to be operated ·pened) when the load applied to them is removed—ie when the boat · raft has been lowered and becomes waterborne.

"On-load" disengaging gear, fitted in lifeboats and rescue boats, is ·signed to be operated (opened) when the boat has been lowered and ·comes waterborne, whether or not a load is being applied to it by the ·tion of the ship or the sea.

·epartment of Transport
Iarine Directorate
ondon WC1V 6LP
·ctober 1986

DEPARTMENT OF TRANSPORT MERCHANT SHIPPING NOTICE
No. **M.1186**

LIFEBOAT WINCHES FITTED WITH A ROLLER RATCHE MECHANISM

Notice to Shipowners, Masters, Safety Officers and Shiprepairers

In 1981 a seaman was injured whilst assisting in the recovery of lifeboat/passenger launch on a passenger cruise liner. The boat wa also badly damaged. The incident occurred as the boat was bein hoisted on board the liner by an electric boat winch. When the bo reached Boat Deck level the launching crew began fitting the tricin pendants as the winch operator stopped the winch. The brak apparently failed to hold the boat and although one of the tricin pendants had been connected to the lower block it could not hold th boat in position. The pendant parted causing the boat to swin violently and drop on the falls into the sea. As the boat swung one the launching crew was thrown into the sea and injured; fortunatel the Chief Officer and the other seamen in the launching crew jumpe clear and were not injured.

When the brake was opened up for examination it was found to b in good condition and working satisfactorily; the roller ratche mechanism within the winch was then suspected as being the princip cause of this incident, but upon subsequent examination it too wa found to be in an apparently satisfactory condition.

Examination of our records then showed that a number of simila winch failures had occurred in recent years but fortunately in none these cases had anyone been injured or any of the boats damaged. I each case however, it was noted that the failure occurred to a winc which had been used for the launching and recovering of lifeboat/passenger launch, a type of boat which is in frequent use o most cruise liners.

In view of these past failures and the knowledge that they coul have led to disastrous results, an extensive programme of winc testing was carried out by one of the country's leading winc manufacturers. As a result of this work it has been concluded that th most probable cause of these winch failures was the weakening of th springs used to retain the rollers in position within the ratchet. Th weakening was due to the frequent use of these particular winches o passenger cruise liners.

Consequently, Shipowners, Masters and Safety Officers ar strongly recommended to ensure that winches used for the launchin and recovery of any boat are regularly checked. In addition:—

(a) Any winch used for any lifeboat/passenger launch fitted o a UK passenger ship or any other highly worked surviva craft or work boat shall be opened up and thoroughl examined every two years. In addition, it is recommende that on those winches which are fitted with roller ratchet the opportunity should be taken to renew the roller retaining springs at these biennial examinations;

(b) Any winches used for a traditional lifeboat fitted on any ship (which is likely to be used less frequently than a lifeboat/passenger launch) should be opened up for thorough examination at intervals not exceeding 4 years as indicated in paragraph 3.11.2 of "Survey of Life Saving Appliances—Instructions for the Guidance of Surveyors". If roller ratchets are fitted to these winches then the roller retaining springs should be renewed during such examinations;

(c) Winches used for any type of survival craft, work boat or launch should, if fitted with a roller ratchet mechanism, have such mechanisms regularly maintained. The ratchet mechanisms should never be packed with grease; a light non-solidifying grease or light oil should be lightly smeared on the mechanisms to assist easy movement and to prevent the onset of corrosion.

It is also recommended that ships fitted with winches incorporating a roller ratchet mechanism carry an adequate supply, or at least one complete set, of spare roller retaining springs.

Department of Transport
Marine Directorate
London WC1V 6LP
August 1985